PACS and Digital Medicine

PACS and Digital Medicine

Essential Principles and Modern Practice

Yu Liu

Jihong Wang

CRC Press
Taylor & Francis Group
Boca Raton   London   New York

CRC Press is an imprint of the
Taylor & Francis Group, an **informa** business

A TAYLOR & FRANCIS BOOK

CRC Press
Taylor & Francis Group
6000 Broken Sound Parkway NW, Suite 300
Boca Raton, FL 33487-2742

© 2011 by Taylor and Francis Group, LLC
CRC Press is an imprint of Taylor & Francis Group, an Informa business

No claim to original U.S. Government works

Printed in the United States of America on acid-free paper
10 9 8 7 6 5 4 3 2 1

International Standard Book Number: 978-1-4200-8365-1 (Hardback)

This book contains information obtained from authentic and highly regarded sources. Reasonable efforts have been made to publish reliable data and information, but the author and publisher cannot assume responsibility for the validity of all materials or the consequences of their use. The authors and publishers have attempted to trace the copyright holders of all material reproduced in this publication and apologize to copyright holders if permission to publish in this form has not been obtained. If any copyright material has not been acknowledged please write and let us know so we may rectify in any future reprint.

Except as permitted under U.S. Copyright Law, no part of this book may be reprinted, reproduced, transmitted, or utilized in any form by any electronic, mechanical, or other means, now known or hereafter invented, including photocopying, microfilming, and recording, or in any information storage or retrieval system, without written permission from the publishers.

For permission to photocopy or use material electronically from this work, please access www.copyright.com (http://www.copyright.com/) or contact the Copyright Clearance Center, Inc. (CCC), 222 Rosewood Drive, Danvers, MA 01923, 978-750-8400. CCC is a not-for-profit organization that provides licenses and registration for a variety of users. For organizations that have been granted a photocopy license by the CCC, a separate system of payment has been arranged.

Trademark Notice: Product or corporate names may be trademarks or registered trademarks, and are used only for identification and explanation without intent to infringe.

Library of Congress Cataloging-in-Publication Data

Liu, Yu, 1963-
    PACS and digital medicine : essential principles and modern practice / by Yu Liu and Jihong Wang.
        p. ; cm.
    Includes bibliographical references and index.
    ISBN 978-1-4200-8365-1 (alk. paper)
    1. Medical informatics. 2. Picture archiving and communication systems in medicine. 3. Medical records--Data processing. I. Wang, Jihong, 1963- II. Title.
    [DNLM: 1. Radiology Information Systems. 2. Electronic Health Records--trends. WN 26.5 L783p 2011]

    R858.L58 2011
    610.285--dc22   2010012790

Visit the Taylor & Francis Web site at
http://www.taylorandfrancis.com

and the CRC Press Web site at
http://www.crcpress.com

To my parents, for their support and guidance; my wife, Hosanna, for her love and encouragement, and my children, Deborah and Samuel, for their understanding and patience.

(Y. L.)

To my wife, Qi, and my children, Christina and Stephen, and my parents. It is only with their support and sacrifices that this book has become a reality.

(J. W.)

Brief Contents

Contents

Preface

Healthcare providers and governments around the world are facing an unprecedented challenge to improve the quality of healthcare while simultaneously reducing costs. Medical information technology plays an important role in meeting this challenge. Today, we have more and better medical information technologies than ever before. However, medical information and records are usually maintained separately by various healthcare providers, often in paper format, preventing their optimal use for the treatment and care of patients. In an effort to both improve healthcare quality and reduce costs, healthcare professionals in all specialty areas are moving from paper-based medical records to integrated electronic health records (EHRs). In addition, clinical departments using imaging equipment are moving to picture archiving and communication systems (PACS) for the management of image data.

As PACS matures and advanced information technologies become available, governments around the world now realize that using information technologies to manage patient medical records (including images) is vital for lowering costs and improving quality and efficiency in healthcare. Over the next decade, health information technologies such as PACS and EHR will be adopted worldwide, providing safer and higher quality healthcare to more people at reduced cost. This transition will involve healthcare organizations such as hospitals, healthcare insurance companies or other payers, and government agencies.

While PACS has been available in radiology for many years, it is only recently that hospitals and clinics have begun to implement large-scale PACS and integrate them with other clinical information systems. These late PACS adopters are implementing state-of-the-art systems for the first time, whereas many early adopters are replacing their legacy systems for the second or third time. PACS is quite different compared to other medical imaging equipment. As radiology PACS matures, its application is extending well beyond radiology into other areas of the healthcare enterprise, such as cardiology, pathology and radiation oncology where other imaging technologies are also used.

This book aims to provide a comprehensive resource for healthcare information technology professionals, including radiologists, PACS administrators, radiology/clinical engineering personnel, radiology/healthcare administrators, and PACS specialists for cardiology, endoscopy, and pathology. Our primary goal has been to present the practical steps for PACS and EHR implementations and their maintenance. Therefore, in each chapter, following the introduction of basic concepts and principles relevant to the topics of the chapter, we discuss practical considerations PACS users may encounter in daily work.

This book covers various components of PACS using state-of-the-art technologies. Unlike the installation of a piece of imaging equipment such a CT or MR, each PACS implementation is unique, requiring customization and workflow considerations specific to the site. Our approach was to discuss introductory and background information about design principles for various PACS components, theoretical and practical issues to consider prior to implementation, post-installation quality control, security and privacy policies, maintenance including upgrade/integration with other information systems, and relevant governing standards. These discussions are supplemented by more than 130 illustrations throughout the book, and case studies of implementation in two institutions.

The book is organized into 12 chapters. Chapter 1 introduces the history and significant milestone contributions by pioneers in PACS and healthcare information technology. Chapters 2 and 3 cover the components of a PACS system, their functionality, and the detailed steps of routine quality control for PACS display monitors. Chapters 4 and 5 discuss IT infrastructures for PACS, essential principles about networks, and various PACS data storage technologies. Chapter 6 covers primary industry standards used in PACS and EHR, such as DICOM, HL7, and IHE, together with their applications for radiology PACS. The use of these standards for nonradiology PACS, EHR, and overall health information technology security is discussed in other corresponding chapters. Chapters 7 and 8 cover practical issues related to implementing radiology PACS and teleradiology, Chapter 9 discusses replacement of legacy systems, and Chapter 10 covers PACS in other medical specialties where imaging equipment is used. Since EHR/PHR will be adopted worldwide, and enterprise PACS is an integral part of patient medical records, Chapter 11 discusses topics unique to implementation and standards of EHR/PHR. Chapter 12 covers security, privacy, and safety regulations, as well as policies related to healthcare information systems including PACS and EHR/PHR.

Given rapid advances in the field of healthcare information technology, the hardware and software we mention in examples will eventually require updating, and standards will continue to evolve. Despite these technology advances, the principles and practical advice presented in this book should remain applicable. Thus, we hope readers will find this book useful, both as a practical guide and as a broad overview of the field, now and in the future.

Yu Liu, PhD, DABR
Jihong Wang, PhD, DABR

Acknowledgments

We want to offer our sincere thanks to staff of Aurora Healthcare for all the support and help, this book would not have been possible without their contributions. Especially we want to thank Jake Nunn, Brian Brennan, Mark Dudenhoefer, Darin Bockman, Julie Melnikoff, Susan McIntosh, Ron Hartig, Wendy Schafer, Greg Ragsdale, Mia Stone, Lisa Egle, Pam Aoys, Mike Whitman, Jim Mills, Jim Jeffery, Kari Rothkopf, and Drs. Robert Breger, David Olson, Dan Bloomgarden from Milwaukee Radiologists, Ole Eichhorn from Aperio Technologies, and Steve Flinn from Wollongon Hospital (South Eastern Sydney and Illawarra Health, NSW, Australia). We also want to thank Kathryn Carnes and Bryan Tutt from the Scientific Publications Department at MD Anderson Cancer Center for editing some chapters of the manuscript. Last but not least, we sincerely thank Luna Han for inspiring the book, Jill Jurgensen and other staff at Taylor & Francis for the production of this book, their efforts and help to make this book possible.

Yu Liu, PhD, DABR
Jihong Wang, PhD, DABR

Authors

Yu Liu, PhD, DABR is currently a senior imaging physicist certified by the American Board of Radiology in the Radiology Department, St. Luke's Medical Center of Aurora Healthcare, Milwaukee, Wisconsin. In 1993, after four years of research at the MRI Laboratory of Mayo Clinic in Rochester, Minnesota, he graduated with a PhD in biomedical engineering from Duke University in Durham, North Carolina. He joined Aurora Healthcare in 1995, and has been involved in PACS implementation and operation since 1997. His expertise and current interests are in PACS and MRI physics.

Jihong Wang, PhD, DABR is currently an associate professor at MD Anderson Cancer Center in Houston, Texas. He graduated with a PhD in medical physics from University of Colorado at Boulder in 1994. After graduation he finished his residency training in medical physics at the Mayo Clinic in Rochester, Minnesota. From 1996 until 2004, he was a faculty member at the University of Texas Southwestern Medical School in Dallas, where he implemented the first major PACS system for that campus. His current research interests are human perception and human–computer interaction in a digital imaging environment, monitor QC in PACS, and functional studies of the brain.

1 Introduction

Over the past two decades, there have been some fundamental changes in the practice of radiology and in medical imaging technology. Besides the broad adoption of advanced imaging technologies such as magnetic resonance imaging (MRI), positron emission tomography (PET), and computed tomography (CT), the complete transition of all imaging modalities to digital format has brought about some profound changes in medical imaging, which have led to paradigm changes in overall patient care. These revolutionary changes include the introduction and broad use of picture archiving and communication systems (PACS) and the adoption of electronic health record (EHR) systems which have dramatically changed the workflow in healthcare organizations and the procedures for image interpretation and communication among healthcare providers and patients. Over the past 10–15 years, the adoption of digital imaging and PACS, along with EHRs and other clinical information systems such as radiology information systems (RIS) and hospital information systems (HIS), has dramatically increased the efficiency of patient care. In this book, the essential principles of PACS, EHRs, and related technological advancements are discussed, together with practical issues concerning the implementation, operation, and maintenance of PACS and EHR systems.

1.1 HISTORY OF PACS: THE TRANSITION TO DIGITAL MEDICINE

Since the invention of computers, people in the healthcare field have been trying persistently to use them to improve the quality and efficiency of patient care. Although it has been a long journey, progress has been made over the past few decades. Stemming from the technological advancements in the computer, semiconductor, and telecommunication industries, digital imaging and information technologies are slowly but surely changing radiology and other medical specialties. Even the practice of medicine itself is undergoing gradual but fundamental changes in areas such as image acquisition, storage, display, and interpretation; data storage and display; and communication among healthcare providers and between patients and their physicians. Yesterday's dream of a completely digital radiology department or a digital hospital has become today's reality. Therefore, it is fitting to briefly review some milestones in the history of the analog-to-digital transition and to pay tribute to the pioneers and visionaries in the field of medical informatics.

1.1.1 From Paper and Film to Computers and Digital Imaging

Computer technologies have progressed tremendously over the past 60 years. In 1959, IBM introduced the first computer, IBM 1620, which was made of 50,000 electronic vacuum tubes, used paper tapes as data input devices, and used FORTRAN and assembly languages as its software. It performed 50,000 operations per second and had a mere 60 kilobyte (kB) of memory. IBM introduced its first transistor-based computer in 1964, which was an important step toward the development of modern computers. The IBM 360/92 was developed with multiuser capability and cost $5 million. In 1965, Digital Equipment Corporation (DEC) introduced the first computer to use a cathode ray tube (CRT) monitor, which was a revolutionary step toward more intuitive human–computer interactions. Later, DEC introduced popular models such as the PDP-11/70 and PDP VAX 11/780 computers. In 1968, IBM introduced the first keyboard with different template overlays for particular applications for the IBM 1092 computer. From the late 1960s through the 1970s, mainframe computers dominated the computer field. They were very expensive and therefore were only used by government agencies and large academic institutions or corporations.

Before the invention and application of computers in medicine, x-ray imaging was the sole medical imaging modality. During the 1960s and 1970s, ultrasonography and nuclear medicine imaging techniques were introduced to medicine. The rapid development of computer technology made possible the invention of advanced imaging techniques for clinical use, such as CT in the early 1970s and MRI in the early 1980s. Unlike earlier imaging modalities, CT and MRI were inherently digital, and computers played an integral part in image formation. These advancements in computer technology also contributed significantly to rapid development in digital radiology imaging from the late 1970s to the 1990s. During the 1980s and 1990s, nearly all imaging modalities in radiology were converted from analog into digital format. This transition was necessary for the success of today's PACS and for the integration of all electronic medical records, including medical images.

1.1.2 Early Attempts to Use Computers to Streamline Medical Records

The first documented attempt to integrate computers into an HIS was carried out in 1964 at a clinic in Michigan. Its goal was to reduce medical errors by reducing paperwork in the process of patient care. However, the effort failed because of the unfamiliar and complicated computer with its inefficient user interface. The other reason for the failure of this experimental system was the elimination of nurses (who were replaced by secretaries), which removed an important quality control step from the patient care process.

After ambitious attempts to develop hospital-wide information systems failed in the 1980s, efforts were made to develop smaller, specific, modular applications. These dedicated modular approaches were better than the centralized approach and had moderate successes in the market-place. In anticipation of the need for information systems specifically designed for radiology, the nonprofit Radiology Information System Consortium (RISC) was formed to develop the first-generation RIS in the 1980s. The effort was spearheaded by multiple institutions including Massachusetts General Hospital, the Hospital of the University of Pennsylvania, Johns Hopkins University, and Washington University in St. Louis.

1.1.3 Early Research, Development, and Pioneers in PACS

Although many people recognized the great potential of the application of computer technology to the practice of medicine, the concept of digitization of medical imaging and digital communication was not introduced and nor did it become popular until the early 1980s. One of the earliest PACS projects using proprietary technology was the U.S. government-sponsored teleradiology project in the late 1970s and early 1980s, which expanded later into the Digital Imaging Network and Picture Archiving and Communication System project and included the participation of commercial vendors Philips Medical Systems and AT&T. Another early PACS research project was funded by the National Institutes of Health and led by H. K. Huang in 1985 at the University of California, Los Angeles. In the early 1990s, the U.S. Army Medical Diagnostic Imaging Support project resulted in several major turnkey PACS installations in Veterans Affairs (VA) medical centers around the country. One of the first installations was done at the Baltimore VA Medical Center, and it has generated many milestone contributions to the clinical application of PACS. This institution's experiences in implementing and using the first clinical PACS have provided much practical guidance for other users.

In most of these early PACS, image data were stored in a centralized system, and there was limited or no RIS integration. Image data were distributed to individual workstations when needed, and various schemes of pre-fetching prior examination data were implemented (often manually for lack of RIS integration). The obvious drawback of this centralized design was its demand on network bandwidth, which was not very high (compared to today's networks).

Some major milestones in the evolution of PACS are:

- 1968: The first attempt to apply teleradiology at the Boston airport walk-in clinic via video link to Massachusetts General Hospital
- 1982: The first international conference on PACS
- 1983: The first EuroPACS conference
- 1983: Start of the U.S. Army teleradiology/PACS research project
- 1983: Introduction of the first commercial computed radiography reader by Fuji Medical Systems
- 1983–1988: Development and establishment of the American College of Radiology (ACR)/ National Electrical Manufacturers Association (NEMA) standard for medical imaging
- 1985: Installation of U.S. Army PACS in the Seattle and Baltimore VA facilities
- 1986–1989: Implementation of PACS in Hokkaido University, Japan, and Utrecht University Hospital, the Netherlands
- 1987: Introduction of the Health Level 7 (HL7) standard
- 1989: Release of the ACR/NEMA 2.0 standard
- 1993: Release of the Digital Imaging and Communications in Medicine (DICOM) 3.0 standard by ACR/NEMA

- 1994: Installation of the first filmless radiology department at the Baltimore VA Medical Center

- 1998: Introduction of the Integrating Healthcare Enterprise (IHE) initiative

There were many pioneers during the development of PACS and digital medical imaging, and it is impossible to compile a complete list of all who have contributed to the successful use of PACS. Some of the most acknowledged individuals who contributed to the early development of PACS are Steve Horii, M. Paul Capp, Joseph Gitlin, Sam Dwyer III, Ed Staab, Heinz Lemke, James Lehr, Gwil Lodwick, Don Lindberg, Bernie Huang, Gil Jost, Hal Kundel, Brent Baxter, Roger Bauman, Don Simborg, Paul Chang, Ron Arenson, and Elliot Siegel, but there are many others whose work helped lay the groundwork for today's PACS.

1.1.4 Formation of Professional Organizations in Medical Informatics

The first professional society for computer users in medicine was the RISC, which was established in the 1980s by 12 academic and private institutional members. The RISC's goal was to develop an RIS with a commercial vendor and make it available to member institutions. In 1989, the Society for Computer Applications in Radiology (SCAR) was created as a subgroup within the RISC and individual membership was introduced. In 1996, RISC and SCAR were merged under the SCAR name and included individual, corporate, and institutional members. Since then, SCAR has been a leading organization for professionals and vendors in the field of clinical applications of medical informatics. In 2006, SCAR changed its name to the Society for Imaging Informatics in Medicine (SIIM) to better reflect the broadened roles of imaging and informatics across healthcare enterprises. Together with the Radiology Society of North America (RSNA), ACR, NEMA, and the Society of Photographic Instrumentation Engineers, SCAR/SIIM has become an instrumental force in the broad adoption of PACS for patient care. On a scale beyond radiology and medical imaging are other initiatives and organizations, such as IHE and the Healthcare Information and Management Systems Society (HIMSS). IHE is an initiative by healthcare professionals and industry partners to improve the way computer systems in healthcare organizations share health information, including medical images and EHRs. IHE primarily promotes and coordinates the use of the HL7 and DICOM standards to address specific clinical system integration needs. Founded in 1961, HIMSS is a professional organization focusing on providing leadership for the optimal use of health information technology and management systems to improve healthcare. In addition to these organizations, other professional societies have contributed to the advancement of PACS and EHRs and to their applications in medicine.

1.1.5 Efforts in PACS around the World

During the early days of PACS, parallel efforts throughout the world were made to develop or implement this new technology, particularly in Europe and East Asia, and most of them were funded either by governments or industry consortiums with government support. Because of differences in healthcare systems and cultures, early models of PACS from Europe and Asia were different from those developed in the United States; however, they all shared the same problems and challenges, such as a lack of standards, inflexibility, and barriers to broad commercial adoption.

Since the late 1990s, when DICOM 3.0 became universally accepted as the international standard, PACS has experienced rapid growth in Europe and Asia as well as in the United States, and currently there are many international PACS vendors who design their products according to the healthcare environment and needs of individual countries or regions.

Today, some Asian governments provide full support for PACS and other EHR initiatives. They support healthcare institutions and the healthcare information technology industry by increasing the reimbursement rate for imaging studies performed using filmless modalities. In Europe, efforts are being made to develop EHR systems that enable patient medical records to fit a particular region's or country's healthcare system while being compatible with the systems of other countries in the region. In the United States, there are similar initiatives by the government to encourage the use of EHRs, with the goal of improving healthcare efficiency and patient safety.

1.2 TECHNOLOGY ADVANCEMENTS ENABLING TODAY'S PACS

Since the 1990s, advancements in information technology (IT) and other industries besides medicine have been instrumental in bringing about fundamental changes in medical imaging and medicine in general, although applications in medicine/healthcare is just a small part of the overall IT market. Today's PACS and EHR applications owe much of their success to the introduction, fast adoption, and expansion of the Internet and the World Wide Web. The following advancements that occurred in the 1980s and 1990s were the building blocks for today's PACS:

- Complete digitization of all radiology imaging modalities

- Dramatic increase in network speed and economical transmission of data since the mid-1990s

- Affordable large-capacity data storage

- Widely available and affordable high-powered computers for workstations and servers

- High-quality, low-cost medical-grade computer display monitors

- Widely available IT expertise and generally computer-literate users

It is also important to note that technological advancements have enabled faster and more intelligent image transfer, making it more practical for clinical use as the size of patient electronic data—including images—grows. Advancements in high-resolution and high-luminance image display devices (including the development of high-resolution graphics card and the capability to support multiple computer monitors) have also been critical in making PACS acceptable for clinical use.

1.2.1 Computer Network Development History

The computer network and Internet technologies used by PACS and EHR systems today are the result of some visionary thinking in the 1960s by people who saw the great value of enabling computers to exchange information. Interestingly, wars or the threat of wars can do wonders for the advancement of technology, and this was exactly the case with the invention of the Internet. Within the geopolitical environment of the Cold War and the space race with the Soviet Union, the original concept was to build a communication system for the United States that could survive a large-scale nuclear attack. The system was to be a decentralized military research network that would allow the military to launch a computer-controlled nuclear counterattack even if some installations were destroyed. In pursuit of this goal, the U.S. Department of Defense funded the initial development of the Internet. J. C. R. Licklider, who then headed the Advanced Research Projects Agency (ARPA) under the Department of Defense, saw the need to broaden the use of computer technology in the military and the need to make computers more interactive and robust. He moved ARPA's contracts to universities and established the groundwork that later became the ARPANET. It was then that the concept of a packet-switched network was proposed, which was the foundation for today's Internet. By 1972, 23 hosts were connected by the ARPANET. In 1973, the development of Transmission Control Protocol/Internet Protocol (TCP/IP) began and was adopted by the Department of Defense as the network protocol for ARPANET, which by then had more than 111 connected hosts.

In the mid- to late 1980s, the U.S. National Science Foundation deployed its T1 line network, and about 1900 hosts were connected. In a few years, the number of users and traffic increased so quickly that the network had to be upgraded. By 1991, the total number of hosts increased to 620,000. In 1992, the Internet Society was chartered, the World Wide Web was released, and the number of hosts reached over 1 million. Since then, Internet use has come out of academic institutions to the general public. Popular use of Gopher, a text-based search tool, followed by the introduction of Mosaic and Netscape, early Web browsers with graphic user interfaces, marked the beginning of commercial Internet use by the general public.

Although Internet use among academic institutions has been available since the 1980s, the communication bandwidth was relatively low until the early-to mid-1990s (primarily 10Base-T systems with a bandwidth of 10 megabits per second were used prior to that time). Internet use by the general public grew exponentially since the mid-1990s, creating a tremendous demand for data communication bandwidth. To meet the need for faster and cheaper data exchange than what was possible with telephone-based Internet systems at the time, communication carriers started to build worldwide fibre-optic networks. Since 1996, most Internet traffic has been managed by commercial communication carriers, and the number of hosts connected is literally countless because it increases every second. It is only in this present climate that PACS and EHR have become practical as just one among many special applications of computer and network technologies in medicine.

1.2.2 Data Storage Technology History

In the 1940s, during the infancy of computer technology, electronic digital data were stored and retrieved on punched paper cards and tapes. In the late 1940s, the first magnetic memory device was developed, which used arrays of magnetic cores with each core storing one bit of data. These were probably the earliest methods to use electronics to store and retrieve digital data. In 1956, IBM introduced the first hard disk drive, marking the beginning of the modern digital data storage era. This hard disk could store up to 4.4 MB of data and could be leased for $3200 per month.

In 1967, IBM invented the monolithic semiconductor memory chip, which allowed fast access to stored data and resulted in the separation of computer memory and data storage. In 1969, the first floppy disk was introduced, which was a read-only 8-in disk that could store 80 kB of data; by 1973, a similar-sized disk could store 256 kB of data with rewrite capability. In 1977, several companies in Japan demonstrated the first optical digital audio disk, which developed into compact discs, a data storage medium that would eventually replace magnetic tapes and floppy disks. Also in the late 1970s, the concept of redundant arrays of independent disks (RAID) was developed.

IBM introduced the personal computer in 1981. As the use of personal computers increased, the dramatic demands on memory and data storage further accelerated advancements in technology and in the standardization of memory and storage interfaces, and companies raced to meet these demands by providing high storage capacity at low cost. In the late 1990s, DVD-ROMs became available to computer users. Today, data storage technologies are much more reliable and affordable than in the past, and a hard disk with the physical size of a small book can easily store several hundred gigabytes of data at a cost of only about $100. All these data storage technologies laid the foundation for computer applications in medicine.

1.2.3 Establishment of Standards for Image Exchange in Medicine

During the implementation of early PACS projects, the lack of communication standards made it nearly impossible or very difficult to transfer digital images (such as MRI images) between imaging systems manufactured by two different vendors. This presented problems for hospitals that installed imaging equipment from various vendors. Even on imaging equipment made by the same vendor, it was very difficult to transmit and display CT images on an MRI workstation. To make systems and devices communicate with each other, researchers spent incredible amounts of time and effort building interfaces between the many devices that constituted the early PACS. Responding to demands from the user community to improve and promote image exchange among different vendors, ACR, RSNA, and NEMA initiated an effort to develop a standardized protocol for data exchange, and these organizations played a pivotal role in the establishment of the DICOM standard in 1983. Under the general direction of the DICOM umbrella, committees were formed to work on standards that were later carefully developed by the community of medical imaging professionals and vendors into a set of standards that enables systems to communicate with each other. The visionary leadership at RSNA in the standardization of digital communication of medical images has proved to be a great success. It is not an exaggeration to say that today's PACS would not exist without DICOM. Since the adoption of DICOM 3.0 in the late 1990s, the communication, interfacing, and integration of devices from various vendors has become easy, and most of the integration can be done via simple software configurations. In addition to DICOM, other important standards used for healthcare information systems include HL7 and IHE, and their history, current status, and other details are discussed in Chapter 6.

1.2.4 Imaging Modality Digitization

Another important step that enabled today's PACS and EHR systems is that by the end of the 1990s, digitized versions of all image acquisition modalities had become commercially available. This complete conversion into digital enabled the acquisition, storage, and transmission of radiological image data in digital format. Even though images on films can be converted into digital data via a film digitizer, this conversion requires additional, labor-intensive steps and can result in loss of image fidelity in spatial and contrast resolution.

1.2.5 Availability of IT Expertise

Just like any equipment used in healthcare institutions and enterprises, PACS, EHR, and other health information systems need trained professionals to operate and maintain them. Thanks to the widespread use of telecommunication and Internet-based software applications, technical expertise in IT has become readily available for PACS and EHR development and implementation. Furthermore, healthcare professionals without IT background and technical training are also becoming more familiar with Web-based software and the general use of computers such as PACS and EHR workstations. These factors all contribute to the wide acceptance of health information systems and their clinical applications in institutions and enterprises.

1.3 DEFINING PACS

The strict definition of a PACS varies, but it is generally agreed that such systems must include image display, data archiving, and data management components. Some PACS also include interfaces (which can be operated purely by software) with image acquisition modalities such as computed radiography/digital radiography, CT, and MRI, and with hard copy output devices such as film or paper printers. Early PACS required special interfaces with image acquisition modalities. These interfaces were developed mostly by the implementing institution, and the process was tedious and time-consuming. Some vendors developed interfacing devices to accommodate these special needs. However, after the creation and gradual adoption of the DICOM standard, interfacing among DICOM-compliant computer equipment and imaging devices became easy, requiring only software configurations. To general users, the most obvious component of a PACS is the image display and visualization workstation, which usually consists of a desktop computer with multiple medical-grade display monitors. This is what most PACS users see and interact with. In fact, a PACS is much more complex than the workstations placed around a healthcare institution. In the following chapters, all major components of a PACS, along with their detailed functionalities, are introduced, and issues related to the implementation and clinical operation of PACS will be discussed.

1.3.1 From Mini-PACS to Enterprise PACS

PACS come in different forms and sizes. Some mini-PACS may consist of only a few workstations and a simple data archiving computer. Mini-PACS are typically found in small independent clinics or outpatient facilities; however, some special-purpose mini-PACS are designed to handle specific tasks in a large institution; for example, an ultrasonography, or PACS in cardiology or other medical specialty departments. The advantages of mini-PACS include the relatively low cost of implementation, limited resource requirements, and high efficiency within the designed realm of operation. In the late 1980s and 1990s, many mini-PACS were installed in institutions across the country. A special-purpose mini-PACS often used proprietary technologies that allowed efficient communication among workstations and archives; these workstations and the archives were usually made by the same vendor and the mini-PACS was used with a specific imaging modality. However, the obvious drawback was the lack of interfacing and integration capabilities with other HIS. On the other hand, large-scale, enterprise-wide PACS may have hundreds of workstations and multiple data archiving centers, which may reside at different geographic locations hundreds of miles apart. Depending on the size of the PACS and the sophistication of the data archive, the requirements for the underlying IT infrastructure may be quite different for various PACS.

1.3.2 Advantages and Benefits of PACS

The foremost benefit of PACS is the increased efficiency of image acquisition, viewing, and interpretation, which can lead to improved patient care. Compared to film-based systems, PACS enable more efficient image data management and archiving, the prevention of film loss, and the avoidance of the costs and environmental hazards associated with film and processing chemicals. PACS also significantly reduce the cost and time of storing and retrieving prior imaging studies. The disadvantages of PACS include high initial capital expenses for equipment, infrastructure, and training, as well as ongoing costs for maintenance and for additional IT staff.

1.3.3 Workflow Changes after Implementation of PACS

Before the broad adoption of PACS, the clinical operation and workflow of film-based radiology departments had remained relatively unchanged for years. In that environment, work orders and information flow were done mostly through paper trails, and medical images were captured on films, usually with one copy per image. Image interpretation and later follow-up consultations with ordering physicians relied on the single set of films, and primary care physicians and patients had only limited access to the actual images unless a set of copies was made, which was not only costly but also inefficient. If the original films were lost, an additional examination had to be performed, adding cost and time to patient care.

1.4 CURRENT STATUS

Today, IT costs consume a substantial portion of capital expenditures by healthcare institutions and enterprises around the world. Although it does not generate any direct revenues, the implementation of PACS, EHR, and other clinical information systems is increasingly recognized as necessary for continued efficiency improvement in clinical operation. Government agencies around the world are also increasing their spending toward the creation of unified patient EHR systems with portability and improved access for the efficient management of care. All these factors have contributed to the rapid development and implementation of PACS and EHR in healthcare systems worldwide.

Currently, there are over a hundred commercial PACS products on the market worldwide. Over the past decade, after consolidations and mergers, well-known imaging equipment manufacturers have emerged as the major players in the PACS market. However, there are still many small- to mid-sized companies specializing in the development of innovative PACS and other EHR products, even though their installation bases are not as large as those of the major medical imaging equipment companies. As is often the case, these smaller companies tend to possess the most innovative ideas and advanced technologies. Since medical informatics is a fast-advancing field, it would be no surprise if one of these companies were to develop a technology that becomes dominant in the near future.

Several business models can be used for PACS implementation and data archiving. In addition to the outright purchase of a PACS, one business model is to use an application service provider. Using this model, a healthcare institution pays a company, based on the institution's examination volume, to provide the hardware, software, and expertise required to manage and maintain the data; the application service provider stores these data off-site and duplicates them for backup purposes. This model saves the healthcare institution the initial capital expenditures for hardware and software and the ongoing maintenance costs, while the application service provider can use its economy of scale to provide relatively affordable services to many smaller hospitals and clinics that lack IT expertise and capital funds.

1.4.1 Prerequisites for PACS Implementation

Although radiology PACS is a mature field and almost all healthcare organizations can find commercial products on the market to meet their needs, in order to successfully implement a system and fully realize its benefits and potentials, several prerequisites must be in place. These prerequisites are critical to the success (or failure) of a PACS implementation.

1.4.1.1 Digitization of All Image Acquisition Modalities

Before implementing a PACS, a healthcare institution needs to confirm that all its image acquisition devices have the capability to transmit image data in standard DICOM format. Although all newly manufactured and installed radiology imaging devices are digital and DICOM-compliant to allow integration with a PACS, it is very likely that there are legacy imaging devices that are not DICOM compliant or not even in digital format. Such devices must be converted to digital or DICOM format, or a means of digitizing their images must be in place before PACS implementation. The selected PACS vendor can often assist this process by providing technical solutions such as interface boxes or film digitizers.

1.4.1.2 Adequate Computer Network Infrastructure

Before the installation of a PACS, the healthcare institution or enterprise must have an adequate computer network infrastructure that is capable of handling the data traffic of the PACS. Since PACS and imaging modality equipment generate a large quantity of data traffic on the institutional network, it is critical to make sure that the network has sufficient capacity for these data to be transmitted at a reasonable speed. Sometimes the PACS vendor can help evaluate the existing network and even install the required network before installing the PACS.

1.4.1.3 Clinical Information Systems

The workflow efficiency improvement functions of a PACS, such as work list management, are very useful in reducing data entry errors concerning patients' demographics. Because such information is provided by an RIS or HIS, a healthcare institution or enterprise should install or upgrade these systems before implementing a PACS, otherwise the full benefits of PACS implementation cannot be realized.

1.4.1.4 Adequate Space and Facility Planning

The implementation of PACS brings new computer equipment into the healthcare institution or enterprise. These computer systems may require additional power and cooling. For instance, the data storage and management system most likely requires a climate-controlled area and additional power supply. Furthermore, depending on the number of PACS workstations in a given area, such as a radiology reading room, it may be necessary to add more computer network and electrical outlets. The area may even require additional cooling. For instance, during some of the early PACS implementations, the amount of heat generated by the monitors of the PACS workstations was so excessive that additional air-conditioning had to be installed after the PACS workstations were installed, causing delays and unexpected costs.

1.4.1.5 Computer and Software Training for Staff

For facilities whose staff members do not have much experience with computers and their applications, adequate staff training before implementation of the PACS or EHR system is necessary for a smooth transition from film to filmless PACS operation or from using paper files to using EHRs in routine clinical patient care. With the extensive use of the Internet both at home and at work, most staff members are now comfortable working with computers and their software, which was not the case for some of the early adopters of PACS in the early 1990s. At that time, a lot of frustration and work slowdowns occurred because radiologists and staff members had to get used to the unfamiliar environment of workstations and associated software tools which although essential to view all PACS images were not user-friendly.

1.4.2 PACS Implementation and Clinical Operation

Today, the decision to implement PACS at a healthcare institution is much easier to make than when it was still in its infancy. In the late 1980s and early 1990s, commercial PACS products were not available and there was no DICOM standard; therefore, pioneers in PACS had to develop home-grown systems. Most of the early sites of PACS installation were academic or university hospitals with strong IT and other research expertise, which enabled them to build some commercially unavailable but necessary software and hardware on their own. The implementation team mostly consisted of computer scientists and medical physicists. Because each system was custom-built for a particular institution, and the fact that each healthcare institution had different imaging equipment and workflow, most of these early, home-grown PACS could not be duplicated at other institutions and were not very scalable either. These systems also required a significant amount of maintenance in order to make them function properly. This was not an issue for a PACS project supported by research funds, but it was impractical for routine clinical use by other nonacademic healthcare institutions or enterprises.

Over the past decade, PACS technology has matured, and almost all imaging acquisition devices are DICOM-compliant now. Since many commercial PACS products are available, most institutions or enterprises choose turnkey products for their PACS implementation. During this type of PACS implementation, the institution selects a PACS product and the vendor delivers and installs it at the institution. Although this model is much more economical and easier to implement than developing a system in-house, it still requires careful planning, configuration, and integration with other existing information systems, such as HIS and RIS. Chapter 8 discusses the processes and considerations involved in the planning, implementation, installation, and maintenance of a PACS.

1.4.3 Toward Fully Integrated Healthcare Enterprises

Since the late 1990s, the RSNA leadership and others in the medical informatics field have realized the importance of a fully integrated health record system throughout the healthcare system in the United States, or a region around the world. To achieve this goal, professional organizations including the HIMSS and RSNA are leading the IHE initiative to improve health information exchange, thus realizing the full potential of the integrated health record. More details and specifics about IHE can be found in Chapter 6.

During the past few years, the U.S. government has introduced programs to encourage the use of EHR products by healthcare institutions and enterprises. These programs aim to streamline healthcare management with EHR to reduce medical errors, improve the quality of care, and reduce the costs of the Medicare program. Various programs provide financial incentives to physician practices that use EHR systems certified by the Certification Commission for Healthcare

Information Technology, and the incentives are based on EHR functionalities for clinical decision support, physician order entry, the ability to capture healthcare quality data, and the ability to support the exchange of clinical data with other organizations. However, there are concerns about privacy and the misuse of sensitive health information with a nationwide EHR system, and people usually trust locally managed systems more than a system operated by a central government. Nevertheless, the broad adoption of EHR has the potential not only to improve the quality of care but also to change the way medicine is practiced and delivered in the United States and around the world.

BIBLIOGRAPHY

Arenson, R. 2008. Building bridges: Centralized versus distributed health information systems in 2008. Opening General Session. *The Society for Imaging Informatics in Medicine 2008 Annual Meeting.*

Augarten, S. 1984. *Bit by Bit: An Illustrated History of Computers.* New York: Ticknor and Fields.

Avrin, D. E. 2009. Storage model. The Samuel J. Dwyer Memorial Lecture. *The Society for Imaging Informatics in Medicine 2009 Annual Meeting.* www.siimweb.org/WorkArea/showcontent.aspx?id=6374. Accessed January 19, 2010.

Dreyer, K. J., D. S. Hirschorn, and J. H. Thrall. 2006. *PACS: A Guide to the Digital Revolution.* Springer, New York, NY.

Duerinckx, A. J. ed. 1982. *Proceedings of Picture Archiving and Communication Systems for Medical Applications. First International Conference and Workshop, Proc. SPIE.* Vol. 318, Parts I and II, Bellingham, WA.

Goldberg, A. 1988. *A History of Personal Workstations.* New York: ACM Press.

Hafner, K. and M. Lyon. 1996. *Where Wizards Stay Up Late: The Origins of the Internet.* New York: Simon & Schuster.

Horri, S. 2009. The Samuel J. Dwyer Memorial Lecture. *The Society for Imaging Informatics in Medicine 2009 Annual Meeting.* www.siimweb.org/WorkArea/showcontent.aspx?id=6176. Accessed January 19, 2010.

Huang, H. K. 1987. *Elements of Digital Radiology: A Professional Handbook and Guide.* Prentice-Hall, Inc, Needham, MA.

Huang, H. K. 2003. Editorial: Some historical remarks on picture and communication systems. *Comp. Med. Imaging Graph.* 27 (2–3): 93–99.

Huang, H. K. 2004. *PACS and Imaging Informatics: Basic Principles and Applications.* Wiley-Liss, Hoboken, NJ.

Lemke, H. U. 1979. *A Network of Medical Work Stations for Integrated Word and Picture Communication in Clinical Medicine.* Technical Report, Technical University, Berlin.

Nie, D., T. R. McClintock, and J. N. Gitlin. 2009. A history of digital imaging in medicine and the implementation of computers in radiology. The Samuel J. Dwyer Memorial Lecture. *The Society for Imaging Informatics in Medicine 2009 Annual Meeting.* Charlotte, NC. www.siimweb.org/WorkArea/showcontent.aspx?id=6178. Accessed January 19, 2010.

PACS History Web site. www.pacshistory.org/. Accessed January 19, 2010.

The Society of Imaging Informatics in Medicine (SIIM). History. www.scarnet.org/index.cfm?id=33. Accessed January 19, 2010.

2 PACS Servers and Workstations

PACS are very complicated systems. However, whether an enterprise-wide PACS for large healthcare institutions or a mini-PACS for a small clinic or independent imaging center, all of them have interfaces with multiple imaging modalities, a server, workstations, data storage, and networks to connect these together. Various imaging modalities generate medical images that are transmitted through a network to a server where they are managed and stored. Images are then sent to or accessed by workstations through a network, and the displayed images are interpreted and diagnostic reports are generated by physicians. These images and reports can be accessed and retrieved later by PACS, EHRs, or other information systems through a network or a digital storage medium. In this chapter, we discuss PACS servers and workstation architecture.

2.1 PACS SERVER

A PACS server is the brain of the entire PACS system. Typical server tasks include receiving DICOM images from various medical imaging modalities: CT, MRI, Digital Radiography (DR), and ultrasound; database management; interfacing with RIS, HIS, and EHR systems; sending DICOM images to PACS workstations upon request; controlling the DICOM image storage archives; controlling the interfaces between the client PACS workstation and other devices; image distribution; and so on. A PACS server may have only a single computer for all server tasks: such small systems are often used in stand-alone imaging centers and clinics where the volume of examinations is relatively low compared to larger healthcare institutions. However, an enterprise PACS system typically consists of a cluster of computers each of which is assigned a specific task such that the total workload is distributed and the system is capable of handling high volumes. Processing power and resource requirements are also high in enterprise PACS. Shown in Figure 2.1 is a single-server basic PACS suitable for independent imaging centers or clinics.

2.1.1 Multiservers for Load-Balancing and Fail-Over

Enterprise PACS dealing with large examination volumes often have multiple servers, each handling specific tasks, such as database management, interfacing with PACS workstations, and other such applications. When a task has its own server, with dedicated Central Processing Unit (CPU), memory, and other processing resources, the quality of performance is vastly superior to that obtained when processing resources are shared with other tasks. To improve performance or reliability, multiple servers may also be assigned to the same task for load-balance and/or fail-over.

2.1.2 Load-Balancing

Load-balancing was originally applied to ensure high availability and scalability for mission-critical application services. Server load-balancing is an approach that divides the amount of work, balancing the load among two or more physical servers and making this server cluster perform like one large server to end users. There are many benefits, but the primary benefit remains scalability and high availability. Scalability means that the resource is capable of adapting dynamically and easily to increased workload, without impacting the existing performance. This is achieved by application service virtualization, which enables the separation of application services from an actual physical server. Adding more servers can scale up application services while remaining seamless to end users. High availability is the capability to continually provide application services and remain accessible and available even during one or more server failure, and so long as not all servers fail at the same time the entire application service remains available to end users.

When service requests are received by a load-balanced PACS server cluster, the requests are routed to the server with the least activity, spreading application service requests to all participating servers. Load-balancing can be achieved using either hardware or software, and it can be obtained as bundled technology from the server itself, or can be bought from a third-party vendor, not necessarily as a PACS vendor-specific application. With load-balancing, either more work can be done in the same amount of time, or all work can be done faster. When multiple servers are assigned for load-balancing, the same types of tasks are spread to all participating servers, improving the performance of the system. Typically, load-balancing is more complicated in a database server cluster, because all activities need to be synchronized among all the participating

Figure 2.1 A single-server PACS for imaging centers or clinics with a directly attached tape library for image storage.

servers. For instance, a PACS database has all relevant information about every patient: demographics, medical history, prior reports, and the location where each imaging study is physically stored. When the PACS is operational, this database is being constantly updated, increasing load-balancing complexity.

2.1.3 Fail-Over

Fail-over is defined as the continuation of application services when one or more components fail. Server cluster fail-over is achieved by hardware redundancy, where the functions of the primary server are automatically assumed by a secondary server when the primary server is not available because of either failure or scheduled maintenance. If hardware or software failure occurs on one server, the function continues and the application services are not interrupted for end users. Server fail over improves fault-tolerance for mission critical PACS, so that when tasks are automatically offloaded from the primary server to a secondary standby server, the procedure is seamless to users at the PACS workstation end. When multiple servers are used for fail over service, usually a secondary "hot spare" server is in standby mode. If the primary server fails, its role is taken by the standby secondary server, which takes over all responsibilities, realizing server fail-over.

2.1.4 Server Load-Balancing and Fail Over Combined

Load-balancing can also be used to implement fail over. When multiple servers are used for combined load-balancing and fail-over services, all servers assigned to the task function, and the workload is spread out to all assigned servers. During normal operations, each server of the server cluster is monitored continuously. When one server is nonresponsive for any reason, the load-balancer is notified and no more application service requests are sent to the nonresponsive server. All application service requests are shifted to the remaining server(s). The performance of the PACS, speed, for instance, may reduce slightly; however, the PACS will continue to function, though impaired, achieving some degree of fault-tolerance. This is a much less expensive and a more flexible approach than a simple fail over, which requires a single "hot spare" standby server paired with each "live" server to take over application services in case of a failure of the "live" server.

Figure 2.2 shows the configuration of a radiology PACS for a large medical center. It includes E1 and E2 load-balancer servers that are clustered, two R servers, and three C servers—the latter five servers are clustered as well. The E servers basically direct traffic to the other five server clusters. They handle frontline requests and distribute processing, communicate with PACS workstations to handle requests from workstations for imaging studies, receiving images from imaging modality scanners or sending imaging studies to workstations based on set rules. The R servers are the "brains" of the operation, and the C servers are the workers—moving images and database requests, and so on.

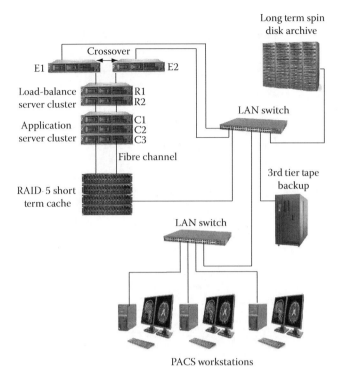

Long term spin
disk archive

Crossover

E1   E2

Load-balance   R1
server cluster   R2

LAN switch

Application   C1
server cluster   C2
            C3

Fibre channel

3rd tier tape
backup

RAID-5 short
term cache

LAN switch

PACS workstations

Figure 2.2 The configuration of the radiology PACS in St. Luke's Medical Center, Aurora Healthcare.

R and C cluster servers run several different services including DICOM, logging, and application-related services. R1 and R2 are central servers that run all the central server services and direct data traffic among all seven server nodes. These two central servers are capable of both load-balance and fail-over, which means they can both handle central services, whether evenly distributed between them or all on one or the other, and in case one fails, the services will run on the other. There is also a monitoring service that tracks the operations of the rest, and it responds to any service going down by attempting to restart it. If a server fails, the system is designed to move processes to another server. An outside server also provides overall monitoring and "call home" capability for the PACS system. It allows vendor service engineers to respond to any immediate problems as well as monitoring logs, disk thresholds, and so on. Shown in Figure 2.3 is the physical server cluster.

C1, C2, and C3 are three archive servers that run load-balanced applications to move data and that are also capable of fail-over. Since there are multiple CPUs, network connections, and Fibre Channels attached from the short-term image storage cache to the C1, C2, and C3 cluster servers, image data can be transmitted and received simultaneously. This server cluster also manages the database, replicating each transaction, so that there is no loss of data or interruption of service in case of one of the three servers shuts down. In such a case, the load is distributed between the remaining two servers. The Integrated Lights Out installed in every server can be accessed no matter what state the servers are in, so that servers can be rebooted through the Internet via a Web application without the administrator having to be physically present at the server cluster to perform the reboot procedure. When the PACS system is interfaced with other information systems such as RIS, HIS, or EHR systems, the database is responsible for the exchange of patient healthcare information among all the systems.

When DICOM images from imaging modalities are received by the seven-server cluster, they are stored on the local short-term cache. This short-term cache, as shown in Figure 2.4, is a spin-disk–based RAID-5 (Redundant Array of Inexpensive Disks) directly attached to the server cluster by a Fibre Channel that has a speed of 10 gigabit/second (Gbps) for fast data access, because it is often the most recent information that is accessed to understand a patient's medical status. This short-term cache can hold 21 terabytes (TB) of raw data, which is approximately two years

Figure 2.3 The actual radiology PACS server clusters (Emageon/Amicas, Birmingham, AL) installed in St. Luke's Medical Center, Aurora Healthcare.

worth of the most recent digital radiology examinations. When the storage size on the short-term storage cache reaches a preset threshold, the data are removed from the short-term cache, in the order that data most rarely viewed or not viewed are removed first. Before a study is removed, the system is designed to verify that the study in its entirety exists on the Long Term Archive (LTA), ensuring long-term storage. The purpose of this server cluster and the short-term image cache is fast, sustained data throughput for image retrieval and forwarding to requesting PACS workstations.

Figure 2.4 A spin-disk based RAID-5 short-term PACS archive attached to the server clusters shown in Figure 2.3.

Figure 2.5 A spin-disk-based RAID long-term enterprise PACS archive located at the data center of Aurora Healthcare. PACS data stored in the short-term archive are also replicated in this long-term archive.

When new studies are stored on the short-term storage cache, the server also sends a duplicate copy off site to the spin-disk–based enterprise PACS LTA located in the enterprise data center. The LTA as shown in Figure 2.5 contains images from various PACS systems of all institutions in a large healthcare enterprise. Compared to the short-term cache, the LTA can hold much more data.

Data from the LTA is then replicated to an enterprise tape backup library, shown in Figure 2.6, which is also in the datacenter for disaster recovery purposes. Detailed discussions of network and data storage technologies are given in Chapter 4 and Chapter 5, respectively.

Most PACS servers run commercially available vendor-developed PACS server software. The hardware components of a server are very similar to a workstation, except that they are more reliable and perform better. Currently IBM (Armonk, NY), Hewlett-Packard (Palo Alto, CA), Dell (Austin, TX), and Sun (Palo Alto, CA) are among the major providers of quality server hardware ranging from entry to high-end level for enterprise mission-critical applications. These servers most likely use Microsoft (Redmond, WA) Windows or UNIX operating systems. These server products are constantly changing when new technologies such as faster CPU and more memory become available.

2.2 TYPES OF IMAGE DISPLAY WORKSTATIONS IN PACS

There are several types of image display and visualization workstations in a full-scale PACS, which are based on the application and functional requirements in clinical settings, and each, as an integral part of PACS, serves a specific function.

2.2.1 Primary Diagnostic Workstation

A primary diagnostic workstation is used by radiologists to make diagnoses. These workstations are usually equipped with two or more high-quality, medical-grade monitors and CPUs, a complete set of image processing and manipulation tools. They are equipped with fast CPUs, large amounts of memory, high-quality graphics card with large amounts of dedicated video memory, and fast network access. They provide high-quality medical image display, fast image load and display, and a wide range of display controls. They are very expensive and therefore there are only a few in a given department. Figure 2.7 shows radiology MRI images displayed on a diagnostic PACS workstation. The primary diagnostic workstations have the most complete set of networking, query, and retrieval privileges in a PACS database. They are often equipped with software that can interface with other medical information systems such as the RIS and speech recognition (SR) dictation systems, in order for radiologists to look up and input their diagnostic results in DICOM structured reports within the same PACS application window. In this case, the primary diagnostic workstation is the centerpiece of a radiologist's workflow.

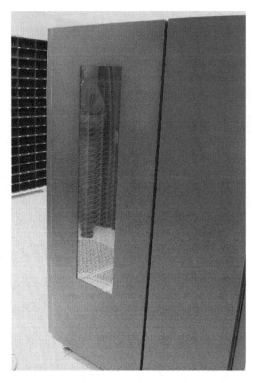

Figure 2.6 An enterprise tape backup system located at the data center of Aurora Healthcare. It provides tape backup for all enterprise data including the long-term enterprise PACS storage shown in Figure 5.5.

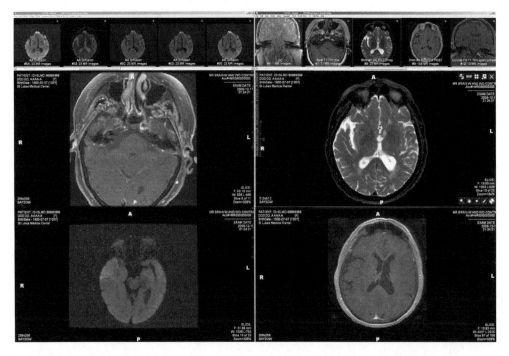

Figure 2.7 Radiology MRI images displayed on a diagnostic PACS workstation (Emageon/ Amicas, Birmingham, AL).

2.2.2 Clinical Review Workstation

Clinical review workstations are usually placed in strategic locations throughout the hospital such as in the Intensive Care Unit (ICU), Emergency Room (ER), and other clinical departments, providing easy access to patient images and other medical record via PACS. Also, these workstations are often connected to image acquisition modality devices such as CT, MRI, Computer Radiography (CR), Digital Radiography (DR), and ultrasound equipment, giving the technologist a view of patient positioning in the modality scanner, and a preliminary assessment of image quality before sending them over to the PACS. These workstations are large in number in a given healthcare institution and are often equipped with one or two consumer-grade monitors or low-cost medical-grade monitors.

2.2.3 Advanced Image Analysis Workstation

Advanced image-processing and analysis workstations are designed to perform post image processing. For example, 3D rendering and functional MRI image analysis are usually processed off-line on these specialized workstations loaded with special image-processing software programs, which are most often made by different vendors and sold separately from the PACS. These workstations are usually expensive and their numbers are much smaller relative to either primary diagnostic or secondary review PACS workstations. After images are processed and saved, they are sent to the PACS server for diagnosis and become part of the patient's medical record. Shown in Figure 2.8 are (a) a three-dimensional CT cardiac image and (b) a functional MRI post-processed

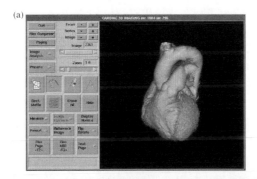

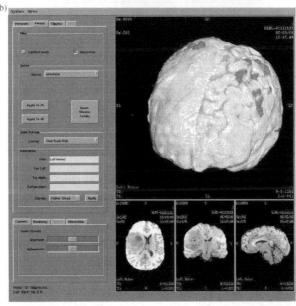

Figure 2.8 Post-processed CT and MRI images using an Advantage Windows workstation (GE Medical Systems, Waukesha, WI): (a) three-dimensional CT cardiac imaging; (b) functional MRI brain mapping.

image using an advanced post-processing workstation. Other advanced image analysis can also be performed on such workstations, which are equipped with high performance CPUs, large amounts of memory, and the best graphics card available, because many software use computer algorithms requiring significant amount of computational power, sometimes even more than a primary diagnostic workstation. However, the display monitor quality of such workstations may or may not be as good as that of a primary diagnostic workstation, for several reasons. For instance, source images for advanced analysis are typically cross-sectional CT or MRI images, and their spatial resolution and data matrix are no more than 512 × 512 (much less than a CR/DR image), and usually the analysis result are sent to the PACS system as DICOM objects. They are reviewed on primary diagnostic workstations and archived by the PACS just as any other source images acquired directly from imaging modality scanners.

2.2.4 QA/QC Workstation

A QA/QC workstation is the platform where the PACS administrator can perform QA/QC tasks. When the PACS administrator logs into this workstation, QA/QC privileges are granted to the user, and image series and demographic information errors can be deleted or corrected as necessary. For instance, if an image series of a patient is accidentally inserted into a different patient's folder, it can be moved to the correct folder. These are very basic QA/QC functions. However, these functions should be restricted and be available only to the administrator to ensure data integrity. Figure 2.9 shows the QA/QC interface of a PACS workstation for the PACS administrator. Administrative procedures, such as merging studies, are shown in Figure 2.10a and correction of DICOM objects or patient demographic data in Figure 2.10b.

2.2.5 Digitization, Printing, CD-Burning Workstation

Before, during, and after the implementation of PACS and the consequent transition to filmless operation in a healthcare institution, there is still a need to digitize the films from a patient's previous examination or from information received from outside hospitals or clinics. Films also need to be printed and CDs burned for a patient or their referring physicians. These processes require a workstation equipped with one or two consumer-grade monitors. Hospitals usually have only a few such workstations that are typically located in a file room or in the film library area. Figure 2.11a shows a film digitizer and Figure 2.11b a digitization workstation that can scan printed films and send them as DICOM images to the PACS system.

Figure 2.12 is a PACS CD and DVD burner, also located in the file room, used to burn CDs and DVDs for patients or referring physicians who do not have access to a clinical viewing workstation or the PACS system. It records DICOM images and diagnostic results to CD and DVD, with an embedded DICOM image viewer included on each disk allowing image review on any PC. Shown in Figure 2.13a and b are images viewed with such a DICOM image viewer, which is a basic

Figure 2.9 QC tools for the PACS administrator to perform QC tasks (Emageon/Amicas, Birmingham, AL).

Figure 2.10 Administrative QC tasks such as (a) merging studies and (b) DICOM objects editing using the QC tools of the PACS.

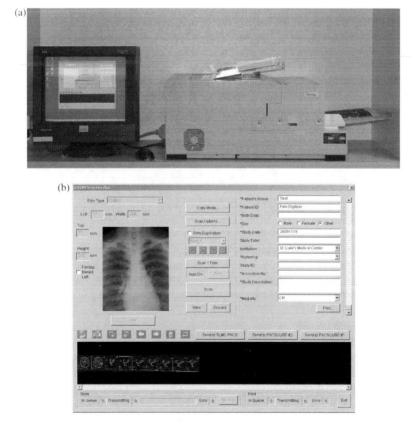

Figure 2.11 (a) A film digitizer and (b) digitization workstation used to scan printed films and send them as DICOM images.

Figure 2.12 A PACS CD/DVD burner (DatCard Systems, Irvine, CA) for storing DICOM images on CD/DVDs. These CD/DVDs can then be sent to referring physicians, replacing expensive film duplication for such purposes.

(a)

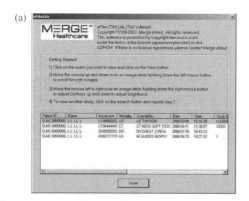

(b)

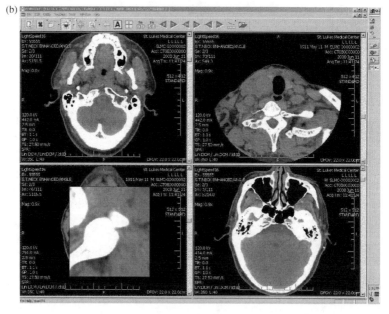

Figure 2.13 (a) A DICOM image viewer software (Merge Healthcare, Milwaukee, WI) embedded in DICOM CD-ROMs; (b) CT images in DICOM format and stored on a CD-ROM.

method of viewing images without many advanced tools and features. When a CD or DVD is burned, the patient's demographics, study information, and a unique ID number for audit trail purposes are also labeled on the disk. Another important function for this workstation is a media import utility, which can easily import studies and images from CDs or DVDs received from outside healthcare institutions.

Replacing film with CD/DVD distribution enables healthcare institutions to communicate with patients better, provide referring physicians with the latest available technology, and realize significant cost savings associated with films.

2.2.6 Office Desktop Workstation

To realize complete and full access to a patient's EHR from anywhere in a healthcare institution or a physician office, today's PACS provide a special basic version of the PACS image display workstation software. These types of desktop workstations are nothing more than a regular desktop computer with a special software program that enables them to access and view images. With the development of thin client and secure network technologies such Secure Socket Layer (SSL) and Virtual Private Network (VPN), any desktop or mobile laptop computer is potentially a workstation capable of reviewing patient images and medical history without too much customization, as long as there is access to the Internet.

2.2.7 Thick and Thin Client

PACS workstations can be categorized as thick or thin clients. They are not totally different workstations, but categorized workstations based on how the PACS application software is loaded. Thick client PACS workstations typically are either primary diagnostic workstations or secondary clinical review workstations. They have all PACS application software installed on the local physical workstation. All functionalities of a PACS workstation are built into the software. These types of workstations usually have stronger processing power, are equipped with large amounts of memory and local storage hard disks, with configuration management tools. PACS application software loading, upgrading, and maintenance on these workstations require either vendor or local PACS administrator involvement. The number of thick client workstations is usually very limited in a healthcare institution, as primary diagnostic workstations are located in the department for which the PACS is specialized, and secondary clinical review workstations are in key areas throughout the healthcare institution. The advantage of thick client workstations is that they have many powerful image-processing functions and are equipped with better medical-grade image display monitors than thin client workstations. However, because of their high cost and complex setup, operation, and maintenance, they are out of reach of most medical staff outside the radiology department.

Thin client workstations, in contrast, are workstations used for image review by clinicians located in all other departments and clinical areas. They are not used for the primary reading that generates diagnostic reports. Compared with thick client workstations, these workstations have less computational power, less memory, and less local storage hard disk space. Very often, they are just plain office desktop personal computers (PCs). The PACS software is mostly Web-based, which means there is no complicated PACS application software installed locally on the workstation. The advantages of thin client workstations are many: they cost much less, software installation and maintenance is embedded in the Web server—there is no need to configure and maintain each workstation individually—and the workstation is fairly simple and widely available across all departments and clinical areas in a healthcare institution. If a computer has a Web browser for surfing the Internet, it can potentially be a thin client PACS clinical workstation. In many PACS teleradiology implementations, remote workstations are thin client workstations using Web access to reach the central PACS image archive.

Although both thick and thin client workstations have their own advantages, they are not likely to replace each other in the foreseeable future, because they have different functional requirements, and complement each other by meeting different clinical needs.

2.3 PACS WORKSTATION ARCHITECTURE

A PACS system consists of many computer hardware and software components, such as image display and visualization workstations, computer servers, data storage devices, networking devices, and so on. In a PACS, the most common device that users interact with is the workstation that allows the user to visualize and manipulate images and other medical information. This workstation is often referred to as the image display workstation (or more loosely as the PACS workstation). In fact, there are many other types of computer workstations that make the PACS work and function properly. In this section, however, the focus is on the particular type of PACS workstation that is primarily used for the display, review, and processing of images and other EHRs. To distinguish these types of workstations, they are referred to as the image display workstations or simply display workstations.

The computer is the key component of a display workstation. Most PACS workstations use a general purpose computer box that includes one or more CPUs, input/output (I/O) controllers, and network communication devices. The computer is usually also equipped with devices for human interaction, such as the keyboard, mouse, trackball, joystick, microphone, and monitor. It includes hard disk and other data storage devices.

In order to display images and other information, computers are equipped with hardware and software components, including the image display devices (monitors) and the controller card (graphics card), which convert digital information from storage devices into analog or digital display signals, such as the image or textual information displayed on monitors. Computers and their software (or computer programs as they are frequently called) allow the access and control of many hardware components to perform desired functions; in addition, there are application programs that can access image data and send it to a display controller in the proper form. One primary difference between a standard computer system and a medical workstation is its associated display interface. The needs of medical imaging often require special display software, high-resolution monitors, and high performance display controllers.

2.3.1 Major Components of a PACS Image Display Workstation

A PACS image display workstation usually consists of one computer box and several image display monitors. The major hardware components of a display workstation include:

- A general purpose computer with one or more CPUs
- Memory and data storage devices, often referred to as hard disk drives
- Communication or network devices
- Graphic controller card or graphics card
- Image display devices (usually liquid crystal display [LCD] monitors)

2.3.2 Hardware

Computers come in various forms and shapes. Computers discussed in this chapter are micro-computers, which are known today as PCs, differing from the mainframe and the supercomputer, which are different classes of computers distinguished by their design and capability.

A computer has some basic functions such as input, output, memory, and central processing. The devices that perform individual functions are connected by a common circuit board often called the motherboard. On this motherboard there are one or more CPUs, which constitute the "brain" of the computer; power supply connections; and many slots where other devices such as input, output, data storage, and information display devices are connected. This board serves as the common platform through which all other devices communicate with each other and perform necessary functions. The number and type of connecting slots on the motherboard determines a computer's expandability.

Just as humans communicate with each other through vision, hearing, touch, taste, and smell, a computer communicates with other electronic devices or human users through input and output devices. Examples of input devices for computers are keyboard, mouse, and microphones. Printers, monitors, and speakers are output devices.

2.3.2.1 CPU

The CPU serves as the central controller, which directs data exchange "traffic" among many devices. Software or preset instructions are needed to properly guide the data "traffic" on the motherboard, and are usually stored in data storage devices, such as the hard disk drive or the memory. The motherboard, along with its many on-board devices, is usually put in an enclosed protective box with many slots in the front or the back. These slots are designed to be plugged in with other input and output devices, such as the network cable and image display devices or monitors.

2.3.2.2 Memory

Computer memory serves a variety of purposes and comes in various forms. It is needed to temporarily or permanently store input and output data as well as software programs that perform specific applications. Random Access Memory (RAM), the onboard memory that is typically used to store data or programs temporarily, is fast but expensive. On the other hand, devices that are typically used for long-term data storage, such as hard disk drives and RAID, are relatively inexpensive. Various data storage devices are discussed in detail in Chapter 5.

2.3.2.3 Communication and Network Devices

Data communication with an individual computer is handled by the circuitry on the motherboard through the CPU. However, to communicate with another computer or devices that are not physically connected to the computer, a network card is required. An oversimplified view of the function of a network card is as a "translator" or enabler of communications between computers. Typically, a wired network card has a connector for a network cable plugged in, such as a 100 Mega bit /second (Mbps) or 1 Gbps network port on the back of a PACS workstation. Different cables such as optical fibre and copper wired network cables use different network connectors on the network card. It should be noted that a network card is rated for a specific transmit and receive mode and data throughput speed. For wireless network communication, the network card does not have a network cable connector. Nowadays, a network card is typically built-in or inserted into one of the slots on the motherboard.

2.3.2.4 Graphics Card

A graphics display control card enables the display of information on the monitor through a combination of hardware and software. It transforms digital values from the motherboard of the host computer to appropriate standard signals for the display monitor.

A graphics controller typically has an on-board processing unit and a special purpose memory that accepts the output of the application program from the computer and quickly converts the signal into a "display-ready" format. The on-board memory (the so-called video memory) of the graphics card supplements the host computer's memory and display signals are processed faster. The digital values in this memory are transformed into video signals, ready for the display device. A computer system may also have driver software that provides an interface for the application to control the contents of the video memory. An application program may change the display by calling driver software that controls the details of updating video memory.

Most of today's display devices accept digital as well as analog video signals in standard formats. The graphics card performs the necessary digital-to-analog or digital-to-digital conversion as the memory is scanned. In the case of analog signals, this conversion creates a voltage called video signal voltage that is proportional to the digital value. The monitor that is connected to the output of the graphics card takes the video signal and performs necessary conversions within its own electronics and finally displays the information.

The graphics card also determines the allowable matrix size on the display device via an automatic synchronization process with the display devices. Additionally, most high-end graphics card have a feature allowing the modification of the so-called look-up-table (LUT) in the graphics memory. An LUT is a conversion table that controls the conversion of digital display signals from input to output. For medical-grade monochrome displays, it typically has a 10- or 12-bit image memory for each pixel. By modifying the proper LUT in the graphics card memory, the grayscale response of the display device can be made to follow the DICOM standard. Some advanced graphics card include a luminance probe and calibration software which can be used to compute the proper LUT.

The methods used to convert display values to monitor luminance are changing with newer display systems, such as flat panel devices, and newer display signal standards, such as Digital Video Interface (DVI). In the newer display system, the graphics card sends a digital signal to the monitor without conversion to analog signal, enabling higher fidelity and better quality in the display. The final conversion to luminance from the digital signal is handled within and by the monitor itself, providing improved performance in several aspects of the display. More details about the processes and steps involved in image display are discussed in Chapter 3.

2.3.2.5 Video Memory

The video memory typically has 8, 16, 24, or more bits per pixel, which is often called the bit-depth of the graphics card. In the case of an 8-bit grayscale controller, up to 256 digital values can be generated. For color images, 3 bytes of storage are used for each pixel (true-color); 8 bits are used for each of the red, green, and blue components of the pixel, resulting in the potential for 2^{24} simultaneous colors. Most graphics card used in medical image display are typically 12- or 16-bit cards, which are more expensive than consumer-grade cards. Primary diagnostic workstations are usually equipped with the best available graphics card and a large amount of dedicated video memory. For thin client workstations, because they are often just regular PCs, sometime there is no dedicated video memory for a graphics card, and instead the video shares memory on the motherboard with the CPU. Especially for thin client and secondary clinical review PACS workstations, one of the most cost effective ways to improve speed is to upgrade the graphics card with a good amount of dedicated video memory and increase the amount of memory on the motherboard for CPU use.

2.3.2.6 Display Monitor

The final hardware element of an image display workstation is the display monitor. A display monitor is the physical unit that generates a visible image from video signals. A workstation may be configured to have two to four monitors. LCD flat-panel monitors are the dominant choice of today's PACS workstation, replacing the cathode ray tube (CRT) which is bulkier, requiring a larger foot print and consuming more power in use. Chapter 3 provides more detailed descriptions of image display monitor technologies.

2.3.3 Software

To make a computer workstation work, software has to be uploaded onto the computer. There are two main types of software on any given computer: the operating system (OS) software and the application software.

2.3.3.1 Operating System Software

The OS software is a specialized program that controls the resources of the computer. It is the fundamental software that controls the proper operation and the functions of the hardware—hard disks, CPU, and I/O devices such as the display monitors and printers. An operating system's software gives the connected hardware detailed and specific instructions to accept input, perform a specific function, produce output, and read and write information from the data storage or network devices. It serves as the backbone of the computer and controls services such as network communications, access control, creation and management of data files, and the execution of application programs.

The most commonly used OS in the medical imaging workstation are Microsoft Windows, UNIX, Macintosh, and Linux. Windows occupies the lion's share of the market, while UNIX and Linux have the promise of being more stable. Each has its advantages and shortfalls. Each of these OS creates an operating environment for the user and for the application programs. An OS may be implemented on many types of computer hardware, and a given hardware configuration may support one or more OS. However, most PACS workstations only have one type of OS loaded, and thus are set up to run only one OS.

Nowadays, all OS provide a graphic user interface (GUI) as well as functionalities that enable easy use of application software such as image viewing and PACS utility software. OS still differ in their way of interacting with users, especially when the user needs to type in the instruction from the command window. They still differ quite significantly in the modes supported, the type and degree of user access control, the type of protection provided between applications, and the services provided by the OS to application programs. Functionally, any OS can support a medical imaging system. Practically, the choice of OS is invisible to PACS users, who do not have to provide the technical support for a PACS system. However, the choice of OS may limit the loading and use of other application software besides PACS. These non-PACS software may include word processing and other medical information system software, which may or may not be compatible with the particular OS under which that the PACS display workstation is operating.

2.3.3.2 Application Software

Application software is software developed for a particular OS, and can operate only on computer systems that run that OS. These are software such as Microsoft Word and Excel that users purchase and install on their computers. A computer is often preloaded with some application software, such as the Web browser. The PACS display workstation software is also an example of such application software.

The application software running on the PACS display workstation has the following main functionalities and features:

- GUI, which provides the user with access to tools and functionalities for image display, visualization, and networking

- Image processing and manipulation tools

- Networking and data management tools

- Worklist management

- User preference configurations, such as image-hanging protocols

- Interface with RIS and HIS

There is a wide variety of PACS programs available for displaying and visualizing medical image data on workstations for all common OS. Some have more features and functionalities than others, but all programs can send or retrieve images via standard DICOM protocols over the network to other workstations or servers. The DICOM standards set by NEMA and ACR have brought tremendous progress in the interoperability of PACS and its realization in the market place. The standard has been developed specifically to aid manufacturers and medical users to seamlessly operate over a computer network a variety of archive, network, and display components made by different vendors. The new initiatives in the standards such as IHE and HL7 address data structures, object and service types, communication protocols, grayscale display, print management, and worklist management. All these standards and their relevance are discussed in detail in Chapter 6.

Some of the programs include tools for more advanced image manipulations, database access and management, and archive querying and retrieval. Almost all commercial PACS software programs support multiple high-resolution medical-grade displays. Other tools such as Region of Interest (ROI) and distance measurement and annotation on selected images are widely available. The latest development in the functionality of these software programs may even include useful features to automatically retrieve and select relevant images from prior examinations of the same patient, and present them in comparison to the most recent images. Additionally, advanced features may include security measures, such as authenticating access and tracking patient data access history, and workflow control.

2.3.4 Typical Features of Diagnostic Image Display Workstation

Although a PACS user has little or no control over the features and functionalities design of the software program for the PACS display workstation, the user should consider them while comparing PACS workstations from different vendors. Shown in Figure 2.14 is a pull-down list of general image display and processing tools built in a PACS workstation. The following is a partial list of desirable features in a PACS display workstation. The latest PACS workstation should have most if not all of the features listed below:

- Networking, archive, query, retrieval, and other DICOM configuration and connectivity functions

- Worklist management tools

- Patient directory and intelligent searching and filtering

- Display screen configuration, image-hanging protocol customization

- Continuous zoom, pan, scroll

- Rotation, flipping, and comparison

- Window and level

- Histogram modification

- Image reverse

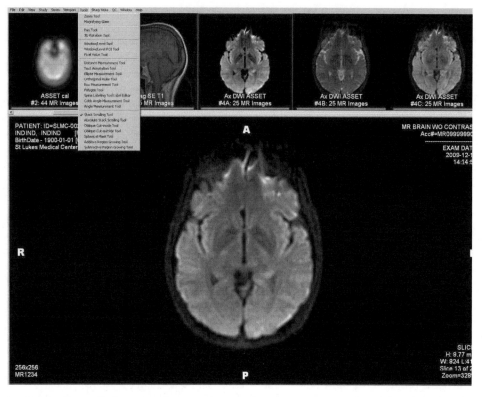

Figure 2.14 The pull-down list of general image display and processing tools built in a PACS workstation (Emageon/Amicas, Birmingham, AL).

- Distance and ROI measurements

- Cine or movie loop viewing

- Advanced image-processing tools such as the 3D processing tools (e.g., surface rendering, Maximum Intensity Projection, etc.)

- Interfacing tools with other health information systems such as RIS, HIS, EHR, and so on.

Although PACS workstations may have some or all of these above-listed functions, vendors tend to differentiate themselves from others by designing the GUI differently. In the following sections, a diagnostic workstation and two stripped-down basic PACS viewer software embedded on burned CDs are used as typical examples.

2.3.4.1 Window and Level

The window and level controls are used most when each study is displayed on the PACS monitor. Window controls the overall contrast of the displayed image, while level controls the overall brightness of the image, using the same digital image pixel values. Because of its frequent use, the control should be very simple. Some systems can set multiple window and level combination presets for various modalities or types of studies performed as a quick default gray scale for image display. From these presets, though, additional adjustment is needed to reach the optimized window and level for a particular study or image. Window and level optimization can reveal subtle details on digital images otherwise invisible on nondigital images, and is very useful in CR/DR diagnostic x-ray imaging, which can reduce repeated scans, which in turn lowers not only the number of exposures to x-ray, but also the resources and costs associated with these additional repeated scans.

2.3.4.2 Pan and Zoom

The pan and zoom function is used to locate the area of most interest to the viewer and display it in a magnified fashion. When an image is magnified, its size can be large compared to the PACS workstation display monitor; therefore, image panning is necessary for the display monitor to show only the required portion of the image. Zoom is usually achieved with image pixel replication or interpolation, sometimes combined with image-smoothing algorithms. Zoomed images can be displayed in two ways: either the entire image is zoomed and displayed in the window resulting in a reduced field of view (FOV), which is reverse proportional to the zoom factor, and the image is changed using the pan or scroll function, or the zoom can be done in a "magnifying glass" fashion, which means only the image portion inside the "magnify glass" is zoomed up.

2.3.4.3 Sorting and Filtering

The sorting and filtering function is that arranges studies together according to their study modalities such as CT, MRI, CR/DR, ultrasound, nuclear medicine, study date, patient ID number such as Medical Record Number (MRU), or patient names in alphabetical order. When studies are sorted or searched by modality, only studies that match the modality selected are displayed. It works just the same for date search, where studies falling into the selected date range are displayed.

2.3.4.4 Worklist

The worklist is a list of work that needs to be done, including imaging studies unread or resident interpreted, or emergency studies for interpretation and reporting. The worklist typically can be incorporated with the above Sorting and Filtering function, so that studies can be arranged in order according to imaging modality, date, or patient name. Different flags are used to flag the status of studies, so that a given study is not worked on redundantly wasting resources, nor are uninterpreted studies missed. The worklist is a very useful tool, especially when the PACS system is interfaced with the RIS, where it contributes to workflow improvement. Shown in Figure 2.15 is the worklist of a PACS diagnostic workstation.

2.3.4.5 Distance and Area Measurement

Distance and area measurement are basic measurement functions for a PACS workstation. In diagnostic imaging reports, the quantitative measurement of length or area might be required. When an area is measured, an ROI is defined first, which can be an adjustable rectangle, circle, or free-hand drawing. The ROI measurement result includes area and pixel values.

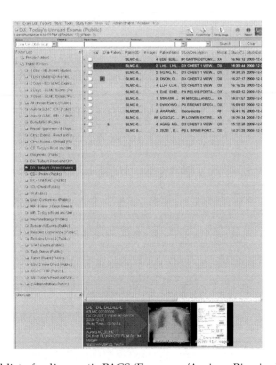

Figure 2.15 The worklist of a diagnostic PACS (Emageon/Amicas, Birmingham, AL) workstation.

2.3.4.6 Hanging Protocol

Hanging Protocol is the default automatic arrangement of the way images from a study are displayed, and is named after the way in which films hang on an alternator and light box. An image reading session may include current and previous studies. Each imaging study can consist of multiple series of different views. When these studies are to be displayed on a PACS workstation with multiple display monitors, how they are displayed on each monitor (such as the number of cross-sectional images displayed on each monitor, window and level preset, on-screen control location and arrangement, etc.) can be tailored to the individual user login, because hanging protocol is a personal preference choice.

2.3.4.7 3D Reformat and Volume Stack

The 3D Reformat and Volume Stack is a very useful tool when viewing large amounts of cross-sectional images in a single series, such as a multidetector CT or 3D MRI acquisition imaging series. When a few hundred images are acquired in one acquisition series, this method allows easier and faster review of all imaging slices at different locations in a single window by moving the mouse around, compared to reviewing each image individually.

2.3.4.8 Movie Loop

Movie Loop or Cine is a tool for dynamic image-sequence display. Images acquired in a sequence such as cardiac MRI Cine acquisitions are displayed in a loop continuously like a movie clip, so that any dynamic changes can be appreciated during such display. Typically, the speed of the movie can be adjusted to a slow or fast motion. The movie may also be paused and resumed.

2.3.4.9 Importing Results of Advanced Image Post-Processing

Ideally, advanced image post-processing, such as computer-aided detection (CAD), functional MRI (fMRI), or cardiac analysis, should be incorporated into the PACS workstation. However, many such advanced post-processing tools are only available on standalone workstations. It is important for these workstations to be able to export the post-processing result in DICOM format, so that the resulting images can be exported to the PACS system and displayed together with other images of the patient during interpretation. Figure 2.16 is an advanced post-processing result from a cardiac MRI imported into the PACS system and displayed on a primary diagnostic workstation.

Although PACS workstations are developed independently by different vendors, there are common basic GUI design principles. To finish the same tasks, the required steps may not be the

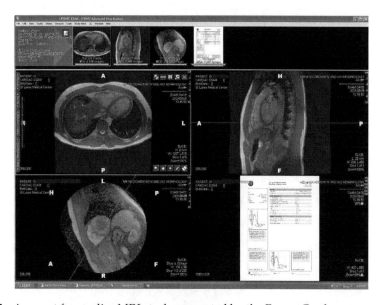

Figure 2.16 A report for cardiac MRI study generated by the ReportCard post-processing software on an Advantage Windows workstation (GE Medical Systems, Waukesha, WI) can be imported into the radiology PACS (Emageon/Amicas, Birmingham, AL).

same, but they can be similar. These features can be evaluated during vendor demonstration sessions, and the user must pay special attention to particular functions.

2.4 ENVIRONMENT FOR PACS WORKSTATIONS

The environment in which medical images are interpreted by clinicians for preparing their diagnostic reports is also very important. Though the environment is not determined by the PACS system itself, it has a direct influence, positively or negatively, on the accuracy and productivity of clinicians who interpret images and dictate reports.

2.4.1 PACS Workstation Reading Room Design

The layout and design of a PACS workstation reading room plays an important role in the efficiency of PACS users. Since the image quality on workstations is affected as well, reading room design should be one of the important considerations while implementing a PACS. A reading room with one or more PACS workstations should be designed in such a way that there is minimum illuminance cross-interference among various workstations. Care should be taken to avoid excessive illuminance on workstations. Finally, heat is generated by the large numbers of computer equipment and monitors, and temperature control in the reading room should be planned ahead of the final implementation of PACS.

2.4.2 The Hardware Configuration of Display Workstations: Number of Monitors

In the early days of PACS implementation, one of the most frequently asked questions related to the number of monitors required in a PACS display workstation. The main reason for such uncertainty stemmed from the fact that radiologists were used to visualizing tens and sometimes hundreds of images on an alternator before the implementation of PACS. At the time, nearly all imaging departments were using automated alternators that could be preloaded with films. So, with a push of several buttons, a user could have eight films at once and select the particular set of images to view. Many sheets of films were displayed and could be scanned and viewed "simultaneously," making the task of browsing through images very easy and efficient. Unfortunately, due to technical and economic constraints, the early PACS workstation was not capable of simultaneously displaying the same number of images as an alternator. The cost was prohibitive, and it was technically difficult to equip each PACS workstation with a large number of monitors. The number of monitors, therefore, was a hotly debated topic.

Earlier studies of PACS indicated that using two monitors instead of one improved radiologists' efficiency dramatically, whereas increasing the number of monitors beyond that did not improve the efficiency significantly. Nowadays, given the PACS software improvements of the past decade, image display and viewing techniques have developed and matured considerably. Today, it is common to have multiple display monitors for each PACS workstation, as shown in Figure 2.17. Most workstations are commonly equipped with two 21-in medical-grade monitors for high-quality medical image visualization, together with one color consumer-grade monitor for other textual information.

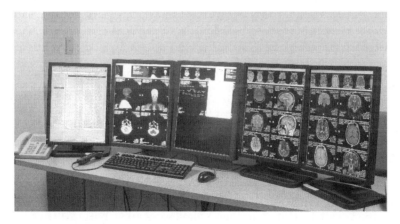

Figure 2.17 A typical five-monitor configuration for a diagnostic PACS workstation: one monitor is for worklist display and four monitors are for image display.

2.4.3 Ergonomic Design of Workstations

Radiologists may spend a good portion of their working hours at the PACS workstation. Besides the reading room design, arrangement of display monitors, the physical characteristics of monitors and the software program, other factors such as ergonomic design of the workstation may also have a direct impact on the quality of the radiologist's work and definitely on the productivity and efficiency. During the early days of PACS implementation, ergonomic consideration in the PACS workstation and the reading room design were mostly afterthoughts. It was not uncommon to have half-a-million dollars worth of PACS equipment put on regular office furniture which was awkward and uncomfortable to work with for a long time. Nowadays, PACS image display workstations are placed on specially designed desks or tables with many ergonomic considerations for users to work around. These desks or benches for workstations are also important not just from the ergonomic standpoint, but for the long-term productivity and efficiency of users. Additionally, today's PACS workstation design includes such features as adjustable monitor angle, height, reduction of acoustic noise from the workstation, and the means to reduce glare or excessive illuminance levels in the reading room.

Factors that are independent of workstation hardware or software design, such as the noise of the workstation and temperature control may have a direct or indirect impact on workflow. Due to the special characteristics of PACS display monitors, which are discussed in Chapter 3, ambient lighting of the reading room is important too, because image quality can be severely impaired by factors such as excess illuminance levels. For users planning to implement PACS in various departments, all the factors discussed above should be considered carefully.

2.5 RADIOLOGY WORKFLOW

Traditionally, disparate clinical information systems function independently from each other in a radiology department. Such clinical information systems include PACS for image review and storage; RIS for managing examination scheduling and order, image acquisition session, prior examination review, and report review; and a report dictation system for speech capture, playback, transcription, and report generation into the RIS. However, recent developments have been in the area of integrating and consolidating these heterogeneous systems to improve workflow for physicians and healthcare professionals.

2.5.1 Film-Based Workflow without PACS

In most radiology departments, the RIS handles all radiology generated text data. The RIS coordinates the department workflow of scheduling, order management, study tracking, film tracking, archiving examination results, billing, and distributing radiology information for orders and results. A typical film-based radiology workflow is shown in Figure 2.18.

2.5.2 RIS and PACS Workflow

A successful digital radiology operation requires seamless integration of RIS and PACS. The role of RIS is that it provides patient information to the PACS, and maintains all examination information, as shown in Figure 2.19. It controls examination information flow from ordering, scheduling, image acquisition, image interpretation reports to posting the final charges, and it also links to systems outside the department such as the HIS. The RIS is thus the centerpiece that connects all clinical information systems and imaging modality scanners within the department. The PACS system receives and stores images acquired by imaging modalities and sends images to workstations for radiologists to interpret and clinicians throughout the healthcare institution or enterprise to review. Therefore, RIS and PACS are the backbone of a digital radiology department.

Radiology workflow is the work process sequence of a radiology department, where staff work with the various equipment to fulfill work goals according to procedural rules. A workflow may be divided into multiple parts each with groups of workflow steps. Every managed radiology workflow starts with the examination order. The order is a service request from another clinical department or healthcare institution. The workflow can be divided into one or more requested procedures.

2.5.3 Filmless Workflow with PACS/RIS

To fully achieve the optimized workflow with the installation of RIS and PACS, it is necessary not only to transform the radiology workflow into a filmless operation but also to integrate the RIS and the PACS, so that the RIS-driven filmless operation integrates the digital image management workflow into an efficient, unified environment. This is also the most challenging and important

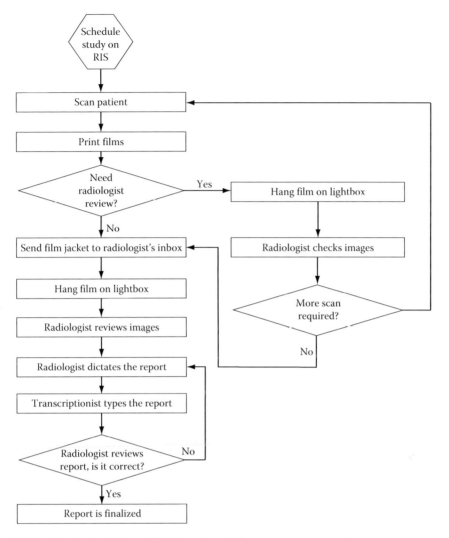

Figure 2.18 A typical radiology film-based workflow.

Figure 2.19 The examination management tool of an RIS system (Cerner, Kansas City, MO).

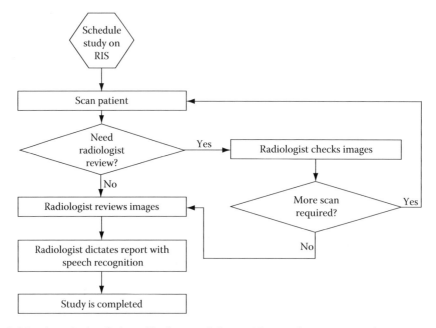

Figure 2.20   A typical radiology filmless workflow with speech recognition for reporting.

step in the implementation of PACS, because both PACS and RIS generate, receive, send, and store patient healthcare information, and proper integration of all these information improves efficiency, accuracy, and utilization of the information. Recent PACS and RIS products offered by various vendors allow data such as examination orders, patient demographic updates, structured reports, and other related information to be exchanged between information systems. A typical PACS and RIS filmless workflow is illustrated in Figure 2.20.

2.6 SPEECH RECOGNITION FOR REPORT DICTATION

Speech recognition, sometimes also known as voice recognition, converts spoken words into machine-readable text. It utilizes proven algorithms and models to recognize human speech—when a person speaks into a microphone—which is digitized thereafter. The digitization can be performed either by the specialized microphone, as shown in Figure 2.21, or by the workstation that the microphone is connected to. The digitized speech signal is then processed by the speech recognition system, which compares these signals to models it stores. When the best match of word signal pattern is found, the word is recognized and converted to text. Speech recognition report dictation allows physicians to dictate medical reports in a way similar to traditional report dictation. The speech recognition system automatically transforms the speech of physicians into draft text that can be edited either by voice command or keyboard.

2.6.1 Benefits of Speech Recognition for Reporting

Speech recognition for medical report transcription is a sophisticated technology. Implemented properly, it can yield results that provide faster and more cost-effective care. The benefits of speech recognition for reporting include:

- Significant reduction in report turnaround time

- Elimination of paper for initiating the dictation process

- Elimination of manual distribution and retrieving patient demographics into the dictation system

- Improvement in referring physician satisfaction

- Department workflow improvement with increased efficiency and productivity

- Reduction or elimination in transcription costs

- Data mining and outcome analysis for advanced quality reporting and operational analysis

Figure 2.21   A specialized microphone (Nuance, Burlington, MA) for dictating radiology reports using speech recognition software.

When the application of speech recognition for reporting is integrated with other information systems, there are additional benefits including:

■ Easy access to automatic speech recognition report dictation from the same PACS workstation

■ Seamless clinical information exchange among PACS, RIS, and speech recognition report dictation systems

2.6.2 Speech Recognition Report Dictation Approaches

There are several approaches to automatic speech recognition that can be utilized to meet the specific requirements and needs of a particular healthcare institution and radiologists. Regardless how speech recognition transcription is used by radiologists, this technology is being adopted by more and more institutions and physicians with various degrees of success.

2.6.2.1 Front-End Real-Time Speech Recognition with Self-Editing

The "Front-End" approach is physician-driven speech recognition, which allows real-time self-edit capability. It allows radiologists to dictate into the speech recognition system and to review the generated text during or at the end of report dictation. Radiologists can use standard word processor software applications with mouse and keyboard, or voice commands and microphone controls to edit the text of the report. If any word has been misrecognized, the user can either choose the correct word from a list or, if the word is not available in the list, dictate it again. Additionally, there is an auto-punctuation feature. Upon completion, the radiologist can sign-off the reports immediately. Under this mode, some products also have batch dictation features, so that the radiologist can hold dictated reports in a queue, and texts of multiple reports are available for reviewing and editing only when all the dictation has been completed. This mode of report dictation requires the involvement of physicians for report editing, but it provides the shortest turnaround time.

2.6.2.2 Rear-End Speech Recognition with Delegated Editing

"Rear-End" speech recognition is speech-assisted digital transcription dictation and has been adopted by many institutions. The report dictation process is illustrated in Figure 2.22. Using this

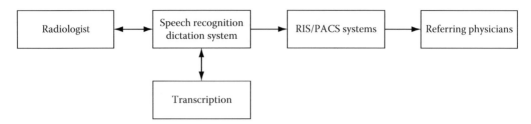

Figure 2.22 The radiology report dictation process using rear-end speech recognition with delegated editing.

mode, radiologists dictate into the speech recognition system, which generates draft reports. These reports are not edited by radiologists; instead they are forwarded, together with the original voice sound files, to transcriptionists for correction and editing. This process is faster compared with traditional medical report transcription, because transcriptionists receive the text of the draft report and only need to perform simple edits to complete the report-generation process. Thus their efficiency increases, and time and costs are reduced. After reports are completed, radiologists can call them up, review, and sign them.

2.6.3 Templates and Shortcuts

Additionally, radiologists can create frequently used templates, macros, and shortcuts for specific procedures, phrases, or imaging modalities to eliminate the need to dictate frequently used normal results or redundant data. Using a properly planned set of templates allows the generated reports to be standardized and consistent, with regular results to be uniformly presented, the number of errors in a report reduced, and the results to be easily communicated by radiologists and referring clinicians. The standard text blocks in templates can also be modified and appended with new dictation for flexibility and optimized productivity. Although it takes time to build such templates, in the long run, building templates allows speech recognition transcription to be easier for radiologists to use, the report generation efficiency of radiologists to improve, and the time required for report transcription to be reduced.

2.6.4 Challenges and Limitations of Speech Recognition for Reporting

A speech recognition report dictation system typically requires a period of training, during which the user trains the system to the speaking patterns and pronunciation of the speaker, and the accuracy of the system improves after it has analyzed many dictated reports. Once the system is trained, it adapts to the user, and may capture continuous, normally paced speech with a large vocabulary with high accuracy. Typical challenges and limitations of speech recognition for report dictation include:

- The speech recognition system is not 100% accurate yet.

- The system needs to work in a noise-free environment; it is especially important to keep it out of heavy traffic areas, where noise from disruptions could affect the quality of transcription.

- Physicians need to speak well and be aware of the surrounding environment to optimize the accuracy of speech recognition, especially in teaching institutions or training facilities.

- It requires radiologists to speak slowly and clearly with a consistent and deliberate speech pattern.

- The speech characteristics must match the training data.

2.6.5 Integrating Speech Recognition Technology into Radiology Workflow

The successful implementation of a speech recognition report dictation system depends on several factors. It is necessary to analyze the workflow of the department and find out what the impact of speech recognition dictation to the existing workflow will be, especially to those procedures that will be changed, added, or even eliminated, and determine the best way to handle these issues. When the new system is installed, these changed policies and procedures should be tested before the system goes "live", so that additional changes can be made if necessary, and all these changes can be incorporated into appropriate user training.

2.6.5.1 Transcription Service

Transcription service should be involved in the establishment of the new workflow, policies, and procedures, because they are an integral part of a speech recognition system implementation, and their knowledge and expertise in the workflow contributes to the choice of optimal setup. Since they are familiar with the process of radiology report generation, they can bring to light issues that otherwise may not be determined. Therefore, radiology transcription service should analyze the new process and set their policies and procedures according to the transcription service's goal.

When PACS, RIS, and speech recognition dictation are all integrated into a single desktop workstation application, radiologists can display the dictated report text immediately when reviewing PACS images. This type of integration eliminates paper work from the dictation process, reduces data entry errors, and provides an efficient paperless workflow.

2.6.5.2 Physician

The physicians group or its representatives should be involved in the decisions relating to implementation of speech recognition, and in all aspects of developing policies and procedures for workflow change. This improves their acceptance and enthusiasm for the new system, because the deployment of such a new report dictation system is very challenging without the physician group's involvement and buy-in.

2.7 NEW DEVELOPMENTS

With the development and integration of EHR (discussed in Chapter 11), PACS, RIS, and other specialized clinical information systems and workstations, the boundary between RIS and PACS is getting blurred. As shown in Figure 2.23, radiology reports and images can be conveniently reviewed on the same desktop, and PACS is no longer limited only to images. Physicians and healthcare professionals need a comprehensive view of patients' clinical data, including HIS, lab reports, radiology images, and RIS from various sources, and they can access this information from a single workstation if the disparate information systems are consolidated, and sometimes

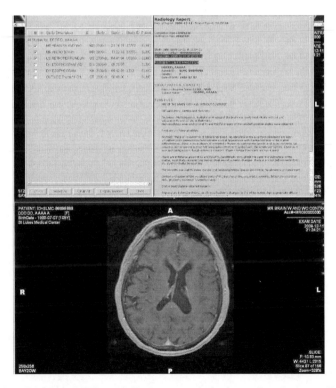

Figure 2.23 When PACS and RIS are fully integrated, the boundary between PACS and RIS is blurred. Both radiology images and reports can be easily accessed and displayed on a PACS workstation.

with a single log on. This integrated approach facilitates the viewing of past and present images of patients, their medical history, and current status in a single workstation or application window. It also reduces the need for duplicate workstation hardware for independent display of images and text.

BIBLIOGRAPHY

Avizienis, A. 1997. Toward systematic design of fault-tolerant systems. *Computer* 30 (4): 51–58.

Boochever, S. S. 2004. HIS/RIS/PACS integration: Getting to the gold standard. *Radiol. Manage.* 26: 16–24.

Branstetter IV, B. F., D. L. Rubin, and D. L. Weiss. 2009. *Practical Imaging Informatics: Foundations and Applications for PACS Professionals.* Society for Imaging Informatics in Medicine (SIIM), Springer, New York, NY.

Cao, F., H. K. Huang, and B. J. Liu, et al. 2001. Fault-tolerant PACS server. *Radiology* 221 (P): 737.

Dreyer, K. J., D. S. Hirschorn, and J. H. Thrall. 2006. *PACS: A Guide to the Digital Revolution.* Springer, New York, NY.

Gale, B., Y. Safriel, and A. Lukban, et al. 2001. Radiology report production times: Voice recognition vs. transcription. *Radiol. Manage.* 23: 18–22.

Gay, S. B., A. H. Sobel, and L. Q. Young, et al. 2002. Processes involved in reading imaging studies: Workflow analysis and implications for workstation development. *J. Digit. Imaging* 15: 171–177.

Hayt, D. B. and S. Alexander. 2001. The pros and cons of implementing PACS and speech recognition systems. *J. Digit. Imaging* 14: 149–157.

Hirschorn, D., C. Eber, and P. Samuels, et al. 2002. Filmless in New Jersey: The New Jersey Medical School PACS Project. *J. Digit. Imaging* 15 (Suppl 1): 7–12.

Horii, S. C. 2009. Workflow and Radiology. *The Society for Imaging Informatics in Medicine 2009 Annual Meeting*, Charlotte, NC.

Horii, S. C. 1992. Electronic imaging workstations: Ergonomic issues and the user interface. *Radiographics* 12: 773–787.

Huang, H. K. 2004. *PACS and Imaging Informatics: Basic Principles and Applications.* Wiley-Liss, Hoboken, NJ.

Huang, H. K., F. Cao, and B. J. Liu, et al. 2002. Fault-tolerant PACS server. *SPIE Med. Imaging* 4685 (44): 316–325.

Integrating Healthcare Enterprise. 2004. IHE radiology technical framework white paper 2004–2005: department workflow.

Langer, S. 2002. Impact of tightly coupled PACS/speech recognition on report turnaround time in the radiology department. *J. Digit. Imaging* 15 (Suppl): 234–236.

Reiner, B. I. and E. L. Siegel. 2000. Effect of filmless imaging on the utilization of radiologist services. *Radiology* 215: 163–167.

Reiner, B. I. and E. L. Siegel. 2003. The cutting edge: Strategies to enhance radiologists workflow in a filmless/paperless imaging department. *J. Digit. Imaging* 15: 178–190.

Weiss, D. L. 2008. Speech recognition: Evaluation, implementation, and use. *Suppl. Appl. Radiol.* December 2008: 24–27.

3 Image Display Devices in PACS

3.1 INTRODUCTION

Image display or presentation is the last yet critical step in the process of medical imaging. As the interface between medical imaging and the viewers who use the images for diagnosis, devices that display medical images with the highest quality possible are just as important as the machines used to acquire the images themselves. The quality and characteristics of the image display directly affect the displayed image quality and thus the viewer's diagnostic accuracy and work efficiency.

3.1.1 Role of Image Display in Medical Imaging

Prior to the broad implementation of the PACS, most medical images were captured directly or printed on films; in the latter case, these films were displayed and viewed on light boxes placed around the hospital, as shown in Figure 3.1. Serving the functions of both image-archiving and image display, the film's characteristics and the quality of film processing thus directly determined the final quality of the medical images. Another consequence of using film was that access to a patient's medical images was limited. Because of the cost of duplicating films, it was common to have only one copy of a patient's images on films, and these films had to be routed to different departments throughout the hospital so that all specialists involved in the patient's care could perform their part in diagnosis or treatment. Patient care was frequently delayed or suspended because of a lack of timely access to medical images, especially if the specialty medical care team had to wait for the films in order to perform their work, thereby reducing overall workflow efficiency. Figure 3.2 presents a diagram illustrating the analog imaging processes in which the image is finally displayed on a light box in a radiology reading room.

PACS and its broad implementation have overcome such inefficiency in the patient care workflow because images now can be accessed by multiple specialists through a PACS workstation (such as shown in Figure 3.3) as soon as the images are acquired and sent into the PACS system. The birth of PACS is a natural result of the progress made in the past few decades in the technology of image acquisition, computer networks, and image display.

A fundamental consequence of digitization of medical images from the point of image acquisition is the separation of image data acquisition and image archiving functions and technologies from image display technology; it is this separation of functions that enables simultaneous display of the same image at multiple locations. This has resulted in improved efficiency in patient care, which is one of the main benefits of PACS.

3.1.2 A Brief History of Digital Medical Image Display Systems

Since the late 1980s and early 1990s, nearly all medical-imaging modalities have been converted into digital mode, although broad implementation of this change did not happen until the late 1990s. Additionally, the wide availability and affordability of computer hardware, increased networking, and easy access to the Internet in the late 1990s resulted in broad adoption and implementation of PACS in hospitals and clinics worldwide. As a consequence, the majority of medical images are viewed on the computer screen instead of film today. As an integral part of PACS, using softcopy display devices (i.e., computer display monitors of various types) to display and view medical images has become common in today's hospital environment.

Unlike on film, the appearance of medical images displayed on PACS workstations can be adjusted for contrast and brightness even after the image has been acquired. This is a fundamental change in the process of medical imaging in that it separates image acquisition from image display/viewing. Take x-ray radiographic imaging, for instance. In the analog world, images are captured by the film/screen system; the films are then developed and placed on light boxes for display and viewing, as shown in Figure 3.1. Films serve as both the image acquisition and display media. In today's digital world of radiography, conversely, images are captured by a digital receptor and converted into digital images consisting of bits and bytes. Unlike those images on films, these digital images can be immediately accessed from any PACS workstations by different individuals at the same time, tremendously improving workflow efficiency and information access and, thus, patient care.

Ironically, throughout the 1980s and early 1990s, because of a lack of adequate digital image display devices (i.e., computer display monitors with adequate spatial and contrast resolution for medical imaging purposes), some of the inherently digital medical images (e.g., those obtained by CT and MRI) had to be printed on films using laser film printers. Computer

Figure 3.1 Radiographic images captured on films and displayed using a light box.

display monitors at that time did not have a high-enough spatial and contrast resolution for primary diagnosis. Prior to the introduction of flat-panel LCD monitors in the late 1990s, CRT-based display monitors dominated the market. Several vendors designed special CRT monitors with high-enough spatial and contrast resolution to meet medical needs. Since the late 1990s, as progress has been made in the LCD flat-panel market, driven mostly by demand in the consumer computer industry, CRT-based display monitors have gradually been replaced by flat-panel LCD monitors, which now dominate the medical imaging market. Previous glitches in terms of quality and grayscale resolution have been overcome, and flat-panel LCD monitors used in today's medical image display have higher resolution and less power consumption. They are lighter, smaller than their CRT counterparts, making them ideal for PACS. Image display devices using newer technologies, such as organic light-emitting diode (OLEDs), that promise brighter images and less power consumption are almost visible on the near horizon. In the foreseeable future, medical-grade display devices will likely become cheaper and have even higher quality than now.

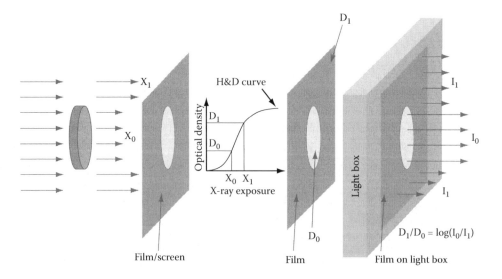

Figure 3.2 Process flow in the generation and display of radiographic images using film both as the image acquisition display media.

Figure 3.3   A PACS workstation for digital image display.

3.2 SOFTCOPY IMAGE DISPLAY SYSTEM

Softcopy image display is usually referred to as the display and visualization of images using such electronic image display devices as computer monitors as opposed to film, which is often referred to as "hardcopy display." A softcopy display system usually consists of one or more monitors (e.g., LCDs) and one or more graphics cards. Figure 3.4 depicts the major steps involved in the image display process when a softcopy image display system is used. From image acquisition to the final step of image display, several conversions are performed on the digital image data; the conversion functions are contained in LUTs. The LUTs are applied to the digital image data before the data are finally presented on computer display monitors as grayscale or color images to be read by human viewers. As a comparison to Figure 3.2 above, Figure 3.4 illustrates the acquisition and data conversion steps in digital image display.

3.2.1 Graphics Card

The computer's graphics or video card is an integral part of the digital image display system. Such cards are typically located inside the computer case and are often called graphics accelerator cards, graphics controllers, video cards, or display adapters. The components' function is to take digital image matrix data and convert them into a video signal output and send it to the display device (e.g., monitor) in which the image is formed. Although newer computer motherboards have graphics cards integrated into their chipsets (i.e., have integrated graphics processors), most PACS medical image displays still use separate graphics cards because of the special graphics requirements of medical image display devices.

The major components of a graphics card are as follows: (1) a graphics processing unit (GPU), which is a dedicated microprocessor optimized for calculations used in 3D graphics rendering and similar functions; (2) video memory, which is the graphics card's own RAM and typically ranges from 128 MB to 4.0 GB; (3) video BIOS, which is the firmware containing the basic program that controls the operation of the graphics card and provides instructions for the interface between the graphics card and the main computer; and (4) output ports permitting the card to be connected to display devices, the most common ports being Video Graphics Array (VGA) and DVI, as shown in Figure 3.5.

3.2.2 Image Display Devices

The most common image display devices in medical imaging are, as noted above, various types of computer monitors, particularly LCD flat-panel monitors. Other not-so-common display devices include high definition TV (HDTV), digital projectors, and products based on other display technologies. Figure 3.6 shows typical display monitors often found in a hospital environment. The function of all these types of image display devices is to convert digital video signals into images. A graphics card combined with the display devices forms the image display system. The working principles of the various display devices are discussed in detail later in this chapter.

3.2.3 Data Flow in Image Display Devices

As can be seen in Figure 3.4, transitioning the digital image data to a displayed image on the monitor requires multiple conversions. The first LUT is applied to data on the graphics card,

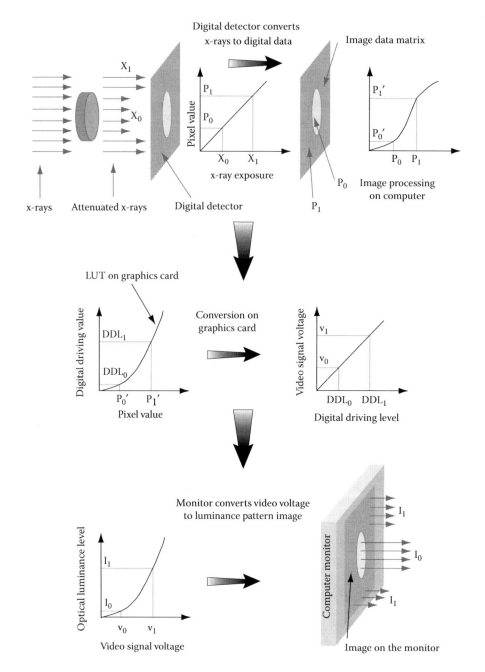

Figure 3.4 The data flow and multiple conversion steps in the acquisition and display of digital images. The lower two panels of the figure illustrate multiple conversions occurring during the image display process.

whereby the pixel values of the image are converted to digital driving levels; these in turn are converted to video signal voltages. The video signal voltages are output from the graphics card via a standard interface (e.g., VGA, DVI, etc.) to the computer display monitor. Within the computer monitor, the input video signal voltage is converted to a luminance level for each image pixel. These pixels form a dot matrix representation of the image, which can be seen by human viewers. After multiple steps of conversion are applied, the relationship between the pixel value of the image data and the luminance level of the pixel on the monitor surface is usually nonlinear.

Figure 3.5 Common connector types (VGA on the left and DVI on the right) for image display devices.

Figure 3.6 Various types of image display devices commonly seen in hospitals.

3.3 OVERVIEW OF IMAGE DISPLAY TECHNOLOGIES

Medical images vary broadly in their characteristics (bit depth, spatial resolution, and color). Their diagnostic quality and accuracy may depend on whether the information contained in these digital images can be displayed with high fidelity (that is, accurately and faithfully). After all, because a human observer cannot "see" the bits and bytes in a digitized medical image, the quality of the display system directly impacts the final quality of these images, just as the quality of film once did. Therefore, computer monitors should be periodically checked to ensure their adequacy for clinical use, just as film/screen and processor systems were.

Medical image display monitors (or medical-grade monitors) require high spatial and contrast resolution, especially those used for primary diagnostic purposes. Prior to the 1990s, most medical-grade computer monitors were CRT-based. At that time, only the CRT-based monitor technology had a high-enough spatial resolution and contrast sensitivity for medical imaging. For those display monitors, special graphics cards and driver software were also required. Because of the relatively small market size, medical-grade display monitors historically were very expensive compared to computer monitors designed for the general consumer market; However, while medical-grade image display monitors remain more costly, the cost differential has declined in recent years.

In the late 1980s and into the early 1990s, LCD display monitors started to enter the consumer market. These monitors became an instant success because of their lower energy consumption (~30% to 50% less than equivalent-sized CRT monitors), brighter images, lighter weight, smaller footprint, and virtually no glare. Although earlier LCD monitors had less spatial resolution than CRT monitors, the deficiency was quickly resolved by new LCD monitors with much higher spatial resolutions and large matrix sizes. Today, LCD monitors have almost totally replaced CRTs for medical image display.

As noted earlier, several other display technologies are on the horizon; while they have potential applications in medical imaging, they are still not fully commercially viable yet. These advanced technologies include those based on OLED technology, micro-mirror, plasma, and head-mounted technologies. LCD displays with high brightness and large pixel array sizes—up to 10 mega pixels (MP)—are now available for use on radiology workstations and have already replaced medical-grade CRTs as the dominant medical display devices. Another interesting technology, the field emission display (FED), is based on the luminescence of phosphors generated by electron bombardment using semiconductors and microelectronics; each pixel element is an individually controllable field emitter. An LCD monitor can be classified as a "transmissive" display device because its pixels alter the transmission of a backlight to the faceplate, whereas FEDs and OLEDs may be classified as "emissive" types, with pixel elements that emit light themselves. Although progress has been made in light-emitter-based image display devices, they have not achieved the promised display quality once predicted. Some emitter-type display devices do not meet the spatial resolution, contrast resolution, or display matrix size requirements of medical imaging displays, even though they may be useful in other applications. Nonetheless, with the rapid development in display technologies, it is likely that they may satisfy the specific needs of diagnostic medical applications in the near future. In the following sections, the fundamental principles of all the display technologies discussed here—historical, current, and future—are briefly reviewed.

3.3.1 History of CRT Monitors

Of the display technologies mentioned above, CRT technology has the longest history. As the first electronic display device, it was developed in the 1920s. As CRTs are still used in many existing imaging modalities as display consoles, and understanding their principles and concepts helps clarify the principles of other display technologies, CRTs are briefly discussed first.

In a CRT, a cathode produces a stream of electrons via a thermal emission process, in which the filament is heated and free electrons are generated at the surface of the cathode. Electrons are drawn from the cathode and moved through the control apertures by a positive electrical potential, accelerated by approximately 25 kV. The CRT screen is coated with phosphorescent materials. When the high-energy electrons strike these materials, a bright dot is generated by the phosphors, emitting visible photons.

To generate a visible image on a monochrome CRT monitor, the single narrow electron beam is moved (or scanned) one line at a time in rectilinear fashion across the face of the phosphor-coated screen. In other words, the bright dot generated by the narrow electron beam moves from one side of the screen to the other and from top to bottom. The scanning electron beam moves so fast (typically the electron beam scans the entire screen over 30 times per second or more) that the human eye is not sensitive enough to perceive the scanning motion; rather, the information is perceived by a human viewer and appears as an image on the screen. The number of times that the electron beam scans the entire screen per second is called the frame rate. Usually, the frame rate is configurable, but it is typically set at 60–70 Hz. When the frame rate is too low, a viewer can see the so-called "flickering effect" of the image.

Monochrome CRT monitors are capable of generating maximum luminance of up to 500 cd/m^2, and these levels are much lower than the luminance levels of typical light boxes on which radiographic films are viewed (about 1000–2000 cd/m^2) or those used for mammography (about 3000 cd/m^2). Therefore, when CRT monitors are used, extraneous light in the reading room has a much larger impact on the viewer's sensitivity to low-contrast pathology. In general, a larger electron beam current leads to a greater luminance level or brightness in a CRT monitor, but it also leads to a larger beam spot size and thus reduces the spatial resolution of the monitor. In addition, larger beam currents and higher image luminance settings reduce the useful lifespan of the display device, because the conversion efficiency of phosphor materials decreases as the number of bombarding electrons grows. This reduction of conversion efficiency is reflected by a reduction in the luminance level or brightness of CRT monitors over time, which is sometimes called an aging process.

The design of color CRT display devices is similar to that of monochrome CRTs, but color devices contain three electron guns instead of one, generating three scanning electron beams. The strength of each of the three electron beams is independently modulated by its own video signal, thus the relative mix ratio of these three basic colors, Red Green Blue (RGB), can be modulated to determine the final perceived color of each pixel. These three phosphor elements in each pixel are called subpixels. When viewed from a normal viewing distance by a human viewer, the subpixels cannot be distinguished and are perceived as one dot with a color determined by the relative mix of the three subpixels.

3.3.2 LCD Monitors

At present, most electronic devices used to display medical images in PACS are flat-panel LCD monitors, which are much smaller in size and weight than their CRT predecessors; consequently, they are sometimes hung on supporting frames rather than standing on the desk or tabletop. The flat-panel LCD monitor is based on active matrix liquid crystal display (AMLCD) technology, which started in the late 1980s in the general consumer computer market. It gradually found its way into the medical display marketplace, benefiting from the consumer market demand for flat-panel monitors which drove production costs down dramatically.

LCD display devices rely on the electro-optical characteristics of liquid crystals (LCs). The polarization properties of LC molecules can be altered by applying an external electrical field. Consequently, the optical characteristics of the LC material can be controlled electronically. More specifically, the polarization orientation of a light beam passing through the LC material can be modified by varying the applied voltage of this electrical field. This modification is called the "electro-optical effect," and it is used in LCD monitors to modulate light transmission. An LCD-based image display device utilizes a large number of such LC cells, with each representing a pixel on the monitor.

Figure 3.7 schematically illustrates the working mechanism of LCD-based monitors. As shown in the figure, the backlight assembly (consisting of light tubes, diffusers, and reflective mirrors) directs light to the front, passing through the first polarizing filter, after which the nonpolarized light from the backlight assembly becomes polarized. Sandwiched between the first polarizing filter and the second polarizing filter is the LC layer (sometimes called the alignment layer in an LCD monitor), which contains pixel elements that are individually controlled via electronics. The LC layer "twists" the orientation of the polarized light emerging from the first filter by changing the voltage across the LC; this twist, the degree of which is controlled by the amount of voltage across the LC layer, and thus controls the transmitted light intensity. After the "twisted" polarized light hits the second polarization filter, the intensity of light exiting the second filter depends on the relative angle between the polarization direction of the light entering the second filter and the filter itself. Consequently, by modulating the voltage applied on the LC layer, the output light intensity can be varied.

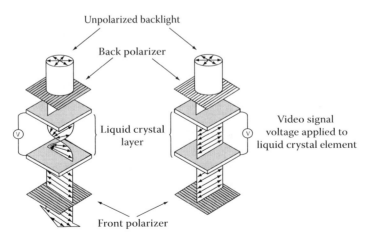

Figure 3.7 LCD working principles.

The maximum amount of transmitted light (i.e., the maximum luminance of the display) is achieved when the polarized light, after passing the first filter and altered or rotated by the LC layer, is parallel to the second polarizing filter. The maximum intensity is determined by the intensity of the backlight, attenuation factors of polarizing filters, transmission of the LC cell in its full ON state, and the transmission characteristics of additional color filters for color displays. For medical-grade monitors, the light output level of the backlight assembly is controlled to be constant over the useful lifespan of the monitor. The minimum luminance of the display is achieved when the polarized light, after passing through the first filter and the LC layer, is perpendicular to the second polarizing filter. The minimum intensity is primarily determined by the maximum opaqueness of the LC cell in its full OFF state, in addition to attenuations of the two polarizing filters. In AMLCD, the switch between ON and OFF states is controlled through voltage changes produced by a thin-film transistor (TFT) array. Shown in Figure 3.8 is a typical cross-section of an AMLCD. For an LCD monitor, the LC layer for each individual pixel element is constructed with this sandwiched structure in order to create an image. While specific LCD monitors vary, the general principle is the same. Further details about specific LCD designs are beyond the scope of this book.

One of the unique aspects of LCD devices is that the light emitted from the monitor surface is non-Lambertian. In other words, the characteristics of the light (e.g., intensity, polarity) emitted from the LCD monitor surface are not symmetrical. For instance, the intensity of light measured at different angles from the central axis of the monitor surface may differ markedly. The two primary reasons for these asymmetries are: (1) the optical anisotropy of the LC cell, which depends on the design of LC pixel elements and the applied voltage; and (2) the effect of polarized light being viewed in a direction co-linear with the polarizing filter. These two effects result in a potentially severe angular dependence of the output luminance level or light intensity, especially for the earlier model of LCD monitors. The angular dependence of LCD monitors can be quite severe and may affect image contrast as a function of the viewing angle. This problem was quite severe in early designs of LCD monitors and delayed the broad adoption of LCD monitors for medical imaging purposes. In the past decade, improved LCD design technologies, such as molecular alignment in subregions within each individual pixel, in-plane switching, and other technologies to compensate for optical anisotropy, have further minimized such angular dependence and have sped up the adoption of LCD monitors. Furthermore, advancement in LCD manufacture technology in recent years has given LCD monitors the very high spatial resolution needed for medical imaging. To date, 2 MP to 5 MP monochrome and color medical-grade monitors are readily available.

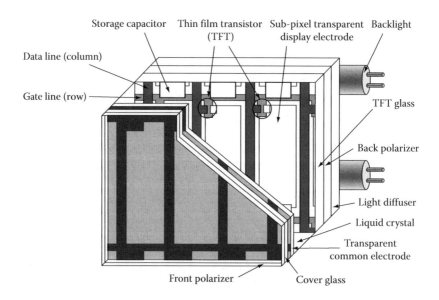

Figure 3.8 Schematic drawing of LCD monitor structures.

3.3.3 FED Devices

FED devices are another type of flat-panel display. FED is similar to CRT technology in that electrons are emitted from a cathode, accelerated, and they bombard the phosphor coating, which emits light. However, FEDs are based on semiconductor microprocessing technology and are only a few millimeters thick. Instead of using the thermo-ionic emission in a single electron gun as in a CRT, an FED utilizes a large array of micro metal tips that emit electrons, as shown in Figure 3.9. These metal tips are positioned behind each phosphor coating dot or pixel and emit electrons when electrical voltage is applied. There are multiple electron emitters behind each pixel; thus the chance of having complete dead pixels, as may occur in LCD monitors, is low.

Additionally, FEDs are more energy efficient, consuming less power than existing LCD and plasma display technologies. They also promise other favorable characteristics, such as better viewing angle independence than an LCD, and potentially higher luminance and higher contrast than both LCD and CRT devices. However, other technical challenges including high-yield manufacturing prevent it from becoming a dominant commercial product in the image display device field. To date, only small-matrix-sized display panels have been produced for demonstration purposes.

3.3.4 OLED Technology

OLED technology was invented in the 1970s. It has the potential to offer brighter and crisper pictures while using less energy than traditional LED technologies. Unlike LCD, OLED screens do not require backlight, and thus have the potential to reduce the screen thickness to several millimeters. OLED is an image display technology based on the principle of electro-luminescence (EL), which directly converts electrical current into light. The luminance of each matrix element or pixel in an OLED display is controlled by the electrical current delivered to each pixel. OLEDs have high light emission efficiency, potentially converting nearly 100% of electrical current into photons of light, thus making this display method potentially very energy efficient. EL-based devices first appeared in the market place in the early 1990s. Since then progress in OLED materials research has resulted in devices that operate efficiently at low voltage. In addition to being superbly efficient, OLED-based technology promises unprecedented brightness of over 1000 cd/m², making it potentially useful in medical image display. This technology is still in the early stages of commercial development for large-matrix display devices.

Other types of display devices (plasma HDTV, projective display, etc.) are not specifically designed for medical use but are appearing in the market for office, desktop, and educational purposes. Although these technologies are being used in medical facilities, their primary purpose is usually not for diagnostic purposes.

3.4 QUALITY STANDARDS FOR DISPLAY DEVICES

Medical display monitors are mostly categorized based on their usage. As described in Chapter 2, a PACS image review workstation may be equipped with one or more monitors. Depending on their primary functions in the clinical setting, monitors at various workstations may differ in quality and cost. Image display devices used in medicine are usually categorized into two classes according to the purpose of their primary usage.

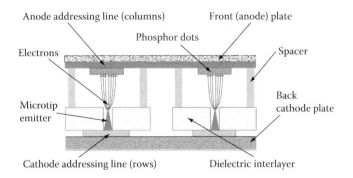

Figure 3.9 FED working principle.

3.4.1 Class 1 and Class 2 Image Display Devices

Class 1 monitors are designated to be used for medical image interpretation and for making primary diagnoses. These monitors are mostly located in radiology departments and other departments where medical images are viewed and primary diagnoses are made (e.g., the orthopedics department). Class 1 monitors typically have high brightness and high spatial resolution. They are usually monochrome or color 3 MP or 5 MP monitors and are equipped with high-performance graphics cards. These monitors require stringent and routine QC checks to ensure they meet the highest performance specifications, and they are mostly located in reading rooms with well-controlled ambient light or viewing conditions, in the radiology department or under the management of the radiology department. Figure 3.10 shows a PACS workstation and its Class 1 monitors.

Class 2 image display monitors are mainly used for nondiagnostic clinical use. They are usually less expensive than medical-grade monitors and mostly use off-the-shelf graphics cards. They are primarily used by general medical staff and medical specialists other than radiologists for patient demographic displays, for clinical image review and image acquisition QC, or as on-board monitors for MRI, CT, ultrasound, and other image-acquisition systems. These monitors may be either monochrome or color. Figure 3.11 shows a QC workstation with a Class 2 display monitor.

Even though these Class 2 image display devices do not have spatial and contrast resolution as high as those of Class 1 monitors, their characteristics, especially their contrast or luminance response characteristics, can directly affect the workflow and quality of medical images. For example, in DR or CR, technologists need to review the images and adjust their contrast on QC monitors before sending them to the PACS for interpretation and diagnosis by radiologists. In the early days of PACS implementation, the luminance response characteristics of QC monitors were often very different from those of the PACS workstations used by radiologists. Radiologists may not be satisfied with image quality and appearance on PACS workstations, while the same images look reasonably good on the technologists' QC workstations.

The primary cause of this difference in quality is the difference in luminance response of contrast LUTs from these two types of display devices. Therefore, Class 2 monitors must also be maintained with some requirements (such as the DICOM-standard LUT) to meet a minimal quality level as a part of PACS. Monitors in both classes must meet the performance requirements of the imaging modality for their intended use. For any given imaging modality, the display system's performance requirements may be more or less stringent than that required by other PACS components. For instance, MRI images are rarely acquired with an image matrix size of more than 512×512, while for CR and DR the acquisition matrix size is at least 2 MP. Therefore, the image display devices equipped with an MRI scanner may be less stringent than those used in CR/DR systems.

Figure 3.10　Class 1 monitors for a PACS diagnostic workstation.

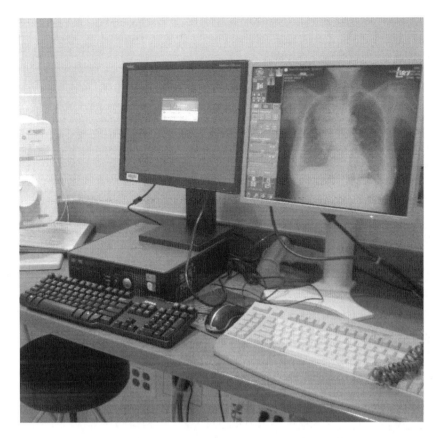

Figure 3.11 A QC workstation and its Class 2 monitors typically used in a radiology department.

Differences between primary Class 1 and secondary Class 2 display devices are mainly in performance requirements on their spatial and contrast resolutions. While selecting the display system during implementing a PACS, the user must understand and establish desired specification and performance requirements based on user needs and workflow in their clinical operations. Requirements must be specified in detail before purchase and clearly communicated between the user and the PACS vendor or the manufacturer of the display system. These requirements become the basis for performance assessment of the display system in acceptance testing and during routine QC procedures once the PACS is installed and in operation.

Ideally, the performance of medical display devices used in diagnostic or clinical modalities should be evaluated and monitored quantitatively on a regular basis, because their performance may deteriorate over time. Routine monitoring is important because such deterioration may be subtle and occur over a long period of time, meaning that the user may not be aware of such changes before it is too late (e.g., before misdiagnosis occurs). Unfortunately, a number of Class 1 medical displays, such as those used in fluoroscopy, digital angiography, or digital subtraction angiography, and some Class 2 displays, including operator console monitors, are "closed" and thus are difficult to assess quantitatively. The American Association of Physicists in Medicine (AAPM) Task Group 18 (TG 18) has developed a series of test patterns that can be used for visual inspection in such a scenario. These patterns have been carefully designed to reveal subtle changes in the performance of display devices. These test patterns can be loaded into the system databases via DICOM, so that even closed imaging systems can be monitored for changes in performance.

3.4.2 Color versus Monochrome Displays

Until several years ago, medical-grade display devices were mostly monochrome monitors. There were many reasons, but the primary one was that luminance level and spatial resolution requirements for medical applications were much higher than consumer-grade monitors could attain. A typical consumer-grade color monitor had a maximum luminance level of 300 cd/m^2 or less, with a spatial resolution matrix of no more than 1920 × 1080. A typical medical-grade monitor, however, usually requires a luminance level of 400 cd/m^2 or higher and a spatial resolution matrix of 1500 × 2000 or more. On the other hand, most of the interfacing software on a PACS workstation and other text-based software such as RIS and HIS programs are in color. Therefore, today's PACS workstations are typically equipped with one consumer-grade color monitor and one or two medical-grade monochrome monitors. The color monitor is typically used for text-based software applications while monochrome monitors are used for medical image displays. In the past a few years, color monitors with high spatial resolution (3 MP and up) and high luminance levels (>400 cd/m^2) are available and are quickly becoming popular for medical use. As more medical images are presented in color, such as those obtained by Doppler ultrasound, functional MRI, nuclear medicine, and PET/CT, medical-grade color monitors with adequate spatial resolution, luminance level, and contrast will find increasing demand in PACS. It is foreseeable that in the future all medical-grade monitors in a PACS workstation cluster will be in color, except in some very special cases.

3.4.3 Image Quality Dependence on the Configuration of Display Devices

As mentioned at the beginning of this chapter, an image display system consists of two components: the image display device or monitor that the user can see and the graphics card(s) located inside the computer case. The characteristics of both these components have a direct impact on the final displayed image quality. In fact, in order to guarantee optimal quality settings for display systems, many medical display device vendors either sell monitors with matching graphics cards or strongly recommend that customers use specific graphics cards that have been configured in a specific way. As discussed earlier in this chapter, there are several steps in the process of transforming bits and bytes of digital image data into a visible image that can be seen by a human viewer. These steps consist of several conversions that change or modify pixel values of the image data; thus, the final image's appearance and contrast are affected by any one of these conversions (the LUTs). In other words, in the display process, the image's contrast or appearance depends on the configuration or adjustment of these LUTs on the graphics card or the monitor. Therefore, it is possible that identical display devices may have different image quality depending on the devices' specific configurations or setups.

Because of the possibility that inadvertent change to display devices' configuration will cause nonoptimal image display quality, most vendors "lock" adjustment functions for display devices, with access to those functions controlled by special passwords or service keys. Additionally, a sudden power surge or corruption of software may cause the configuration to change. Therefore, it is imperative for users to routinely perform a QC on all image display devices throughout a PACS to ensure their consistency in quality and performance. AAPM TG 18 have developed detailed guidelines and recommended procedures for routine QC of image display devices. DICOM Part 14 specifies a standard luminance conversion table or curve. The DICOM LUT or DICOM luminance conversion table (the DICOM Gray Scale Display Function [GSDF]) is a standard with which all medical display devices must comply.

3.4.4 Factors Determining Display Device Quality

The technical specifications or engineering parameters of display systems are vital to the final quality of images displayed and seen by the viewer. Sometimes PACS users face the task of specifying and selecting appropriate image display devices for their PACS or their EHR systems. At such times, users have to evaluate and compare specific parameters to ensure that the display systems' quality is adequate for their designated uses. To help users understand specifications that are critical for display devices, some important factors are discussed below. The most relevant specification may vary depending on the display device's intended use; for example, some color characteristics of a display device may not be as critical for medical-imaging applications as they are for printing or advertising use. Sometimes, the outside appearance of a medical-grade display monitor may not be too different from that of a consumer-grade display monitor.

The parameters and specifications discussed in the following sections apply mostly to medical imaging. They should not be thought of as minimum requirements; rather, they should be used as references or points of consideration when evaluating and comparing different display devices.

3.4.4.1 Physical Dimensions

The physical dimensions of a display device are its height, width, depth, and weight. Although a display device's physical dimensions do not directly affect image quality, these characteristics are relevant to space planning and PACS workstation installation. For medical use, the most popular size of monitor is 20–22 in diagonally, which is approximately the same size as a conventional chest x-ray film. Even though larger display systems are available, at the normal viewing distance of 18 in or so, 20–22 in monitors seem to be optimal, allowing the viewer to browse through the entire display area without moving his or her head too often. Today's display systems are mostly LCD-based flat-panel monitors that occupy much less space and weight than traditional CRT-based medical-grade monitors. These flat-panel monitors are lightweight and can be hung on moveable arms, thus making more ergonomic workstation design possible.

3.4.4.2 Input and Output Signal Types

As mentioned earlier in this chapter, two main standard input types are used in today's display systems: VGA and DVI. VGA was introduced with the IBM PS/2 line of computers in the late 1980s and has been adopted as the analog computer display standard, with connection to the workstation/PC via a 15-pin connector. As the analog image source, that is, the graphics card's analog output transmits each horizontal line of the image, it varies its output video voltage to represent the desired brightness. In a CRT device, this process is used to vary the intensity of the scanning beam as it moves across the screen. VGA is the last graphical standard introduced by IBM and has been superseded by several slightly different extensions to VGA, such as Super VGA. DVI is another video interface standard designed to maximize the visual quality of display devices such as flat-panel LCD monitors. DVI was developed by an industry consortium and is designed for digital video data being carried to a display device from the graphics card. Shown above in Figure 3.5 are a VGA and a DVI connector on a PACS workstation.

The DVI interface utilizes a digital protocol in which the desired luminance levels of the pixels are transmitted as digital binary data. When the display device is driven at its native resolution, it reads the data and applies a brightness value to each pixel. Therefore, each pixel in the output buffer of the graphics card corresponds directly to one pixel in the display device, whereas with an analog signal the appearance of each pixel may be affected by its adjacent pixels as well as by electrical noise and other analog distortions. Unlike DVI, earlier input types (such as the analog VGA) were designed for CRT-based devices and thus do not use discrete time display addressing, as DVI does. DVI cable connectors are designed so that users cannot connect the cable in an incorrect position or orientation. DVI has many advantages over analog input types and has therefore become widely used in newer display devices.

3.4.4.3 Matrix Size

Matrix size describes the number of independently addressable pixels in a display device, as indicated by the number of pixels running along its horizontal and vertical axes. It specifies the maximum number of individual pixels in both the horizontal and the vertical directions that can be accepted by the display device from the graphics card. Current medical display devices are capable of providing a maximum matrix display size of up to 10 MP. Most of today's consumer-grade display monitors can support a matrix size of approximately 1900 × 1200 pixels, or 2 MP, and they cost much less than the medical-grade monitors. Matrix size, together with the active display area, determines a display device's nominal pixel size, which affects its spatial resolution. The nominal pixel size equals the dimensions of the active display area divided by the number of pixels in each direction. It should be noted that nominal pixel size is not the only factor defining the actual spatial resolution. Rather, spatial resolution is a function of the nominal pixel size that is defined by the matrix pixel size, the active display area, and the luminance profile of the pixels displaying the image. For most LCDs, if the display device is running at its native display matrix, the spatial resolution most likely equals the nominal resolution. In general, the spatial resolution may be equal to or lower than the nominal spatial resolution when display monitors are running at their native matrix sizes.

3.4.4.4 Display Area

Display area specifies the physical dimensions of the active image display area. Conventionally, the size of an image display device is defined as the diagonal length of the active display area. For

CRT displays, the diagonal measurement is specified as the outside dimension of the faceplate. Usually, the useful display area is less than the specified dimension. For instance, a quoted 21-in display device may only have 20 in of active display area. For LCD flat-panel displays, the specified and actual display areas are almost the same. Today's image display devices may specify both the horizontal width and vertical height of the display area along with the conventional diagonal dimension.

3.4.4.5 Pixel Size

Display pixel size refers to the nominal physical dimension of the smallest addressable, light-emitting element of the display device. The nominal pixel dimension in the horizontal direction equals the horizontal dimension of the active display area divided by the horizontal matrix size. Similarly, the pixel dimension in the vertical direction equals the vertical dimension of the active display area divided by the vertical matrix size. Theoretically, smaller pixel sizes should yield better resolution.

For CRT-based display devices, the actual pixel size is usually larger than the nominal pixel size. The actual pixel size is defined as the area of light emission of the phosphor upon excitation by the electron beam. The industry standard is to specify the actual pixel size at the 50% point of luminance energy of its luminance profile. The ratio between this value and the nominal pixel size is called the resolution-addressability ratio. For medical-grade monitors, resolution-addressability ratio is approximately 0.9–1.1. Additionally, the size of the CRT pixels in the horizontal and the vertical directions may be different. In the horizontal direction, the electron beam spot size increases on the phosphor screen at larger deflection angles from the center of the CRT; this process causes a larger pixel size and, consequently, poorer spatial resolution at the edges of the CRT monitors. The pixel size is, therefore, usually larger near the corners and edges of the screen than in the center, reducing spatial resolution in those areas. Furthermore, in a CRT display device, as the electron beam current increases, so does the luminance level of a pixel, but the spot size also increases. Therefore, spatial resolution is specified at particular luminance levels.

In LCD-based display devices, the pixel size is determined by the dimension of each microcell of the liquid crystal matrix and is independent of the luminance level of the backlight. It should be noted that, typically, in LCD display devices each pixel usually consists of multiple subpixels. As shown in Figure 3.12, three primary color subpixels are present in each pixel. For medical-grade monochrome display devices, the color filter for each subpixel is eliminated; thus, each pixel consists of three or more grayscale subpixels. The subpixel design affects characteristics of the overall display image quality, such as the angular dependence of the LCD display device.

3.4.4.6 Refresh Rate

The refresh rate defines the frequency with which the entire display frame is updated. In CRTs, refresh rates are expressed as the frequencies of the vertical and horizontal scans. The vertical scan frequency, sometimes also called the frame rate, is usually between 55 Hz and 150 Hz. A refresh rate that is too low generates a flickering effect that can be perceived by the human eye, and it may result in lower user performance and cause user fatigue. In the past, Class 1 CRT-based display devices were recommended to have a minimum refresh rate of 70 Hz. In LCD-based flat-panel display devices, because of data persistence resulting from the relatively slow speed of switching from one polarization state to another in the LC material, the flickering effect is less prominent, which allows these monitors to run at a lower refresh rate. Display devices with higher frame rates, however, have better dynamic characteristics, which is very useful when displaying fast-changing images such as dynamic cine movie files.

3.4.4.7 Luminance

Luminance level usually refers to the output brightness of display devices. A regular consumer-grade desktop color display monitor has a maximum luminance of approximately 100–300 cd/m^2, while a high-luminance display device can have a maximum luminance level up to 700 cd/m^2. These luminance levels may subject to change during the device's lifetime because of the aging of the backlight in LCD devices. High-quality medical-grade display devices have internal, dedicated control circuitry to stabilize luminance levels, which is very important for reliable and consistent quality among display devices over time. However, low-cost consumer-grade display devices usually do not have a stabilizing circuitry.

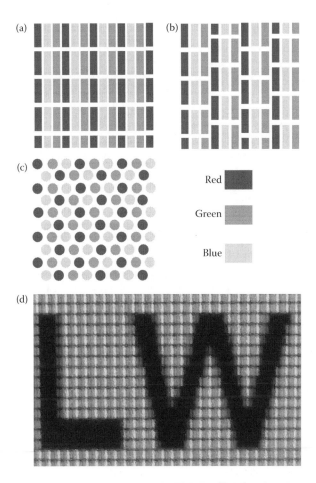

Figure 3.12 Schematic drawing of various types of subpixels layout in color monitors of (a) LCD; (b) Television CRT; (c) computer monitor CRT; and (d) a magnified photographic view of the actual subpixels and the letters "LW" on a color LCD monitor. Three subpixels of RGB form pixels of any color.

Maximum and minimum luminance levels are both important parameters in medical display devices. Vendors frequently advertise their monitors' ratio of maximum to minimum luminance, that is, the contrast ratio or luminance ratio. Usually medical-grade display devices have a higher maximum and a lower minimum luminance level than consumer-grade products, thus giving the former a higher contrast ratio. For a medical-grade display device, the contrast ratio requires regular recalibration in order to comply with the DICOM LUT standard. In some medical-grade display devices, the luminance level itself may be adjustable. It should be noted, however, that there is a balance between achieving higher maximum luminance level while maintaining a lower minimum luminance level. That is, a medical-grade monitor can be too bright, causing fatigue and strain on the user's eyes; thus, the contrast ratio should be considered in the purchase and configuration of display devices.

3.4.4.8 Luminance Uniformity

Luminance uniformity refers to the degree of variation in luminance over the entire active display area when an image with a uniform background is displayed. A uniform output luminance over the active display area is important for the monitor to achieve uniformity or consistency in contrast and spatial resolution. For CRT-based display devices, uniformity tends to degrade at the edges of the active display area because of the increased deflection angle of the electron beams. In LCD flat-panel display devices, nonuniformity is primarily due to nonuniform luminance output of the backlight assembly behind the LC panel and variations among individual pixels. The uniformity characteristics of LCD-based displays are generally better than those of CRTs.

3.4.4.9 Surface Antireflection Treatments

To reduce undesirable ambient light reflected on the surface of the image display device, most medical-grade monitor manufacturers apply a coating of materials with antiglare/antireflective properties. The antireflective coating is applied to the surface of the display device, and has a thickness equals to $\frac{1}{4}$ of the wavelength of the ambient light. In consumer-grade displays, the antireflection treatment is often done by making the glass surface rough through some chemical processes or by applying a spray coating. Such treatments diffuse incident light, thus reducing specular reflections, but they also degrade the displayed image. For practical purposes, the user should check the reflective characteristics of display devices in the evaluation process to make sure there is no excessive reflection.

3.4.4.10 Bit Depth

The bit depth of an image display system refers to the maximum number of gray or color levels that the system can display simultaneously. An 8-bit display device theoretically can simultaneously display 256 distinct gray levels, whereas a 10-bit system can display 1024. The display graphics card usually determines the bit depth of a display device. At present, most medical-grade graphics cards support at least 10 bits. Users should realize that even though, theoretically, the number of shades of gray is claimed to be high, the actual number of gray levels distinguishable by a display system may be less because of limitations in the device's capability. Furthermore, the adjustment of LUTs on the graphics card, made to comply with the DICOM standard GSDF, may significantly reduce the actual number of distinctive luminance levels that a display system can produce.

Bit depth is also very important to the display image quality. In fact, one reason graphics cards for medical use are much more expensive than those for consumer use is that the former requires much higher bit depths.

3.4.4.11 Angular Dependence

The luminance, contrast response, and chromaticity of LCD display devices are usually angle-dependent; that is, the measured luminance output and contrast resolution of the display decrease substantially as the viewing angle to the surface of the devices deviates from normal. This angular dependence was much more severe in the early days of LCD displays. The viewing angle specification of an LCD display device indicates the angular range within which the contrast ratio of the device is maintained within a certain range. Viewing angle is usually separately specified in the horizontal and vertical directions. Today's LCD designs have dramatically improved the angular dependence. However, even within the specified viewing angle range, an LCD display may still exhibit marked variation in luminance, contrast, and chromaticity as the viewing angle increases. Generally speaking, the larger the viewing angle range is, the better the quality of the display device.

3.4.5 DICOM Standard GSDF

A DICOM GSDF is a conversion table that dictates the relationship between each input pixel value and the output luminance level of a display system. This table covers the entire range of input pixel values that the display system can sustain and their corresponding output luminance levels, from the darkest to the brightest. It can be considered as the combination of all conversions that occur in the graphics card and display monitor.

The GSDF was selected by the DICOM standard committee because it is believed that when all image display devices satisfy these conditions (or "are in-compliance to the DICOM standard GSDF"), identical images displayed on various devices will have a similar appearance (i.e., the image will be "equalized" on all display devices that conform to the standard LUT). The DICOM GSDF promises an identically perceived relative contrast at all luminance levels that a display device is capable of producing, or "perception linearization."

One important aspect of compliance with the DICOM GSDF standard is that the GSDF is dependent upon both the minimum and the maximum luminance levels the image display spans, or the dynamic range of the display system, which also determines the contrast ratio described above. However, the contrast ratio may be changed depending on the ambient light conditions in which display systems are located. For instance, a system may need to be recalibrated if it is moved to a different location with different ambient conditions.

Most high-end medical-grade display monitors are calibrated to conform to the DICOM GSDF either at the factory or at the customer's site on demand. The monitors should be verified regularly and recalibrated if needed to ensure compliance to the DICOM standard GSDF. More details about the DICOM GSDF are covered in Chapter 6.

3.5 QC AND QA FOR IMAGE DISPLAY DEVICES IN PACS

Just as in the traditional film-based radiology department, where regular QA and QC on film processing and light box luminance were important for the quality of a patient's imaging studies, QA and QC processes on all PACS workstations are vital to ensure the highest quality and consistency in a patient's medical images display. Since image display is the last, but a critical, step in the imaging chain, display device quality may have a direct impact on diagnostic accuracy and patient care.

3.5.1 QC Requirements for Medical Image Display Devices

Digital image display devices or monitors deteriorate gradually over time, mainly because the brightness of these display devices gradually decreases as they age. This causes a reduction in the contrast resolution. Without a routine QC check, the deterioration may not be noticed. If low-contrast lesions or other pathological conditions are present in an image but the image display device is of low quality or is not configured properly, these abnormalities may simply be invisible to the physician, potentially resulting in misdiagnosis. In addition, the quality of digital images on PACS display devices is strongly affected by the ambient environment or viewing conditions (e.g., excessive light in the reading room area where the workstation is located). Therefore, it is critical to check the displayed image quality in these devices' actual ambient environment to ensure optimal image quality.

Furthermore, in multicenter clinical trials that include digital medical imaging as a screening or diagnostic tool, it is important to ensure consistency and optimal settings of digital display devices at all participating institutions. In that scenario, a centralized QC and QA tool is highly desirable. A centralized QC database containing information about the image display devices' settings, contrast sensitivity, and similar data is useful to ensure high quality of the imaging studies. Such a database also makes it easy to compare and manage quality disparities between institutions, so that frequently occurring mistakes can be documented and corrected. A centralized system is also applicable outside of multicenter clinical trials. For instance, it can be used to optimize telemedicine and teleradiology applications. Service providers can use such systems to validate and assess image display quality before allowing images to be read remotely.

AAPM TG18 has recommended QC procedures involving tests and the use of sophisticated equipment to measure the physical characteristics of display devices. To make the tests more practical for routine QC checks in the clinical environment, the TG18 report also provides test pattern images that test particular aspects of device quality.

The ultimate purpose of routine QC and QA for image display devices is to detect subtle and gradual deterioration in the quality of displayed images in clinical settings, so that corrective action can be taken before misdiagnosis or other medical errors occur.

3.5.2 Image Display Device QC and QA Procedures

The following tools are needed to perform QA and QC using the AAPM TG18 guideline:

- Luminance meter or photometer that is calibrated to the national standard

- Calibrated illuminance meter

- Colorimeter

- Test pattern images (e.g., test patterns provided by AAPM TG18)

There are two major categories of routine QC procedures: routine qualitative checks and less frequent quantitative checks. Quantitative checks require the use of sophisticated measurement tools such as calibrated luminance meters, illuminance meters, and colorimeters. AAPM TG18 has published detailed guidelines and descriptions for these test procedures (see the bibliography at the end of this chapter). Among these QC tests some need to be done more frequently than others. Although there is not yet a specific regulatory requirement for softcopy display systems QC in

PACS, as a good clinical practice and in the spirit of the Joint Commission on the Accreditation of Healthcare Organizations (JCAHO) guidelines, users of PACS should perform necessary QC and implement QC measures regularly to ensure high-quality patient care. This following section specifies aspects of an image display device that the authors believe should be assessed quantitatively and/or qualitatively, with recommended guidelines for the frequency of such testing, and aspects of the testing and results that PACS users should pay attention to.

3.5.2.1 Ambient Lighting Level

Digital display devices are particularly prone to be affected by their ambient environment, mainly because of their low luminance output levels (350–500 cd/m^2) relative to light-boxes (1500–3000 cd/m^2). The main effect of improper ambient environment is a reduction in contrast resolution directly resulting from excessive light in the area where the image display devices are located. Therefore, it is critical to maintain a well-controlled, low-light reading room environment. Artifacts and loss of contrast sensitivity (and thus loss of the ability to detect low-contrast objects) caused by reflections from the display device surface can be minimized by controlling the ambient environment. A general rule is that the ambient lighting level in reading rooms for primary diagnosis in radiology should be maintained at 10 lx or lower for x-rays and mammography and below 60 lx for cross-sectional images from imaging modalities such as MRI, CT, nuclear medicine, and ultrasonography. Also, display devices should be positioned so that no light falls directly on the display surface. If the viewer's reflection can be seen on the display device surface, the reading room's lighting should be rearranged or the display monitor relocated. The position of the display device should also be verified or corrected before any QC test is performed.

The lighting condition in a reading room should be measured, recorded, and routinely checked with an illuminance meter, as shown in Figure 3.13. Some display monitors are equipped with photocells for ambient light detection, and allow monitors to self-adjust the output luminance level and luminance response according to ambient light changes. This feature, however, should be used with extreme caution because the self-adjustment may cause noncompliance to the DICOM standard GSDF. Therefore, it is critical to strictly follow the vendor's guidelines.

Tool Required
Calibrated illuminance meter.

Procedures
Measure and record the illuminance levels at the same locations everytime in the reading room.

Figure 3.13 An illuminance meter (left) and a luminance meter (right).

Recorded Values

Illuminance level (in lx).

QC Check Frequency

Quarterly or every time there is a change in the layout of the reading room or lighting conditions.

3.5.2.2 Minimum and Maximum Luminance Levels

The minimum and the maximum luminance levels are important characteristics of a display device because they are part of the determining factors for contrast sensitivity and the luminance response curve. Therefore, they should be measured and recorded regularly and monitored over time for drift or degradation. Most of the medical-grade monitors have internal control mechanisms to maintain the maximum luminance constant. This test can also verify the working status of such mechanisms.

Tool Required

Calibrated luminance meter, as shown in Figure 3.13.

Procedures

To measure luminance levels, the user needs to use the luminance meter and AAPM TG18 test patterns TG18-LN12-01 for minimum luminance and TG18-LN12-18 for maximum luminance measurements. Shown in Figure 3.14 are TG18-LN12-01, TG18-LN12-09, and TG18-LN12-18 patterns for testing. At the center of each of these test patterns is a square patch. The luminance level of this patch is measured and recorded to define the minimum and the maximum luminance values. When using a near-range luminance meter, the meter should be placed directly over the patch and be in contact with the display device's surface, but no pressure should be exerted on the surface itself. When a telescopic luminance meter is used to measure luminance levels, the measured area (as indicted by the area seen through the meter's viewfinder) should be centered on the square patch.

Recorded Values

Minimum and maximum luminance levels, which are compared with the previous QC test results. Watch for gradual drift and any trends, and make sure the values still fall within the specifications or the AAPM TG18 guidelines.

Recommended QC Check Frequency

Monthly at the beginning of use, then quarterly if no changes are observed.

3.5.2.3 DICOM Standard GSDF Compliance

Tools Required

The TG18-LN, TG18-CT, and TG18-QC patterns and a calibrated luminance meter.

Figure 3.14   From left to right, AAPM TG18-LN12-01, TG18-LN12-09, and AAPM TG18-LN12-18 patterns for luminance measurements. Luminance response curve or LUT of monitor can be obtained by measuring luminance level of the center square of multiple TG18-LN patterns using a luminance meter.

Procedures

There are several methods of checking for DICOM GSDF compliance. For a quantitative measurement, the user can use the calibrated luminance meter to measure the luminance levels of the center patches of all the eighteen TG18-LN patterns. Luminance levels are then plotted against the input pixel values corresponding to the patches to obtain the measured LUT, which is then compared with the DICOM standard GSDF.

To qualitatively check for compliance, the user can bring up the TG18-CT pattern after recalibrating the monitor according to the vendor's specifications and then evaluating the test pattern. Half-moon features should be visible in all 16 patches. Shown in Figure 3.15 is the AAPM TG18-CT low-contrast test pattern for qualitative evaluation of display devices' luminance response.

In another qualitative check, the user can review the TG18-QC test pattern shown in Figure 3.16. Each letter of "QUALITY CONTROL" should be visible in all three patches, as well as all small squares in each of the 18 patches at the center of the test pattern.

Recorded Values

For quantitative measurements, the deviation of the measured LUT should be within the control limits specified by the AAPM TG18 report. For qualitative measures, the number of visible objects should be documented, compared with the value recorded at acceptance testing, and any decrease in the number of visible objects should be noted.

QC Check Frequency

The user can evaluate the display quality using test patterns at least weekly or as needed. Quantitatively the LUT should be measured annually. Most medical-grade monitor vendors

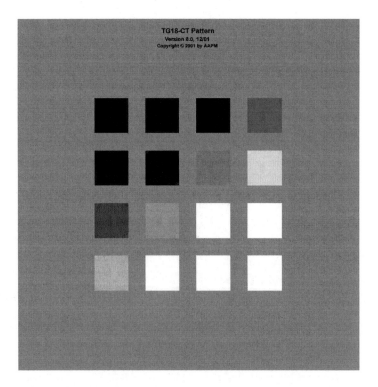

Figure 3.15 AAPM TG18-CT low-contrast test pattern that can be used for the qualitative assessment of display systems' luminance response. On this test pattern, there is a circular object containing two half-moon features in each of the 16 squares. All 16 half-moon features should be visible if the LUT conforms to the DICOM standard.

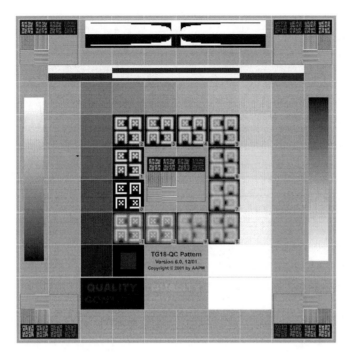

Figure 3.16 The AAPM TG18-QC test pattern. It can be used for daily QC overall quality checks. On the lower part of this pattern, inside the three rectangles there are low-contrast words "QUALITY CONTROL". For a Class 1 medical grade monitor used for primary diagnosis all those letters should be visible. The pattern can be used for other quality measurement such as geometric distortions for the CRT monitor and spatial resolution. Additionally, the two grayscale strips on the sides of the test pattern should appear smooth without appreciable steps.

provide an optional tool for DICOM LUT self-calibration. This can be very helpful in the quantitative assessment of the LUT.

3.5.2.4 Contrast Resolution and Noise

Tool Required

The user can visually assess the display noise using the TG18-AFC test pattern. The pattern shown in Figure 3.17 is contrast-enhanced to illustrate the features of low-contrast objects.

Procedures

The TG18-AFC test pattern is brought up on the display device to be tested. The user then counts the total number of visible squares that contain two low-contrast objects. For a square to be counted as visible, both the target at the center of the square and the one at the corner must be visible. Alternatively, one can count the number of targets visible in five big squares, one at each corner and one at the center of the test pattern. In theory, if the noise level of the monitor is high, the number of visible squares will be less.

Recorded Values

The user should record the number of visible squares or targets and keep track of its trend over time. A downward drift may occur due to a loss of contrast resolution and an increase in noise. The user should understand, however, that this test pattern has been found to have limited sensitivity for deterioration in contrast resolution and spatial noise in digital display devices. A recent study showed that a more sensitive test may be needed in order to assess the contrast-resolution characteristics of the monitor.

QC Check Frequency

Bi monthly or as needed.

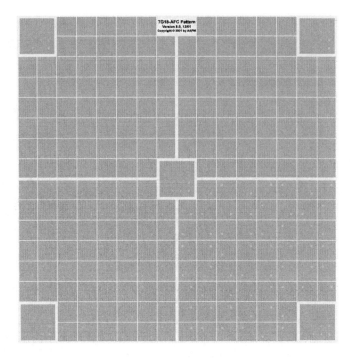

Figure 3.17 The AAPM TG18-AFC test pattern can be used for visual assessment of display system's contrast-noise characteristics. There are two dots inside each square (the contrast of the picture shown in this book has been enhanced in order to show the features). Both dots must be visible to be counted as a visible square when counting the total number of visible squares in the qualitative assessment of monitor's quality.

3.5.2.5 Spatial Resolution

Tool Required
TG18-QC test pattern.

Procedures
The user brings up the TG18-QC comprehensive test pattern image shown in Figure 3.16 on the display device to be tested. The spatial resolution of line pairs at the four corners and at the center of the test pattern is checked. All line pairs should be clearly resolved.

Recorded Values
Pass or fail.

QC Check Frequency
Monthly or as needed.

3.5.2.6 Geometric Distortion

Geometric distortion is mainly an issue for CRT monitors. Geometric distortion is minimal for LCD flat-panel display monitors, so it may not be necessary to test geometric distortion on LCD devices.

Tool Required
TG18-QC pattern and a soft flexible ruler.

Procedures
The user brings up the TG18-QC pattern on the display device, and then measures the horizontal and vertical dimensions of the center square and several vertical and horizontal lengths on the

square. Obvious distortions of straight lines in the test pattern indicate that corrective action is needed.

Recorded Values

Record and document measured values.

QC Check Frequency

Annual or as needed.

3.5.2.7 Display Reflection

Tool Required

TG18-AD test pattern.

Figure 3.18 shows the TG18-AD test pattern for visual evaluation of a display's diffuse reflection response to ambient light. The pattern is brightened and contrast enhanced to illustrate its features.

Procedures

The user brings up the TG18-AD test pattern image on the display device to be tested. The room lighting should be set at the ambient conditions present during actual clinical use. The test pattern is observed at a normal viewing distance, and the number of squares with a visible line-pair pattern is counted.

Recorded Values

The number of visible squares is recorded and compared with previous QC check results. The user should also watch for degradation or downward drift between QC checks.

QC Check Frequency

Semi annually or as needed after cleaning the image display device.

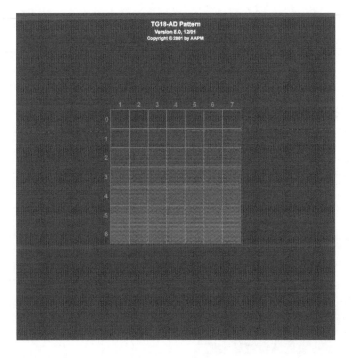

Figure 3.18 The AAPM TG18-AD test pattern used for evaluating the antireflection coating on the monitor and its response to ambient light. The user should count the number of squares inside which all the lines are clearly resolvable. The contrast of the picture shown here has been enhanced to demonstrate the features contained in the test pattern.

3.5.2.8 *Luminance Uniformity*

Tool Required

TG18-UN test pattern images, calibrated luminance meter.

Procedures

The TG18-UN images are displayed one at a time, as shown in the Figure 3.19, on the display device to be tested. The luminance meter is then used to measure the luminance levels at the four corners and at the center of each test pattern.

Recorded Values

Measured luminance level values are recorded and nonuniformity of the luminance level can be calculated from those values.

QC Check Frequency

Quarterly, during the initial months of use, and then annually or as needed thereafter. Note the uniformity is usually not an issue in LCD monitors.

3.5.2.9 *Angular Dependence*

Regular QC for angular dependence is not required; however, the angular dependence of the display device should be checked during acceptance testing. Detailed procedures for angular dependence measurement can be found in the AAPM TG18 report.

3.5.2.10 *Color Temperature*

As shown by several studies, color temperature is not as critical to image quality as the other factors mentioned. However, some viewers dislike it if the color temperature of one display device is obviously different from that of others. The only qualitative way to ensure the match of color temperature is to manually match display monitors as closely as possible (when used at a single PACS workstation) or to request replacement display monitor(s).

Tool Required

Colorimeter.

Procedures

The user will load an image with a uniform background and then use the colorimeter to measure the color temperature.

Recorded Value

Color temperature.

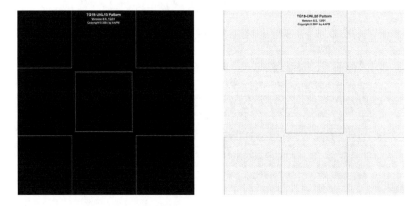

Figure 3.19 The AAPM TG18-UN10 (left) and TG18-UNL80 (right) test patterns for luminance uniformity check. The luminance level at the four corners and the center of each test pattern can be measured and the percentage variation of the luminance across the display area calculated.

QC Check Frequency

As needed.

3.5.3 QC Using Third-Party Software and Vendor-Provided QC Tools

Some third-party, vendor-independent PACS display device QC program tools are available; these tools are designed for ease of use and have an automated documentation feature for QC management. These program tools generate test patterns following the general principles of AAPM TG18 and may also interact with the user to get the user's response as for the visibility of the test targets. Additionally, most high-end medical-grade image display device vendors provide their proprietary QC tools as an option along with their hardware. However, such vendor-specific QC software is not a replacement for a general QC program implemented by the user because multiple vendor products may be used at the healthcare institution's clinical sites. Nevertheless, vendor-specific QC tools can make the overall QC program easier and less labor intensive.

3.5.4 Frequency of QC for Display Devices

The following are recommended QC schedules for PACS display devices:

Daily QC

- If possible, use the AAPM TG18-QC pattern to look for main features as a quick qualitative check or use an interactive QC software tool to quantitatively assess the quality of display devices. The user should stop using the monitor for primary diagnosis if the main features on the test pattern are not visible.

Monthly QC

- Clean the display monitor surface following the vendor's recommendation.
- Visually check the ambient environment in the reading room.

Annual QC

- Review the previous QC record of the display device.
- Check the ambient environment and positioning of the display device.
- Measure and document the illuminance level of the ambient environment.
- Measure and document the minimum and maximum luminance of the display device and compare them with previous QC test results. If changes are found, the display device should be recalibrated.
- Check for spatial resolution using test patterns.
- Check for contrast resolution by measuring the detectability of low-contrast objects in the test pattern.
- Verify the LUT either by measuring it following the procedure discussed earlier in this chapter or by taking the post-calibration LUT provided by the display device vendor and verifying its conformance to the DICOM standard GSDF.
- Load the TG18-QC test pattern on the display monitor for a final review after LUT recalibration, making sure all important features are still visible.
- Document the QC test results.
- Analyze and compare recent QC test results with those of previous tests and with the results obtained during acceptance testing.
- Check the reflective characteristics of the display devices.
- Check for excessive noise.
- Check for bad pixels or defects in monitors using a magnifying glass.

PACS users need to understand that the recommended frequencies for QC checks shown above are not regulatory requirements, but are based upon good clinical practice. PACS users should decide which specific items to include in QC checks and how frequently to perform such tests based upon their own unique situations and available resources.

BIBLIOGRAPHY

American Association of Physicists in Medicine (AAPM) Task Group 18. 2005. *Assessment of Display Performance for Medical Imaging Systems*. AAPM On-Line Report No.03. www.aapm.org/pubs/ reports/OR_03.pdf. Accessed January 20, 2010.

American Association of Physicists in Medicine (AAPM) Task Group 18. *Test Patterns*. www.aapm. org/pubs/reports/OR_03_Supplemental. Accessed January 20, 2010.

Arenson, R. L., D. P. Chakraborty, and S. B. Seshadri, et al. 1990. The digital imaging workstation. *Radiology* 176: 303–315.

Barten, P. G. J. 1992. Physics model for the contrast sensitivity of the human eye. *Proc. Soc. Photo Opt. Instrum. Eng.* 1666: 57–72.

Barten, P. G. J. 1999. *Contrast Sensitivity of the Human Eye and its Effects on Image Quality*. SPIE Press, Bellingham, WA.

Blume, H. and B. M. Hemminger. 1997. Image presentation in digital radiology: Perspectives on the emerging DICOM display function standard and its application. *Radiographics* 17 (3): 769–777.

Brem, M., C. Bohner, and A. Brenning, et al. 2006. Evaluation of low-cost computer monitors for the detection of cervical spine injuries in the emergency room: An observer confidence-based study. *Emerg. Med. J.* 23: 850–853.

Cook, L. T., G. G. Cox, and M. F. Insana, et al. 1998. Comparison of a cathode-ray tube and film for display of computed radiographic images. *Med. Phys.* 25 (7 Pt 1): 1132–1138.

Doyle, A. J., J. Le Fevre, and G. D. Anderson. 2005. Personal computer versus workstation display: Observer performance in detection of wrist fractures on digital radiographs. *Radiology* 237: 872–877.

Dwyer, S. 1997. Soft copy displays and digitizers. *Proceedings of 1997 AAPM Summer School: Expanding Role of Medical Physics in Diagnostic Imaging*, Eds. G. D. Frey and P. Sprawls. p. 381, Medical Physics Publishing, Madison, WI.

Flynn, M. J. and A. Badano. 1999. Image quality degradation by light scattering in display devices. *J. Digit Imaging* 12: 50–59.

Hangiandreou, N. J., K. A. Fetterly, and J. P. Felmlee. 1999. Optimization of a contrast-detail-based method for electronic image display quality evaluation. *J. Digit Imaging* 12: 60–67.

Hendee, W. R. and P. N.T. Wells. 1997. *The Perception of Visual Informatics*. New York: Springer.

Horii, S. 1995. Quality assurance for picture archiving and communication systems (PACS) and PACS technology applications in radiology. *J. Digit Imaging* 8 (1): 1–2.

Krupinski, E. A. and H. Roehrig. 1999. Influence of monitor luminance and tone scale on observers' search and dwell patterns. *Proc. Soc. Photo Opt. Instrum. Eng.* 3663: 151–156.

Krupinski, E. A. and M. Kallergi. 2007. Choosing radiology workstation: Technical and clinical considerations. *Radiology* 242: 671–682.

LeGrand, Y. 1957. *Light, Colour and Vision*. Chapter 5. New York: Wiley.

Li, M., D. Wilson, and M. Wong, et al. 2003. The evolution of display technologies in PACS application. *Comp. Med. Imag. Graph.* 27: 175–184.

Liu, Y. 2007. Implementation of PACS QA using AAPM TG-18 in a clinical healthcare center. *Med. Phys.* 34 (6): 2327.

Liu, Y. 2008. Monitoring quality. *Advance for Imaging and Radiation Oncology,* http:// imaging-radiation-oncology.advanceweb.com/Article/Monitoring-Quality.aspx. Accessed January 20, 2010.

Pilgram, T. K., R. M. Slone, and E. Muka, et al. 1998. Perceived fidelity of compressed and reconstructed radiological images: A preliminary exploration of compression, luminance, and viewing distance. *J. Digit Imaging* 11: 168–175.

Reiker, G. G., N. Gohel, and E. Muka, et al. 1992. Quality monitoring of soft-copy displays for medical radiography. *J. Digit Imaging* 5 (3): 161–167.

Riggs, L. 1965. *Vision and Visual Perception.* Chapter 1. Ed. C. Graham. New York: Wiley.

Roehrig, H. 1999. Image quality assurance for CRT display systems. *J. Digit Imaging* 12: 1–2.

Roehrig, H. and E. A. Krupinski. 1998. Image quality of CRT displays and the effect of brightness on diagnosis of mammograms. *J. Digit Imaging* 11 (3 Suppl 1): 187–188.

Samei, E. 2005. Assessment of display performance for medical imaging systems: Executive summary of AAPM TG18 Report. *Med. Phys.* 32 (4): 1205–1225.

Wang, J. 2006. Using a gradient plot of luminance response curve as a metric for assessing and comparing softcopy display system quality. *Med. Phys.* 33 (4): 1182.

Wang, J. and Q. Peng. 2002. An interactive method of assessing the characteristics of softcopy display. *Proc. Soc. Photo Opt. Instrum. Eng.* 4686: 189–197.

Wang, J. and S. Langer. 1997. A brief review of human perception factors in digital display for PACS. *J. Digital Imaging* 10: 158–168.

Wang, J., J. Xu, and V. Baladandayuthapani. 2009. Contrast sensitivity of digital imaging display systems: Contrast threshold dependency on object type and implications for monitor quality assurance and quality control in PACS. *Med. Phys.* 36 (8): 3682–3692.

Wang, J., K. Compton, and Q. Peng. 2003. Proposal of a quality-index or metric for soft copy display systems: Contrast sensitivity study. *J. Digit Imaging* 16 (2): 185–202.

Weirecht, M., G. Spekowius, and P. Quadflieg, et al. 1997. Image quality assessment of monochrome monitors for medical soft copy display. *Proc. Soc. Photo Opt. Instrum. Eng.* 3031: 232–244.

Weiser, J. C., K. T. Drummond, and B. D. Evans, et al. 1997. Quality assurance for digital imaging. *J. Digit Imaging* 10: 7–8.

4 Networking Technologies

A network can be defined as groups of transmission links and the associated equipment that allow information exchange among the users connected by these links. Its application in PACS has improved radiology significantly. For healthcare professionals, connectivity has never been as important as it is today, and this connectivity is dependent on the quality of hospital computer networks. Not every aspect of computer networking is covered in this chapter; only such technical components that are a part of the healthcare institution network are discussed with a view to understanding their applications and limitations.

4.1 ETHERNET LOCAL AREA NETWORK (LAN) AND TCP/IP

Ethernet is a family of computer network technologies for a LAN. It defines a number of wiring and signaling standards for the physical layer of the network through network access at the data link layer and a common addressing format. Because Ethernet is currently the most widely used type of LAN network in the world, especially in PACS/EHR systems, this section describes Ethernet, the very commonly used protocol; TCP/ IP; and the environments in which they operate.

4.1.1 Open Systems Interconnection (OSI) Model

The OSI reference model defines a seven-layer model that can be used for any data communication. The layered approach is the foundation for flexibility through defined interfaces. Interfaces allow some layers to be changed, while others remain unchanged. Each layer of the model provides a set of functions to a layer and to the layer above it, and relies on functions provided by the layer below it. Each layer can also communicate directly with other connected nodes on the same layer, as illustrated in Figure 4.1.

4.1.1.1 Physical Layer

The physical layer is the lowest layer responsible for data transmission over the communication link. It is the physical medium where mechanical, electrical, and functional standards are accessed.

4.1.1.2 Data Link Layer

The data link layer is where data transfer function and protocols are provided, together with error detection and correction.

4.1.1.3 Network Layer

The network layer establishes the network connection, including information routing using protocols such as IP.

4.1.1.4 Transport Layer

The transport layer is the layer where data transfer is provided, such as TCP.

4.1.1.5 Session Layer

The session layer organizes interactions among devices and applications.

4.1.1.6 Presentation Layer

The presentation layer is used for user or system data representation.

4.1.1.7 Application Layer

The application layer is the highest layer, with which the end user is most concerned. One example is the PACS application or DICOM image viewer software, which the end user uses to view medical images.

All network devices can be categorized in terms of which OSI layer they operate on, and the amount of hardware or software involved.

4.1.2 Ethernet

Ethernet LAN is relatively easy to implement, maintain, and manage, compared to other network technologies, and Ethernet products are also readily available. Currently, Ethernet has a variable data-carrying capacity (i.e., it has bandwidth limitations), and uses different physical network

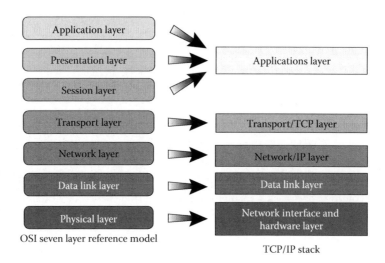

OSI seven layer reference model

TCP/IP stack

Figure 4.1 The OSI seven-layer model and TCP/IP five-layer stack.

cables. In the past, Ethernet used coaxial cables, but today's Ethernet for the most part uses fibre optic cables for the backbone, and twisted pair cables to connect all the workstations of the PACS/EHR to the network. Using these cables, together with switches and routers, reduces installation costs, improves network reliability, and increases the data-carrying capacity of the network. Although different generations of Ethernet use different physical layers of hardware, they can communicate with each other because of the use of the same standard. Most current PACS/EHR workstations have built-in Ethernet twisted pair network interface cards on their motherboards; however, PACS/EHR servers using optical fibre Ethernet network connections have separate optical fibre network cards installed. Ethernet with optical fibre cables are frequently used in healthcare institutions' backbone network infrastructures. The 1 Gbps optical fibre Ethernet is currently the most commonly used backbone network, with 10 Gbps optical fibre Ethernet getting more popular, and 100 Gbps on the horizon. PACS/EHR servers with very high bandwidth requirements can be connected to the optical fibre Ethernet using fibre-optic network cards installed inside servers. Optical fibre Ethernet has the advantages of high bandwidth and non-electrical interferences compared with twisted pair copper wire cable Ethernet. However, optical fibre Ethernet is not usually directly connected to PACS/EHR workstations, mainly because of convenience and cost concerns.

4.1.3 Function of Ethernet

All nodes, such as the various servers and workstations on the Ethernet LAN, transmit information randomly and retransmit it when collision occurs. In Figure 4.2, a workstation first listens to see if anyone else is transmitting on the network. If the network is busy, the workstation waits until the current transmission stops before sending out its own data in units defined as frames, ranging from 64 to 1516 bytes. Because the network has a physical length, it takes time for the data

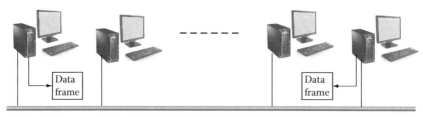

Ethernet LAN

Figure 4.2 In an Ethernet LAN, workstations send out data packets as frames. If any workstation transmits its data when there is traffic on the network, a collision occurs.

frame to reach from one workstation to another. During this time, if any workstation thinks there is no traffic on the network and proceeds to transmit its data, a collision occurs and is detected by all workstations on the network. If there is a collision, all workstations involved in the collision need to wait before retransmitting their data. The longer the distance the network extends, the more workstations there are on the network and the greater the likelihood of a data collision.

4.1.4 Transmission Control Protocol/Internet Protocol (TCP/IP) Layers

The main design goal of TCP/IP is to connect networks, so that servers and workstations on different networks, even separated by long geographical distances, can communicate with each other. It provides data communication services that run between the programming interface of a physical network and user applications, such as the PACS applications, for instance. TCP/IP has become the IT industry's standard method for connecting networks, servers, workstations, and the Internet. Its power and simplicity have led to its becoming the single network protocol of choice in the world. Within TCP/IP there are five layers. As shown in Figure 4.1, the first layer is the physical network interface and hardware layer, which can be a twisted pair or optical fibre cables; the second layer is the data link layer, interfacing to the physical network hardware; the third layer is the IP network or Internet layer, separating the physical network from the layers above it; the fourth layer is TCP for data transport, providing flow control and error correction for reliable exchange of data; and the fifth and final layer is the application layer, where application software is run. TCP/IP is widely used in healthcare institutions because it is relatively inexpensive, able to merge different physical networks while providing a common suite of functions to users, allows interoperability for various vendor products on different platforms, and can access the Internet. Today's Internet consists of regional, national, and international backbone networks on a large scale. Use of the Internet itself in the general consumer market has grown exponentially, with much potential to be realized in the applications of PACS/EHR in the future.

4.1.4.1 TCP

TCP is a network transmission control protocol. It is a transport protocol that sends data in a stream of digital bytes. TCP provides the sender with delivery information about the data it sends by using acknowledging messages and sequence numbers. If the data are lost or if an error occurs during data transmittal from sender to recipient, TCP will retransmit until either the data are successfully sent or the message is timed out. By using flow control, it can also coordinate the data transmission speed between the sending and receiving nodes.

4.1.4.2 IP

IP is a high-level protocol communicating with a logical IP address. The IP address is used for network devices to communicate over the network. Such devices include servers, workstations, switches, routers, printers, and so on. Every device in the network is assigned an IP address that allows it to be identified and each is connected to the network through a network interface card with a unique hardware address. Network devices communicate with each other through this hardware interface at the physical layer. When a network device with an IP address A communicates to a workstation with an IP address B, the logical addresses of A and B are translated into hardware addresses for network interface cards to understand. When a PACS server has more than one network interface card, each interface card can have a unique IP address.

An IP address consists of two parts: network number and host number. The network number, its Internet identification, is assigned by an Internet assigned numbers authority. The host number part of the IP address is assigned by the local institutional network administration. Some IP addresses are globally unique; some are only locally unique. The IP address is used for communication, and it is not, therefore, always a unique identifier for a network device. Currently, IP addresses are defined by Internet Protocol version 4 (IPv4). Because only 32 data bits are used to define each address, the total number of unique addresses that can be defined by 32 bits is $2^{32} = 4.3 \times 10^9$. The IP address is usually in the form of four decimal numbers separated by dots, for example, 10.22.20.56. Each decimal number has 8 bits, ranging from 0 to 255, as described in Figure 4.3.

There are three classes, A, B, and C, of IP address networks. Class A network is intended for very large networks and has only 8 bits for the address field. Class B has 16 bits, and Class C, 24 bits for the address field. For Class C, there are only 8 bits left for the host field in the entire 32-bit address field; so there is a limit of 1024 hosts per Class C network.

Class	First octet range	Number of addresses	IP address format
A	0–127	$2^{24} = 16,777216$	**Net ID** ____ **Host ID** Network . Host . Host . Host 1 Octet ____ 3 Octet
B	128–191	$2^{16} = 65,536$	**Net ID** ____ **Host ID** Network . Network . Host . Host 2 Octet ____ 2 Octet
C	192–223	$2^{8} = 256$	**Net ID** ____ **Host ID** Network . Network . Network . Host 3 Octet ____ 1 Octet

Figure 4.3 Class A, B, and C networks with their corresponding IP address formats.

4.1.4.3 Internet Protocol Version 6

The above IP address structure is IPv4. The problem associated with IPv4 is that, with the explosive use of the Internet and mobile devices, the number of addresses available for future use is going to be exhausted very soon. The next generation address, Internet Protocol version 6 (IPv6), is 128 bits long and is intended to replace IPv4. There are $2^{128} = 3.4 \times 10^{38}$ unique IP address when the 128-bit address scheme is used. That is $2^{96} = 7.9 \times 10^{28}$ times more than can be provided by IPv4. The purpose of IPv6 is not to provide a unique IP address for every possible network device, but to allocate large blocks of addresses for various purposes. Also, with the abundance of available IPv6 addresses, there is no need to employ or develop complicated address-conserving schemes. An IPv6 address is in the form of eight groups of hexadecimal numbers separated by dots, for example:

2009:0ce8:82b1:17a3:2119:5c2d:1003:6214

Using a subnet mask, IPv4 and IPv6 can divide each IP address into network address and host address parts, and create a subnet.

4.1.4.4 Static IP Address

There are two ways to assign an IP address to a network device. The first uses the same IP address assigned by the network administrator each time it is connected to the network; this is called the static IP address. Typically, PACS/EHR servers and PACS diagnostic workstations are assigned with static IP addresses. The second uses a dynamic IP address, which means a computer's IP address may often change. This is typical for computers that log onto the Internet through residential Internet service providers, or general office computers, which log onto the company's LAN for retrieving e-mails and browsing the Internet. Sometimes clinical PACS (nondiagnostic)/ EHR workstations may also employ dynamic IP addresses.

4.1.4.5 Dynamic Host Configuration Protocol (DHCP)

For a LAN in a large healthcare institution, or broadband Internet access by an Internet Service Provider (ISP) for smaller clinics, dynamic IP addresses are usually assigned to computers or workstations using DHCP servers. In a healthcare institution, the number of clinical workstations or PCs in various departments can be large, and they are constantly changing. Using a dynamic IP address significantly reduces the burden on the administrator to assign an IP address to each and every computer or workstation in the institution. With the increased use of Web-based PACS, any computer that can access the Internet with a dynamic IP address may potentially become a clinical PACS workstation. The dynamic IP address assigned by DHCP generally changes, although sometimes it can stay the same for a long time. The network administrator can also configure the DHCP to always assign the same IP address to a particular computer, and not use this IP address for any other network device. In this way, the administrator can remotely assign static IP addresses to workstations. This is useful when the total number of PACS/EHR workstations is very high, or in a teleradiology setup, where a static IP is also desirable. In contrast, static IP addresses are usually assigned by the administrator to PACS/EHR servers, PACS diagnostic workstations, and high-end printers.

4.1.5 Network Cables

The combination of twisted pair versions of copper wire cables connecting end systems to the network, along with the optical fibre cable versions to be discussed later, is the most widespread wired LAN technology. Ethernet over twisted pairs refers to the use of cables that contain insulated copper wires twisted together in pairs at the physical layer of an Ethernet network, in which the Ethernet protocol provides the data link. The cable in Figure 4.4a illustrates four pairs of insulated copper wires, each pair twisted, forming a twisted pair cable. There are several different standards for the copper wire-based physical medium. The most widely used in healthcare institutions are 10Base-T, 100Base-T, and 1000Base-T, all using the same 8-position modular connectors, sometimes also called RJ45 connectors, as shown in Figure 4.4b. For these designations, the number refers to the theoretical maximum transmission speed in mega bit per second; the "Base" is the baseband, indicating that there is no frequency-shifting modulation or multiplexing in use, each signal has a single frequency on wire; and the "T" means twisted pair cable, where each pair of wires for the signal is twisted together to reduce radio frequency interference and crosstalk among different pairs. These three standards support both full-duplex and half-duplex communications. Full duplex means that data can be transmitted and received simultaneously, whereas half duplex means that data can either be transmitted or received, but not at the same time. There are various categories and classes of cables such as Category 3 for legacy systems, and Category 5, Category 5E, Category 6, and Category 6A, each rated to different maximum bandwidth for data throughput.

4.1.6 Network Bandwidth

Ethernet LAN has evolved from the original 10 Mbps to the so-called Fast Ethernet of 100 Mbps, and now to the 1 Gbps known as Gigabit Ethernet and thence to 10 Gbps, and beyond.

4.1.6.1 100 Mbps

Though the Fast Ethernet is ten times faster than the original 10 Mbps Ethernet, with 100 Base-T cabling used the most, it is no longer considered fast. However, it is still widely used in today's PACS/EHR workstations. The data framing structure of the Fast Ethernet is the same as the original 10 Mbps Ethernet. Because of the higher bandwidth and unchanged framing, the length of the network is significantly reduced to avoid collision issues. When used with switches providing duplex mode operation, workstations can send and receive data simultaneously, effectively doubling the connection bandwidth. This operation mode requires at least Category 5 Unshielded Twisted Pair (UTP) network cables, with two pairs of wires used for transmitting and two pairs of wires for receiving data.

4.1.6.2 Gigabit and 10 Gigabit Ethernet

The Gigabit (1 Gbps) Ethernet is ten times faster than the 100 Mbps Fast Ethernet. Since the framing remains unchanged from the original 10 Mbps Ethernet, the network length is actually

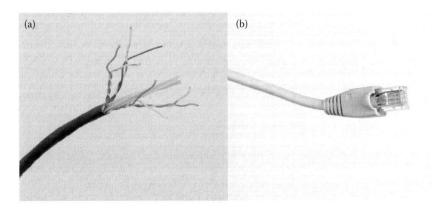

Figure 4.4 A Category-5 network cable with (a) inside four pairs of twisted and insulated copper wires and (b) an eight-position modular connector.

reduced. Gigabit Ethernet uses either a UTP cable, or a fibre optic cable. When single-mode fibre is used, the network length can reach 3.5 miles or 5 km. In healthcare institutions, Gigabit Ethernet is mainly used for high-speed backbones, or used in most PACS servers, but its use in PACS workstations has just begun. In 2005, the 10-Gigabit Ethernet (10 GbE) standard was introduced, which is ten times faster than the Gigabit Ethernet.

4.1.6.3 Line Speed and Data Throughput

It should be noted that users need to distinguish between nominal line speed and actual data throughput. Line speed, which is the raw transmission speed of the network interface card of a workstation and the port on a switch, is quite different from the actual rate at which data are exchanged across the network. If the network interface card and the switch port it connects are both 1 Gbps, the true data throughput speed is not 1 Gbps. All network technologies and applications have overheads, though their impact depends on specific situations. However, the final true throughput speed of useful data, after considering all protocol and application overheads for packaging and delivering data across the network, is always much less than the raw transmission speed.

Besides, the network bandwidth requirements for various applications are not similar. In a healthcare institution, the application that runs the HIS for patient registration does not require the same bandwidth as a radiology PACS application, which involves a large amount of image data transmission. Therefore, while network design is determined by the applications it currently supports, network bandwidth design must consider current applications as well as the types of applications the network is likely to support in future.

4.1.7 Microwave Wireless Ethernet

Gigabit microwave wireless Ethernet provides a backbone link to extend LANs between buildings at full 1-Gbps Ethernet speed. The advantage is easy installation and cost savings for inter-building LAN connections. Since the technology uses high-frequency electromagnetic waves in the microwave range, its performance, including link distance and reliability, is sensitive to rainfall. Its link distance is in the range of a few hundred meters. For large healthcare institutions with multiple buildings in a campus, or if buildings are not too far away, this may be used as an alternative backbone LAN extension among buildings, or as a redundant network path for primary optical fibre or copper wire Gigabit Ethernet LAN.

4.2 WIDE AREA NETWORK

Wide Area Network is a network that connects computers in a much larger area, usually covering various regions, cities, or coutries. It is mainly used to connect networks that are geographically separated. As an example, the Internet is the largest (and most widely known) WAN.

The purpose of WANs is to link localized LANs. Using WANs, computers belonging to LANs remotely separated can communicate easily. The requirements and choices of WAN technologies are different from LAN technologies. The main difference is that WAN technologies are usually a subscribed service provided by carriers. Often leased communication lines, which can be expensive, are used to build WANs. WANs also differ from LAN technologies in the area of speed. For WANs, the data transmission rate is lower than LANs. Just as in LANs, TCP/IP is the most used protocol for WANs. Typically telephone lines, satellite channels, and microwave links can be utilized for WANs.

4.2.1 ATM

Asynchronous Transfer Mode (ATM) is a data link layer service protocol running over a physical network layer. It encodes data traffic into fixed size cells of 48 bytes of data and 5 bytes of header information, so that it can be reassembled when it is received. This method is in contrast to IP or Ethernet, which is based on variable length packet transmission. The ATM protocol has been used widely in WAN set-ups; however, its use in LAN is very limited. Though ATM will remain useful for some time, it is very likely to be replaced by Gigabit or higher-speed Ethernet for new WAN implementations.

Table 4.1: Data Transmission Speeds of Various Leased Lines

Europe			United States		
E0	E0	64 kbps	DS0	DS0	64 kbps
E1	32 E0	2 Mbps	T1/DS1	T1/DS1	1.544 Mbps
			DS1C		3 Mbps
E2	128 E0	8 Mbps	T2/DS2	4 T1/DS1	6.3 Mbps
E3	512 E0	34 Mps	T3/DS3	28 T1/DS1	45 Mbps
			DS3C		89.5 Mbps
E4	2048 E0	140 Mps	T4/DS4	168 T1/DS1	274 Mbps

4.2.2 Leased Lines

Leased lines are intended for mission-critical applications. They are widely used in long-distance data communication to form WANs, and to connect business LANs at various geographic locations. The most commonly used leased lines are T1 and T3 lines. The T1 standard for data communication has been around for several decades. A T1 line is a digitally conditioned four-wire full duplex, bi-directional circuit. Twenty-four 64-Kbps channels form a data transmission rate of 1.544 Mbps for a single T1 line. A T3 line aggregates 28 T1 lines or circuits, including 672 channels of 64 Kbps each, yielding a total data transmission rate of 44.736 Mbps. A T3 line is also referred to as a DS3 line, well suited for network traffic of large businesses, or large file transmissions such as PACS images. Listed in Table 4.1 are various leased lines and their data transmission speeds in the US and Europe.

Listed in Table 4.2 are common optical carrier (OC) line speeds.

Since PACS/EHR applications are mission critical to any healthcare institution, leased lines are the primary means of interconnecting various hospitals and clinics enterprise-wide. The only concern with using leased lines may be the cost—they are expensive. A dedicated T3 leased line can cost a few thousand US dollars per month.

For WAN setup using leased lines, cost-effective VPNs using readily available Internet access is a powerful competitor. Institutions can have remote network access using VPN technology at a much reduced cost.

4.3 NETWORK TOPOLOGY

Network topology describes the arrangement of network elements of nodes and links at both physical and logical levels. The infrastructure of a LAN and the data flow among nodes of the LAN determine, respectively, the physical and logical network topology, which may be the same or different.

4.3.1 Basic Network Topologies

The elements of networks can be arranged in ways similar to a few basic topologies, or their combinations. As illustrated in Figure 4.5, the basic topologies include (a) Bus, (b) Star, (c) Ring, (d) Mesh, and (e) Tree. Network topologies can be either physical or logical. Physical topology indicates how the nodes of a network are connected; logical topology is used for data flow.

4.3.1.1 Linear Bus

A linear bus topology network is also frequently called a network backbone. In this network, all network devices/nodes connect to a main transmission medium, which starts at one end point and finishes at another, as shown in Figure 4.5a.

Table 4.2: Data Transmission Speeds of Various Common Optical Carriers

Optical Carrier	Bandwidth
OC1	51.85 Mbps
OC3/STM1	622 Mbps
OC12/STM4	1.244 Gbps
OC48/STM16	2.5 Gbps
OC192	9.6 Gbps

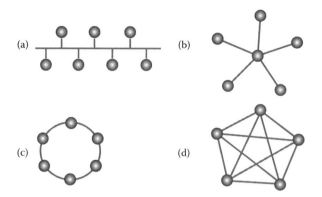

Figure 4.5 Four basic network topologies: (a) linear bus; (b) star; (c) ring; and (d) mesh.

4.3.1.2 Star

Figure 4.5b depicts a star topology network, where there is a central switch through which all network nodes or devices transmit or receive data. A single network node failure does not bring the network down (only that single node is affected). However, the failure of the central switch brings down the entire network.

4.3.1.3 Ring

In ring network topology, as illustrated by Figure 4.5c, each node of the network is connected sequentially, with the first and the last node connected directly forming a closed ring loop. In this network topology, a single network node failure results in failure of the entire network.

4.3.1.4 Mesh

In mesh network topology, as shown in Figure 4.5d, each node connects to all other nodes directly. The advantage is that there is more than one path between any two nodes. In case of a link failure, a redundant path can be found. This type of topology is practical only if there are not too many nodes in the network. If there are many nodes in the network, usually a partially connected mesh topology is deployed. This reduces the complexity and cost and maintains some redundancy offered by a fully connected mesh topology network, with some nodes connected directly and several other nodes in a point-to-point link. Mesh topology is commonly used in networks where a high degree of availability is required.

4.3.2 Hybrid Network Topologies

Hybrid network topology is a network using any combination of at least two of the above standard topologies. In a large healthcare institution, a single topology network is used only rarely. Most often there is a hybrid topology network with features of several topologies meeting the specific requirements of the institution. For example, a star bus topology may be used, with several star topologies linked to a backbone bus topology.

4.3.3 Logical Topology

Logical topology describes the digital data flow path of the network. The definition of a topology is the same as a physical network topology. The logical topology of a network is generally determined by a network protocol in contrast to physical topology which is determined by the physical layout of network cables. Logical network topologies can be configured using switches and routers.

4.3.4 Routing and Switching

Network switches and routers play an integral part in most Ethernet LANs. A large LAN, such as in a large healthcare institution, can contain a number of linked and managed switches and routers. There are common features and differences between switches and routers.

4.3.4.1 Router and Gateway

A router is a device connecting groups of networks and is an important component in an IP network. A router works in the network layer. It checks data packet information relevant to the network layer, then forwards the data packets using setup rules. Typically, it checks more information than just the data link layer information in a packet and has much more processing power for information exchange between different groups of networks. A router understands addressing structures associated with the network protocols it supports and makes decisions as how to exchange data packets. Routers are able to find the best data transmission paths and optimal data packet size. Dedicated routers can even provide much more sophisticated routing functions than basic minimum IP routing. When a powerful CPU and a large amount of memory are used, a router can check information contained in a data packet at a higher layer than the network layer, functioning as a firewall, which is discussed later in this chapter. Since a router is very powerful, capable of making decisions based on protocol and network information such as IP addresses, and has sophisticated redundancy features, its configuration is complicated and requires great expertise. Sometimes a router is called a gateway; the term is simply a historic name in the IT community.

4.3.4.2 *Switch*

Today, most computers and network devices on a LAN are connected through switches. A switch is an electronic device with many ports where network cables end and are connected to the LAN. A switch guides data traffic from a source computer, connected through a cable on one port, to the destination computer, connected through another cable on another port of the same switch. A switch functions at the data link layer through hardware implementation, making it very efficient. Switches are used to segment a large network into smaller groups. Many switches allow network administrators to login from the LAN to configure switch settings and administer network connections. The LAN infrastructure of a large healthcare institution may have many switches located in various communication closets throughout the institution. Shown in Figure 4.6 is a rack-mounted network switch in a communication closet located in a physician's office building, which connects many clinical workstations. The switch has optical fibre network cable ports for connecting optical cables to the fast speed LAN backbone, while many ports are for copper wire cables, each connecting to a wall network port in the physician's office.

4.4 NETWORK DESIGN CONSIDERATION

Ethernet has been around for many years because of its low cost, simplicity, widely available products, and the proliferation of Internet use, and it will be used for many years to come. For healthcare institutions, a network design should consider the following areas.

Figure 4.6   A rack-mounted network switch with fibre optical cable ports and copper wire network cable ports.

4.4.1 Standards

Open standards should be used by all components for building the network. This ensures flexibility and is cost effective. It is very likely that various equipment from different vendors are to be interconnected into the network.

4.4.2 Scalability

The network designed should be scalable, so that it can accommodate new PACS/EHR servers and workstations without requiring a complete redesign of the network topology.

4.4.3 Reliability

When the operation of a healthcare institution increasingly relies on HIS, RIS, PACS, and EHR, the reliability of the network becomes extremely important. A network capable of a few seconds response time is meaningless if it frequently and unexpectedly goes down. Network redundancy and reliability should be incorporated into design cost before the network is implemented.

4.4.4 Security

Network security is another important issue, especially because the healthcare institution's network is most likely interfaced with the Internet. Taking care of basic known security problems and security risks is essential during network design.

4.4.5 Performance and Cost

Both data throughput, which is the amount of data sent in the shortest time possible, and response time, which is the time a user needs to wait before a result is returned from the system, are performance measures to be considered. It is a challenging task for the healthcare institution to balance cost and requirements at the same time.

4.4.6 Network Planning Considerations

Building a network for digital data exchange in healthcare institutions requires much planning. Network infrastructure that is well planned not only provides reliable and fast data exchange between HIS, RIS, and PACS, but is also able to adapt to changes when these systems are expanded or upgraded. There are many aspects to be considered in the process of building a healthcare institution's network, including

- Evaluating available technologies
- Selecting the LAN technology to be used inside the healthcare institution building complex
- Selecting the WAN technology to connect geographically separated sites
- Designing a plan for network implementation
- Determining components needed for building the network
- Determining scalability and a modular design for the network
- Determining the budget and cost
- Planning network management and administration

In planning a network, in addition to many other issues, the lower three layers of the OSI model need to be carefully analyzed. Before a decision is made to select a particular technology, it is important to understand at which layer of the OSI model the network equipment function. After their implementation, these equipment need to meet the demands of various healthcare information system applications and perform tasks as designed.

Since it is easy and simple to use TCP/IP for device interconnection, it is not uncommon for networks to be set up without much careful planning. For a small network, this may not be a serious issue. However, for large healthcare institutions, problems can arise when changes need to be made to the network at a later time. As an example, networks not carefully planned may have problems in IP address management such as address allocation, static or dynamic addresses, centralized network coordination, and so on. Figure 4.7 shows complicated cable arrangement in an enterprise data center after careful planning. Network design and planning should take place before any actual implementation is done. The network design and infrastructure should also be frequently reviewed, because network application demands are frequently changed; at the same

Figure 4.7   Network cable arrangements in the Aurora Healthcare enterprise data center.

time, the infrastructure of the network itself may also undergo constant change. When a healthcare institution's network is upgraded, it is usually done in several phases. In some phases, there is a need to connect new technology with legacy networks and equipment.

Understanding the limitations of both new and current technologies is necessary to avoid creating new problems during the process of upgrade. Detailed documentation of the network for future reference is definitely a necessity for any well-planned network design, which results in smooth network implementation with few surprises.

It must be realized that there is no perfect network. The best designed and planned network based on current demand and technology may not address future requirements of the healthcare institution. Healthcare operation is itself constantly changing in this age. Expectations and requirements from medical staff and healthcare professionals are also changing, which in turn changes requirements from the network and the underlying infrastructure supporting the network. The best that can be done in network planning is to incorporate scalability and adaptation to changes into the network design. Before a healthcare information system such as PACS application is deployed for clinical use, much planning must be done to analyze network infrastructure, server deployment, security policies and procedures, and types of workstation requirements. All of these should be well coordinated.

4.5 OPTICAL FIBRE COMMUNICATION

Optical fibre network communication systems are increasingly being used in healthcare institutions' network infrastructure. The idea of using light to send messages and of glass fibre carrying optical signals for communication is not new. Light is actually an electromagnetic wave, and optical fibre is a waveguide. Flexible optical fibre cables can be used for data communication and networking. Though optics and electronics are closely related, optical fibre network communication is quite different from electronic communication in many aspects. The former uses light and optical fibre as a media to transmit information between two places. Information is modulated and carried by light as opposed to electrical current in copper wire. Optical fibre communication has many advantages over classic electrical transmission, mainly because of light propagating through the optical fibre with low signal attenuation, low interference, and high network capacity. Because of these advantages, optical fibre network technologies are extensively used in today's LANs and WANs. Optical fibre and associated device development for communication started in the 1960s and continues even today. Building an optical communication system infrastructure is relatively complex and expensive; however, its use is increasing because of the continuing drop in cost and increase in network capacity.

4.5.1 Basic Optics

Before discussion of optical fibre communication, it is helpful to review some basic optics principles. The nature of light is complicated. Analogies of rays, waves, and photons are all useful in understanding what light really is in different situations from various viewpoints. In geometric optics, light is a ray that is refracted and reflected when projected at mirrors and prisms. This model can be used when geometric dimensions involved in optical devices are much larger than the light wavelength used. In optical communications, when the optical fibre core diameter is less than about ten times that of the wavelength of the propagating light, it is best to treat light as an electromagnetic wave. In such cases, more complicated electromagnetic wave equations should be used for analysis. However, in this chapter geometric optics principles treating light as a ray are used to simplify discussion, especially for light propagation in multimode optical fibre as a waveguide.

4.5.1.1 Refraction and Reflection

When light is projected to the boundary of two materials, the portion that passes through is called refraction, and the amount that bounces back is called reflection. The refraction index of a material R_1 is used for characterizing light speed in the material. Light travels the fastest in a vacuum at approximately 300,000 kilometers (km) per second. The refraction index of a material is defined as the speed of light in a vacuum divided by the speed of light in this material. Therefore, by definition, a vacuum's refraction index is 1. The larger the refraction index of a material, the slower light travels in this material. For a typical optical fibre core with a refraction index value of 1.48, the digital light signal is transmitted at a speed of approximately 200,000 km per second.

4.5.1.2 Critical Angle

When light is projected at the boundary from a dense material to a less dense material with a different refractive index, the larger the angle to the normal, the smaller the refracted portion of light passing through the boundary, until total internal reflection occurs; the angle reached is called the critical angle. The critical angle θ_c can be calculated as

$$\theta_c = \sin^{-1}\left(\frac{R_2}{R_1}\right),$$

where R_2 is the refraction index of the less dense medium, and R_1 is the refraction index of the dense medium. In the case of an optical fibre cable, R_2 is for the cladding and R_1 is for the optical fibre core.

When light meets a boundary at an angle greater than the critical angle while traveling in a dense medium, the light is totally reflected and, just like a diver looking upward from underwater, the water surface appears like a mirror. This effect is used in optical fibre communication to confine light propagation to the core of an optical fibre. When light travels in a multimode optical fibre, it bounces off from the core boundary back and forth many times, as shown in Figure 4.8a.

Since the light must hit the core boundary at an angle larger than the critical angle, only light that enters the optical fibre core with an angle within a certain range can travel down the optical fibre core without leaking out. This range of angles is called the acceptance cone of the fibre and is a function of the refractive index difference between the optical fibre core and the cladding. The larger the range of angles, the lesser the precision required to splice and work with the optical fibre, which is the case for larger core diameter multimode optical fibre. The coupling of light into fibre is more efficient when a light ray is both close to the fibre core axis and can transmit down the fibre at various angles. However, this coupling causes higher dispersion because light rays at different angles have to travel various path lengths inside the optical fibre core, which takes different times to propagate across its entire length.

4.5.2 Characteristics of Optical Fibre Communication

An optical fibre is a cylindrical waveguide transmitting light through its core by internal reflection. An optical fibre for data communication consists of a core with a cladding layer surrounding it. The refractive index of the core is greater than that of the cladding to restrict the optical signal within the core. In digital optical communication systems, digital data bit information is transmitted as pulses of light: the light represents "1," and dark represents "0." However, because of the characteristics of optical fibre communication discussed in the following sections, the digital optical signal received may not be the same as that transmitted, causing data error.

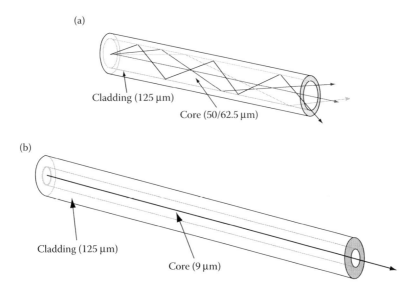

Figure 4.8   In a multimode optical fibre (a) an optical signal travels through the fibre in multiple paths. In a single-mode optical fibre (b) the optical signal travels through the fibre only in one path.

4.5.2.1 Multimode Transmission

When a light source sends light into an optical fibre, and if the fibre core is large enough, there are multiple paths the light can travel down the fibre, as shown in Figure 4.8a. This propagation of light transmission is called multimode transmission. However, whereas the diagram is two-dimensional, true multimode optical transmission is three-dimensional. Many light propagation modes are spiral and oblique, and never cross the optical fibre axis when they travel down the fibre. In multimode fibres, the fibre core is larger, between 50 micrometer (μm) and 62.5 μm, thus connections among fibres are simpler and lower-cost electronic devices can be used. The problem for multimode optical transmission is that some paths are longer than other paths; in other words, light arrives at the destination at different times because of the lengths of the various paths it takes. The result is that light pulses disperse after propagation through the fibre. To reduce this dispersion effect, multimode optical fibres are used typically for short-distance communication links and for applications where high optical power is transmitted. The light gathering capacity of multimode optical fibre is also higher than single-mode fibres. Typically LEDs with a wavelength around 850–1300 nanometer (nm) can be used in multimode fibres, which also have less information carrying capacity.

4.5.2.2 Single-Mode Transmission

If an optical fibre core diameter is very thin compared to the wavelength of the light used for optical transmission, the light can only find one path to travel down the optical fibre in one propagation mode. This is termed single-mode transmission. There is no way for the light to disperse over multiple paths as it does in thicker multimode optical fibre. Single-mode fibres use more expensive laser sources with a wavelength of 1310–1550 nm. In the single-mode, laser light transmission does not depend on reflection from the boundary between core and cladding. Rather, the laser light as an electromagnetic wave is held tightly to travel down the optical fibre. Optical attenuation for single-mode fibres is much lower than in multimode fibres. To be accurate, a single-mode optical fibre is quoted with a cut-off wavelength, because when a wavelength is shorter, the optical transmission becomes multimode.

A single-mode optical fibre has a typical core diameter of 8–10 μm, with a cladding diameter of 125 μm, as shown in Figure 4.8b. Single-mode optical fibres are used mostly for communication links with a distance longer than 1800 ft or 550 m. Since tighter tolerance is required for coupling of light into single-mode optical fibres with a thin core diameter, transmitters, amplifiers, receivers, and other components for single-mode optical fibre are usually more expensive than those of multimode optical fibres.

4.5.2.3 Scattering

When light travels through an optical fibre core, it relies on total internal reflection of the light in multimode transmission. However, any irregular or coarse surfaces even at very small scales at the molecular level of the glass or plastic core of the optical fibre can cause the light to be reflected in many directions randomly. This type of reflection is termed light scattering, and is dependent on the wavelength of the light. Light scattering is the main reason for optical attenuation.

4.5.2.4 Dispersion

Dispersion occurs when a light pulse spreads out during transmission. A short duration light pulse gets longer and mixes with the light pulse that follows. This optical phenomenon is similar to that in a rainbow, which is the spatial spreading of white color light into different colored light as a prism. This makes optical data bit-stream recovery at the receiving end impossible. In optical fibre light transmission, the distance the light can travel before any relay or amplification is required is determined by dispersion. In multimode fibre intermodal dispersion is caused by the nature of the transmission itself. Single-mode fibre is dominated by chromatic dispersion, which is caused by light with different wavelengths bearing slightly different indexes in the same optical fibre material. Dispersion makes optical pulses spread out as they travel along the optical fibre, so that the digital optical information of "1" = light and "0" = dark is not distinguishable; hence, signal loss occurs. The longer the fibre length, the more the dispersion effect in the fibre. Therefore, when the fibre is used for larger bandwidths (i.e., bits of light pulses are getting closer), the distance it can reliably transmit the optical pulse signal is reduced. Light dispersion is also related to the spectral width, which is the range of wavelengths of the light source. Usually light produced by a light-emitting diode (LED) has a spectral width of 30–150 nm, while typical semiconductor lasers have a spectral width of 1–5 nm, much narrower than light generated by LEDs. Because of their spectral width differences, laser has much less dispersion than LEDs, which is why laser is used for single-mode long distance transmission, and LEDs are frequently used for multimode shorter distance LAN applications.

4.5.2.5 Optical Attenuation

When light travels in an optical fibre, there is transmission loss and the light signal intensity is reduced. Optical attenuation is caused by light scattering, lack of uniformity in the optical fibre density, impurity of the fibre core glass, splicing of optical fibre, and absorption by the optical fibre material itself. Among these contributing factors, the amount of scattering and the rate of light absorption are dependent on the characteristics of the particular glass or plastic core and the wavelength of the light used. The rate of optical power loss with respect to the length of the optical fibre is defined as the attenuation coefficient, typically expressed as decibels per kilometer (dB/km). The lower the attenuation coefficient, the lesser the attenuation of the optical fibre. Amplifiers are needed if the transmission linkage is long. Nevertheless, attenuation in modern optical cables is much less than in electrical copper cables.

4.5.3 Principles of Optical Communication

There are four basic steps taken in modern optical fibre communication systems: (1) a series of data bit streams is sent in electrical form to a modulator to be encoded for optical fibre transmission; (2) this encoded signal is fed to a laser or LED to generate a light input into the optical fibre, finishing the transmitter's process of converting electrical digital signals into optical signals for transmission; (3) light travels down the optical fibre, reaches a detector, and is converted back to electrical signals; (4) finally, the electrical signal is amplified and the sequence of state changes are decoded back into the original timed sequence of data bit streams, ready to be used for application devices, thus completing the process of the receiver converting the received optical signal back into a digital electrical signal.

4.5.3.1 Transmitter

The transmitter is where the electrical digital signal is converted into optical pulses for transmission in an optical fibre network. In optical communication systems, two types of light-generating devices are used. One is an LED, which is relatively inexpensive and mainly used for multimode applications; the other is a laser diode, which costs more and is used typically for single-mode implementations. LED transmitters generate incoherent light, in contrast to the coherent light generated by lasers. Of the LED incoherent light, only a very small amount of input signal power is converted into optical power for transmission in the optical fibre. LED transmitters cost less due to their simple nature and are therefore widely used in LAN infrastructures. Lasers used in optical transmitters are semiconductor lasers; that is, they are laser diodes. The output laser is directly controlled by the electrical current. The output power is larger than that of a conventional LED, typically in the range of 100 mW. Laser can be used in high bandwidth applications because of its narrow optical spectrum width. Since laser is directional compared to LED, approximately 50% of the output power is converted into optical transmission.

4.5.3.2 Amplifier

When a light signal travels in an optical fibre, the light pulse is gradually attenuated because of dispersion and fibre attenuation. To send the light pulse a long distance, it must be either relayed or amplified. Traditionally, the relay is performed by repeaters, which convert optical pulses into electrical signals, which are then converted back into stronger optical pulses for continued transmission. Now, however, optical amplifiers are used when relay is needed because of their simple design. Optical amplifiers use lasers to directly amplify an optical signal without converting it into an electrical signal.

4.5.3.3 Receiver

An optical receiver is the device at the end of the optical communication chain that converts the signal-carrying optical pulse back into a digital electrical signal. The major component of the receiver is a photodetector, which converts the light it detects into electrical current or voltage, depending on the operation mode. The photodetector can be a photodiode, or an avalanche photodiode similar to an LED but working in reverse. Unlike LED-emitting light photons, photodiodes take light photons as input and generate electrons as an output electrical signal. The photodiode is similar to the ones used in CT detectors, but is specially designed for optical communication. It is capable of accurate measurement of the light signal pulse intensity. Often an avalanche photodiode, which can also magnify the signal, may be used as a photodetector. In the receiver there are also other circuits for various digital signal-processing, because the strength of the light signal pulse is attenuated and fidelity reduced when it reaches the receiver after traveling in the optical fibre. Figure 4.9 shows an optical transceiver used to convert between electrical and optical signals for transmitting and receiving optical signals over optical fibres.

4.5.3.4 Wavelength-Division Multiplexing

Multiplexing is the process of transmitting and receiving multiple signals across the same physical connection at the same time. Wavelength Division Multiplexing (WDM) lets several lasers at different wavelengths carry different optical signal pulses of various data streams simultaneously over a single physical optical fibre. WDM greatly increases the bandwidth capacity of the optical fibre network system. WDM requires a WDM in the transmitter, and a wavelength division

Figure 4.9 An optical transceiver used to convert between electrical and optical signals for transmitting and receiving signals over optical fibres.

demultiplexer, similar to a spectrometer, at the receiver end. Since each data stream uses a different optical wavelength, different protocols and data bit rates can be used at the same time.

4.5.3.5 Dense Wavelength-Division Multiplexing

Dense Wavelength Division Multiplexing (DWDM) is a technique combining data from many different sources together on a single optical fibre. Each signal is carried at a different wavelength but at the same time. This is achieved to increase conventional optical fibre data-carrying capacity by dividing it into many channels, each at a different optical wavelength, or color of light. Thus, each channel is capable of carrying a digital signal at any bit rate not exceeding the upper limit defined by the electronics, which is about 10 Gbps. Essentially, DWDM is similar to WDM, but with the additional benefit of supporting much more data, increasing the bandwidth capacity of an optical fibre significantly. Since each optical channel is demultiplexed back into the original source at the receiving end of the optical transmission, another benefit of WDM and DWDM is that digital data in different formats, transmission protocols, and data rates can be transmitted all together.

4.5.3.6 Multiplexer and Demultiplexer

In the operation of WDM and DWDM, multiplexers are internal components. Their function is to take various input optical wavelength signals from input optical fibres, converge them into one beam of light that is composed of multiple wavelengths. This composite light beam is then transmitted through the optical fibre to the receiving end of the WDM and DWDM. When an optical signal composed of multiple wavelengths is received at the destination, a reverse process is performed. The composite light beam goes through the demultiplexer first. Demultiplexer devices are required to split individual wavelength optical signals, because photodetectors are not capable of distinguishing different wavelength optical signals. Therefore, after a ray of combined wavelength light enters the device through a special lens, one wavelength of light is separated, or demultiplexed, after which the separate individual wavelength optical signal is ready for input into a photodetector in the receiver. Usually, a demultiplexer can also operate as a multiplexer.

4.5.4 Optical Fibre Cables

All the devices discussed in the previous sections are located at the transmitting and receiving ends of a complete optical communication system. Optical fibre cables are a medium linking these two ends. Different optical cables are used in fibre optic data transmission over long distances at large bandwidth, and in shorter-distance digital network LAN systems. When light enters one end of the optical fibre it is confined inside even though the fibre may be bent around corners. It travels throughout the entire length without much loss along the fibre, and leaves the optical fibre at the other end.

4.5.4.1 Optical Fibre Cable Structure

An optical fibre is a very thin, long strand of glass or plastic cylinder with unique characteristics. As illustrated in Figure 4.8, an optical fibre cable has a small cylindrical strand of glass or plastic as the core. The cladding surrounds the core so that light is confined in the core of the optical fibre. The core has a refraction index around 1.5, slightly higher than the cladding, so that the boundary between the core and cladding behaves like a mirror. The cladding itself is then surrounded by a buffer layer, which is further covered with a plastic jacket layer. These layers not only protect the core and the cladding of the optical fibre, but also strengthen the optical fibre cable. Optical fibre cables used in healthcare institutions are typically flexible. However, care should be taken not to bend optical cables too much because it can increase optical signal loss.

As discussed earlier, there are two optical communication modes: single-mode and multimode. There are also two types of optical fibres currently in use: one is single-mode optical fibre, the other is multimode optical fibre. Single-mode optical fibre usually has a smaller core diameter, typically 8–10 µm, which requires more precise connection and equipment. Multimode fibres have larger core diameters. For example, the current widely used multimode fibre 50/125 has a core diameter of 50 µm and a cladding diameter of 125 µm. This type of fibre can be used for Ethernet LAN applications from 10 Mbps to 1 Gbps. Low-cost LED transmitters can be used because of the larger core diameter. Optical fibre network can be best used for the network backbone infrastructure in a healthcare institution, or for network applications requiring very high bandwidth, such as between the radiology department and the PACS server, which demands the highest bandwidth in a hospital.

Multimode fibre usually has more attenuation and is more expensive. Because of the larger diameter of the fibre core, greater than 50 μm, less precise and inexpensive equipment can be used. Multimode fibre also has multimode distortion; therefore, the data bandwidth and distance of the fibre link without relay is less compared with single-mode fibre. However, the fibre length for the link may not be a concern for backbone LAN infrastructure in buildings, because typical transmission bandwidth and corresponding distance limits are around 100 Mbps up to 2 km, 1 Gbps up to 500–600 m, and 10 Gbps up to 300 m. These transmission distances are sufficient for most LAN backbone infrastructure or for PACS applications.

Connectors, transmitters, and receivers associated with single-mode fibres usually cost more than those of multimode optical fibres, though single-mode fibre itself costs less. In a single-mode optical fibre, only a single ray of light is carried. A single ray of light consists of light within a small range of wavelengths. The ray of light travels along the length of the fibre, but the electro-magnetic wave of the light actually vibrates in a direction perpendicular to the length of the fibre. There is no dispersion in single-mode optical fibre. Dispersion is a phenomenon similar to what happens in a rainbow, which separates the white light into different colors, and causes the fidelity of the signal carrying light pulse loss. Without dispersion, higher data bandwidth capacity can be achieved and the light signal can travel longer distances before relay is required.

4.5.4.2 Cable Joining

Joining two separate pieces of optical fibre is much more complicated than joining electrical wire or cable. Two pieces of optical fibre can be connected together by splicing. When optical fibres are spliced, the ends first need to be cleaved carefully. Cleaving involves deliberately breaking the optical fibre under control using a cleaving tool to create a perfectly flat end-face which is perpendicular to the longitudinal axis of the fibre. A good cleave is required for a successful splice. Then two cleaved ends are spliced together either by fusion with an electric arc to melt the fibre ends together, or by mechanical means.

4.5.4.3 Cable Termination

Optical fibres can be terminated with connectors, with the optical fibre end held securely, precisely, and ready to be connected to devices or other optical connectors. A typical optical fibre connector is a solid round barrel covered by a sleeve holding the barrel in a mating socket. The connector is installed with the optical fibre end prepared and inserted into rear of the connector. Usually, a quick-drying adhesive is used for the connector to hold the optical fibre end securely. When the adhesive is set, the optical fibre end is carefully polished to a mirror-like finish. Figure 4.10a is a multimode optical fibre cable cleaved and polished, installed with an end connector in a communication closet, and Figure 4.10b are two multimode optical fibre cables with different connectors.

4.5.4.4 Cable Bending

In a straight or slightly bent optical fibre with a diameter of at least 20 cm, a light signal travels through the core by internal reflection, which means light is reflected from the interface between the core and the cladding material back and forth without almost any signal loss. When the optical fibre is bent, there is complete internal reflection only when the light incidence angle to the axis of the optical fibre is low. When the bent angle is large up to a point, some light starts to escape and

(a)   (b)

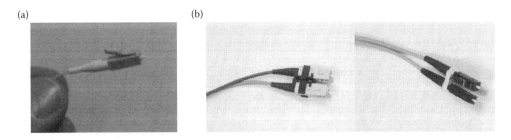

Figure 4.10   A multimode optical fibre is (a) cleaved, polished, and installed with an end connector in a communication closet; and (b) two multimode optical fibre cables with different connectors.

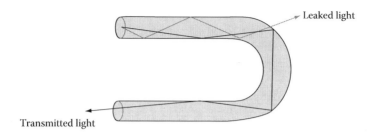

Figure 4.11 When an optical fibre cable is bent too much, some light starts to escape and leaks out of the core into the cladding.

leaks out of the core into the cladding, as illustrated in Figure 4.11. When the bent angle is too large, the light leaves the optical fibre and gets lost. Though it is much more likely in multimode fibres, this effect also applies to single-mode optical fibres. Generally speaking, the maximum bend for most optical cables is around 4 cm in diameter. The maximum allowed optical fibre cable bend is determined by the refraction index difference between the core and the cladding. Therefore, optical fibre cables used in any PACS server, storage device, or workstation should not be bent too much and must be handled carefully.

4.5.4.5 Water Proofing

Contrary to the common assumption that optical fibres are immune from water damage because no electricity is conducted in the fibre, water is the worst enemy of an optical system, and water-proofing for optical fibre is even more important than for electrical cables. Glass from which the fibre core is made can gradually pick up hydroxyl ions if immersed in water, contributing as a major source of optical attenuation due to light absorption. Also, water can cause microcracking in the optical fibre glass, which not only causes light scattering, but also weakens the optical fibre significantly.

4.5.4.6 Optical Cable Handling

Though an optical fibre cable is very strong under tension stress along its length, stress can cause significant increases in optical attenuation. An optical fibre can be broken easily under lateral sheer stress when handled roughly during installation or usage. Therefore, optical fibre cables should be kept away from any sort of stress.

Glass optical fibres are very thin, rigid, and sharp. It is very easy to get cut by unclad optical fibres when fibre joints are made or connectors are fitted. During these processes, leftover short lengths of glass fibres, which are very sharp, like glass needles, and hard to see, can easily pierce into skin and break off, leaving minute pieces inside. Such situations can be avoided by immediately disposing of small pieces of glass fibres into a designated, specially designed container.

4.5.5 Benefits of Optical Transmission

Optical fibres have many advantages and require much less maintenance once installed compared to copper wire cables.

4.5.5.1 Electromagnetic Interference

Optical fibre is free from electrical interference arising from crosstalk between signals in different cables, which limits the bandwidth of an electrical network cable. Since there is no electrical connection, there is no electromagnetic interference, which is a major source of noise in electrical network cable systems, especially for high bandwidth applications. It is no longer a problem if optical fibre cables are placed close to elevators, power lines, emergency generators, and so on inside a healthcare institution, where electrical network cables are affected by electromagnetic interferences.

4.5.5.2 Ground Loops

Since there is no electrical connection in an optical fibre, the possibility of ground loops caused by voltage potential difference among "ground" at various locations is eliminated. Ground loops can pose serious problems for LAN networks using electrical wire cables.

4.5.5.3 Bandwidth

When upgrading a LAN network to a higher bandwidth using electrical cables, the cabling infrastructure of the network often needs to be replaced too, because the various categories and classes of unshielded or shielded twisted pair cables all have different maximum bandwidth limit. They may not function properly in handling higher bandwidth requirements after other network equipment are upgraded. Optical fibre communication systems have enormous potential transmission bandwidth to which it is relatively easy to upgrade, because installed optical fibre cable has a very large theoretical maximum capacity. Research has shown that bandwidth of up to 100 Gbps can be achieved for a single channel in an optical fibre. When dozens of channels are multiplexed for DWDM, the total data bandwidth is potentially several hundred times higher than can be handled by the electrical wire cables typically used for a LAN network. Current fibre optical communication network systems are limited by the bandwidths of the electronics used for transmission and receiving. When new optical communication technology and equipment are available and installed, the bandwidth capacity of the optical fibre can be increased without changing the optical fibre cable itself.

Though the theoretical bandwidth is very high, in practice not that many optical channels have yet been multiplexed in a single optical fibre because it may be more economical to use multiple optical fibre cables than to use very complicated multiplex technologies to incorporate many channels into a single optical fibre cable. However, as already mentioned, optical fibre can be upgraded for much higher data bandwidths when new cost effective technologies become available. . In addition, for network applications inside a building, such as a LAN in a healthcare institution, optical fibre saves space in cable ducts because of the much higher data-carrying capacity each cable has compared to an electrical cable.

The bandwidth demands for various specialized PACS system is ever-increasing, because of demands for more data and faster speed for transmission of images from modalities to PACS servers and workstations. Optical communication networks are becoming the technology of choice to meet these requirements.

4.5.5.4 Fibre Channel

Fibre Channel is a technology mainly for used data exchange among computer servers and storage devices. It is replacing the Small Computer System Interface (SCSI) as the interface for data transmission between clustered storage devices and servers. Though it uses mainly optical fibre connections, electrical cables are also used for some of the relatively lower speed Fibre Channels. Since it is widely used for Storage Area Network (SAN), much of the details are discussed in Chapter 5, where PACS storage systems are covered.

4.6 NETWORK SECURITY

Network security is very important for any business, especially for healthcare institutions handling private, confidential, healthcare information. Some applications on the network do not have built-in security features, or have not implemented standard security mechanisms. The following sections discuss various security measures for different purposes such as authentication and data integrity.

4.6.1 Data Encryption and Decryption

Digital data encryption is a method of altering a clear text message into an unreadable form, so that its meaning is hidden, keeping the message secure. Decryption is the reverse transformation, which retrieves the original clear text message. With technological advances in this area, there are many ways to encrypt data, each having procedures that transform data based on methods that are difficult to reverse. Essential to strong encryption is the difficulty of reverse engineering. Strong encryption indicates that, without knowing the proper keys used for data encryption, it is practically impossible to retrieve the clear text message with conventional computation resources. Data encryption and decryption can be achieved with mathematical algorithms as the cipher. Current data encryption and decryption algorithms use a parameter called "key," and the "key" used determines the security strength of these algorithms. The longer the key length, the stronger the data encryption.

4.6.2 Authentication and Data Integrity Check

Authentication in data communication over networks is a method of verifying that the message sender is truly who one claims to be. Authentication can tell if an individual is an intruder if one pretends to be someone else. A data integrity check is a process to verify that during data transmission over network, the data are not altered. Data tampering by an unauthorized intruder is detected by the data integrity check. Both authentication and data integrity check are essential components of network security.

4.6.3 Digital Signature

A digital signature is essentially data encryption with a public key in reverse. For data transmission over an unsecured network, a properly implemented digital signature provides the recipient reason to believe that the data have been sent by the claimed sender. A sender digitally signs a file for transmission on the network using a private key, while the recipient of the file then uses the public key of the sender to verify that the "signature" is truly from the claimed sender. Properly implemented digital signatures are more difficult to forge than handwritten ones. A typical digital signature-generating scheme including the following steps is illustrated in Figure 4.12.

For the sender:

- Use special software to obtain a hash which is a mathematical summary and the digital signature of the message file

- Use a private key previously obtained from a public–private key authority to encrypt the hash

 For the recipient:

- Create a hash from the received message

- Use the sender's public key to decrypt the received hash

- Match the two hashes to authenticate the validity of the received message

The main purpose of the digital signature is to authenticate the source of the message, which means that the message file really comes from the individual claiming to be the sender.

4.6.4 Digital Certificate

A digital certificate is an electronic document issued to individuals by a certificate authority to prove their or a Web site's identity. Digital certificates are trusted identification cards in electronic form that bind a Web site's public encryption key to their identity for the purpose of public trust. A digital certificate is issued by an independent, recognized, and mutually trusted third party that guarantees that the Web site is what it claims to be. This third party is called a certification authority. An individual or Web site applies for a digital certificate from a certificate authority; the certificate authority then issues an encrypted digital certificate to the individual or Web site. Usually the name of the digital certificate holder, a serial number, a copy of the certificate holder's public key for message encryption and digital signature, the valid period of the digital certificate, the digital certificate authority's digital signature for a recipient to verify if the digital certificate is real, and so on are contained in the digital certificate. An individual's identity is tied to a public key. The certificate authority makes its own public key readily available. When a recipient receives an encrypted message, the certificate authority's public key can be used to decode the digital certificate attached. Then it can be verified that the certificate is really issued by the certificate authority, and the sender's public key can be retrieved to encrypt the reply message file.

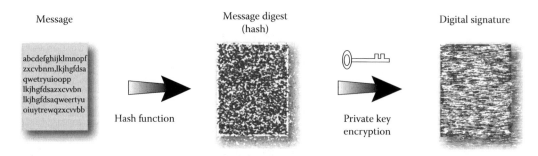

Figure 4.12   A typical digital signature-generating scheme.

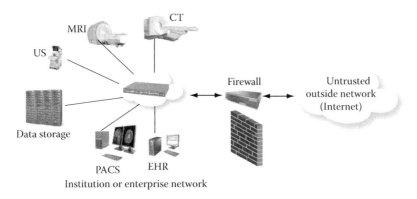

Figure 4.13   A network firewall allows computers and devices on an institution's intranet to access outside networks, while keeping unauthorized outsiders from reaching the institution's intranet.

4.6.5 Firewall

When computers from a healthcare institution's internal network are connected to outside networks such as the Internet, there is always a challenge in terms of maintaining safety for the internal network to not be attacked by outside intruders, while at the same time meeting the need for accessing outside networks for business purposes. Theoretically, whenever the institution's network is connected to the Internet, outside intruders can break into the institution's network, causing damage to the institution's network, computers, and data, including patients' confidential information. To solve this problem, computer network firewalls are built, as shown in Figure 4.13, to allow computers located in the institution's intranet to access outside networks, yet keep unauthorized outsiders from reaching the healthcare institution's intranet.

A firewall is a mechanism that enforces a network security policy, separating an institution's secure internal network (intranet) from an unsecured network such as the Internet. It is simply a device and software, restricting communication and helping to improve security. Firewalls are also commonly used to segregate anything else that is considered high risk.

4.6.5.1 Rules to Build a Firewall

Depending on various network security goals to be achieved and the specific network environment of the healthcare institution, there are many ways to build a network firewall. However, some rules are common, such as:

- Only allow safe network traffic in and out

- Track firewall activities and keep thorough auditing and logging records, so that if an institution's internal network is compromised, the records can reveal the root cause and help prevent damage

- Setup the firewall and describe exactly what network traffic is allowed; anything not explicitly permitted is denied

- Provide warning if there are suspicious activities

- Keep safe private network information

4.6.5.2 Firewall Components

A firewall is composed of hardware and software installed at the entry point where the secure internal network meets the unsecured external network. A firewall may include one or more software running on one or more hosts. The hosts can either be a general computer system, a specialized system such as routers, or a combination of these systems. This firewall can then decide what information or services are accessible from outside, and who is allowed to use such information and services from outside.

4.6.6 Secure Socket Layer and Transport Layer Security

In addition to firewalls, Secure Socket Layer (SSL) is a security protocol providing a private channel to parties, including the server and the client, communicating over an unsecured public

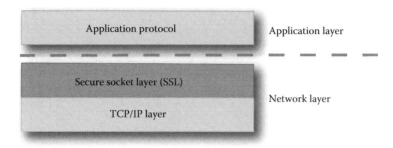

Figure 4.14   The SSL is between the TCP/IP layer and the application protocol.

network such as the Internet. The term "socket" refers to the method of passing data back and forth between a server and a client application across the network layer. SSL uses the public and private key data encryption system and a digital certificate to ensure the authenticity of the parties, and the privacy and integrity of the data communicated. SSL is designed to use TCP as a secure and reliable communication layer. Its relationship with TCP and the application supported is explained in Figure 4.14. The network connection thus established is authenticated and secured so that data are protected during transmission.

Using SSL, both client and server can be authenticated. SSL provides standard public key encryption techniques for authenticating the parties involved in the communication. SSL is not a single protocol, but a set of protocols including the SSL Record Protocol, the Alert Protocol, the Change Cipher Spec Protocol, and the Handshake Protocol. During an SSL session, data cannot be tampered with either intentionally or unintentionally. In addition, data are fully protected and can only be read by the intended recipient, and cannot be intercepted by anyone else. Since SSL is an alternative to the standard TCP/IP socket that is implemented with security, any TCP/IP application can potentially run in a secure way without requiring a change of application. The importance of SSL for healthcare institutions is that it is used in many teleradiology PACS systems to ensure secure data communication over unsecured networks.

4.6.6.1 Objectives of SSL

Though SSL is used mainly for secure Internet access, the SSL protocol can be used to protect any TCP/IP service transmissions. SSL authentication has both server authentication and optional client authentication, allowing both parties involved in the communication to be verified. Server authentication enables the application client, such as a remotely located PACS/EHR workstation, to go through the following steps to authenticate a teleradiology server's identity:

- Check the validity period of the server certificate.
- Check if the certificate is issued by a trusted certificate authority.
- Check if the public key validates the issuer's digital signature. This public key is maintained by the client and is from an issuing certificate authority such as VeriSign.
- Confirm that the domain name in the server's certificate is actually the domain name of the server, and the server is located at the same address as listed on the server certificate.

For client authentication, the server goes through the following steps:

- Checks the validity of the user's certificate.
- Checks if the user's digital signature can be validated by the public key in the certificate.
- Checks if the certificate authority in the user's certificate is trustworthy.
- Checks if the digital signature of the user or client can be verified by the certificate authority's public key.
- Verifies if the user or client is authorized to access the requested resources on the server; even the identity of the user or client is verified. For example, a remotely connected PACS workstation may be authorized to use the PACS application software on the PACS teleradiology server, but may not be authorized to access other resources located on the server.

4.6.6.2 SSL Certificate

An SSL certificate includes a public and a private key. Information encryption is done with the public key, so data are scrambled and invisible to anyone else on the network, and the private key is used to decipher the data received. Both the server (i.e., a Web server) and the client (i.e., a Web browser) are authenticated when the browser points to the server. Secure data transmission can start after a data encryption method is established with a unique session key. Data encryption is the strongest when a 128-bit SSL certificate is used. Although an SSL certificate is issued by a trusted certificate authority, many certificate authorities simply verify the domain name and then issue the certificate. VeriSign is one of the recognized certificate authorities verifying the ownership of an Internet Web site, the existence of the business of the Web site, and the applicant's authority to apply the certificate.

4.6.6.3 SSL Considerations

The SSL protocol is well designed; however, it is not concerned with the processes and procedures an individual user or institution must go through to acquire a certificate. The SSL protocol depends on a trusted third-party authority. The assumption is that parties involved in the communication over a secure channel trust the identities of SSL certificate holders, which are issued by a third-party certificate issuing authority. The SSL protocol does not address a few critical issues associated with authenticating an individual or institution before a certificate is issued. Therefore, the process lacks authentication standards: sufficient proof for identity, various levels of proof and their representation in certificate issuance, and consistent standards for different third-party certificate-issuing authorities. These issues are all related to the nontechnical aspects of issuing certificates.

4.6.6.4 Purpose of Transport Layer Security

The goal of Transport Layer Security (TLS) is similar to that of the SSL. It is also a cryptographic protocol providing communication security over TCP/IP networks and data integrity at the network transport layer. It runs on layers beneath application protocols such as Hypertext Transfer Protocol (HTTP) and above a reliable protocol such as TCP. TLS can also be used as a tunnel for an entire network used in a VPN. Additionally, TLS supports the more secure mutual authentication connection mode, which means both parties involved in the communication are assured the true identity of the other party. In this case, of course, the TLS client too is required to hold a certificate. TLS has inherent advantages in firewall administration when there are many remote access users. Thus, TLS has been adopted by the DICOM standard in its Basic TLS Secure Connection Profile as a secure channel for DICOM image and object exchange.

4.6.6.5 Difference between SSL and TLS

SSL is the predecessor of TLS. Although SSL and TLS differ in several aspects, in terms of security, they are considered equal. Both SSL and TLS are protocols providing data encryption and authentication in applications where an unsecured network is used for data exchange, and both enable identity verification of the parties involved in the communication. The main benefit of using TLS is that it is designed as an open standard, and is likely to be supported by other Internet standards in the future. Compared to SSL, TLS has additional features, including interoperability—when the application is enabled with TLS, exchange of TLS parameters can be initiated by either party involved without the other party having to know the detailed TLS implementation—and expandability—future extensions can be easily built with new cryptographic techniques based on the TLS framework. Currently, SSL is the most frequently used method for secure communications on the Internet. However, it is anticipated that TLS will become a recognized Internet communication security standard.

4.6.7 Virtual Private Network

A VPN is an institution's intranet extended over the public Internet, establishing a private secure connection to wherever the Internet reaches. The Internet has become a widely available low-cost public network infrastructure. Its universal reach provides various institutions and organizations an opportunity to construct a VPN. The Internet can be safely used when VPN and other network security technologies are deployed. The most common Internet access methods are discussed next.

4.6.7.1 *Virtual Private LAN/WAN Services*

A VPN is a communication network tunneled through another network, and can be dedicated for a specific network such as a LAN or WAN. The most widely used VPN application makes the Internet a part of a secure enterprise network, as illustrated in Figure 4.15.

For users such as healthcare institutions, a VPN can securely exchange confidential healthcare information over the Internet, connecting remote clinics, physicians, and healthcare professionals into an extended healthcare enterprise network. For example, a teleradiology or telemedicine workstation located outside the premises of the hospital may use the Internet to connect to the hospital's telemedicine servers through a VPN and exchange PACS and EHR data. Thus, Internet access services provided by ISPs can be a cost effective alternative to healthcare institutions' expensive leased network communication lines. Additionally, in many places around the world, Internet access may be the only option to accessing other computer networks.

The advantage of VPN is not its high performance, but the convenience of allowing a remote PACS/EHR workstation to be connected to a server using an Internet network infrastructure. However, attention should be paid to the security of the VPN, and security should not be sacrificed for convenience.

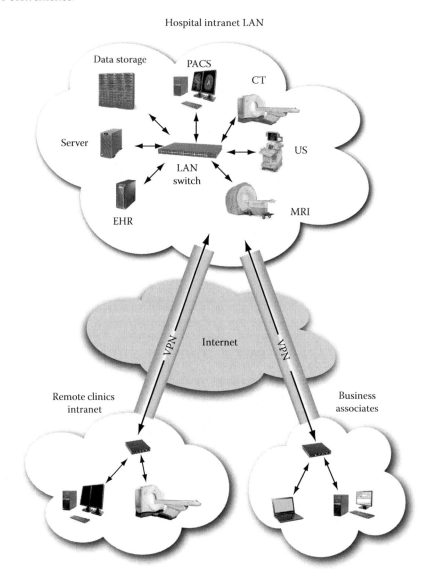

Figure 4.15 VPN is widely used to securely exchange data over the unsecured Internet.

4.6.7.2 VPN Security Models

The core of VPN technology is data encryption and user authentication; therefore, an institution's private network can be embedded in the public Internet backbone while still being able to maintain the security and safety associated with a traditional private institutional intranet. From the standpoint of security, a VPN must enforce security mechanisms within itself, or the underlying delivery network upon which the VPN operates must be secure. IP Security (IPSec) is the most-used VPN protocol for IPv4, and it is a mandatory part of the IPv6 standard.

4.6.7.3 Authentication for VPN Connection

Several methods of authentication can be used individually or in combination in a VPN connection. These include biometrics, passwords, cryptographic methods, access gateway, or firewall. Some of these measures require user interaction, whereas others are built into the PACS/EHR workstation. Authentication is strong if several methods are used in combination.

4.6.7.4 VPN Implementation

A VPN is an economical way of connecting remote PACS/EHR workstations to the PACS/EHR networks located inside a healthcare institution. However, there are some issues to be kept in mind before a VPN is put in place. Security for the VPN, as has been mentioned, is a major concern because of the public nature of the Internet network infrastructure. To use VPN securely, either the physical delivery network in which the VPN is embedded should be secured, or the VPN itself should have security measures. For users of remote PACS/EHR workstations, authentication is always needed to access the VPN.

Another consideration for healthcare institutions when a VPN is implemented is the ability to upgrade the VPN as needs change, or when TCP/IP technology develops. Usually, vendors supporting a wide range of VPN hardware and software products are flexible in meeting client needs. Current VPN solutions are suitable mostly for an IPv4 environment. It is important to keep in mind that the VPN solution used should be upgraded to maintain interoperability when IPv6 is widely deployed. For smaller clinical or imaging centers without their own IT department to administer VPN servers, there are ISPs who can provide such services. They can provide remote workstations with secure VPN connections to other healthcare institutions' internal PACS/EHR networks, and may even include other security services for those remote workstations.

Successful VPN solution implementation involves more than technology. A VPN should be designed and operated under well-defined institutional security policies. When network access goes to clinics, to homes of clinicians, or to healthcare professionals where there is no professional network administrator, security must be maintained as transparently as possible for end users.

4.7 WI-FI AND WIRELESS LOCAL AREA NETWORK

In addition to these wired networks, which use either copper wired cables or optical fibre cables as the data transmission medium, wireless network technologies are also frequently used in healthcare institutions. Generally speaking, Wireless Fidelity (Wi-Fi) refers to any wireless LAN (WLAN) technology defined by the Institute of Electrical and Electronic Engineers (IEEE) 802.11 standards, including 802.11, 802.11a, 802.11b, 802.11g, and 802.11n, and so on. It defines the interface for over-the-air wireless communications. Compared to wired networks, wireless network implementation, especially LAN for data communication within a healthcare institution, is increasing because of its unique advantages, especially for EHR applications. The implementation involves many antennas installed throughout the building as WLAN access points. Since each access point only provides a small area of coverage, there are many access points connecting wired communication closets. Though it may cost much less to connect many computers wirelessly compared to installing physical cabling to each computer, WLAN poses several challenges. It is possible for an individual to access the wireless network and intercept sensitive confidential healthcare data communication. Also, wireless communication speed is slower than physically cabled network technology. Because of these drawbacks, wireless network is not going to totally replace wired network in healthcare institutions any time soon. However, understanding various wireless network technologies and their limitations helps in making informed decisions when selecting them for applications appropriate in healthcare environments.

4.7.1 Wireless Application Protocol

Wireless Application Protocol (WAP) is the standard for Internet communications and services on mobile digital devices. This protocol enables secure access to applications and services on the

Internet and an intranet using wireless digital devices. However, because of the nature of these devices and the wireless environment, there are limitations in both digital devices and wireless network environments. For devices, the limitations include less processing power of the CPU, less memory, power consumption restrictions, and limited interfaces, such as display size and input devices. Limitations in the wireless network environment are limited network bandwidth, limited network stability, and longer latency. When using these wireless devices for any medical image display, such as PACS or EHR applications, these limitations should all be taken into account, in addition to data privacy, integrity, and authentication considerations mandatory for dealing with confidential healthcare information.

4.7.2 Wireless 802.11 Family Standards

The core of WLAN is the 802.11 family of IEEE standards. These standards define operations at 2.4 gigahertz (GHz) and 5 GHz frequencies. The same basic protocol is used for various over-the-air electromagnetic wave modulation techniques. The first wireless network standard is 802.11, followed by the amendments of 802.11a, 802.11b, 802.11g, and 802.11n. Almost all mobile laptop computers used in healthcare institutions for various applications employ some 802.11 family techniques.

4.7.2.1 IEEE 802.11a

IEEE 802.11 is already obsolete. IEEE 802.11a operates in the 5 GHz band with a maximum data rate of 54 Mbps, although a realistic achievable net data throughput rate may be much lower. It uses the same original 802.11 standard data link layer protocol and frame format. Since the 5 GHz radio band is used much less compared with the 2.4 GHz band, it is less likely that the data transmission will be interrupted by other devices operating at the same frequency. However, the shorter wavelength of the electromagnetic wave of the relatively high 5 GHz frequency also has its own disadvantages. The effective wireless network link distance and penetration of 5 GHz 802.11a is not as good as those operating at a lower 2.4 GHz frequency band.

4.7.2.2 IEEE 802.11b

IEEE 802.11b operates in the 2.4 GHz frequency band, using the same media access method defined in the original 802.11 standard, with a maximum raw data throughput rate of 11 Mbps. When the wireless signal strength is low, the data speed is reduced, and at a lower speed, 802.11b has a higher link distance range. Since many electronic devices, such as microwave ovens, cordless phones, and Bluetooth devices, all operate at 2.4 GHz frequency, 802.11b wireless devices suffer from occasional interference from these electronic devices. There is no surprise that when full wireless coverage is implemented in a new healthcare institution, it may be found that the wireless connection is intermittently interrupted, and the root cause is the microwave ovens used in employee lounges. In this instance, additional electromagnetic wave shielding is necessary for those microwave ovens.

4.7.2.3 IEEE 802.11g

Following 802.11b, IEEE 802.11g also operates in the 2.4 GHz frequency, but uses the same transmission scheme as 802.11a. The maximum data bit rate is 54 Mbps, the same as that of 802.11a; its hardware is back compatible with that of 802.11b, but suffers from similar interference from other electronic products operating in the same 2.4 GHz frequency.

4.7.2.4 IEEE 802.11n

IEEE 802.11n is a significant improvement in speed and link distance range compared to 802.11g. It operates in 5 GHz or 2.4 GHz, with a maximum theoretical data bit rate of 600 Mbps, with a realistic throughput around 100 Mbps. This is achieved by simultaneously transmitting three and receiving two data streams using separate nonoverlapping channels. Also, more data is packed into each transmitted data packet for improved efficiency. Because of the high data throughput, 802.11n opens the way to far more applications using WLAN.

Since various 802.11 family wireless devices are used in healthcare institutions, understanding their advantages and limitations helps to better utilize the technology and to take the right decisions. It is anticipated that newer standards and speeds will continue to emerge in the future.

4.7.3 Wi-Fi Operation

The operation of Wi-Fi is based on nonswitched Ethernet. A WAP broadcast has Service Set Identifier (SSID) beacon packets every 100 ms. Hotspots or access points are defined as all areas covered by this WAP beacon. When the client wireless network interface setup is configured for connection to the SSID, the data encryption level of WAP is acceptable, and the detected WAP wireless signal is strong enough for the wireless client or user to be connected to the network. When large areas are to be covered wirelessly, multiple WAPs may be deployed to overlap individual hotspots or access points using the same SSID. In case multiple WAP signals are detected, the user should choose to connect to the WAP showing the strongest signal, because signal strength is related to bandwidth and reliability of wireless connection, which are limitations of wireless networks.

4.7.4 Wireless Equivalent Privacy (WEP) and Wi-Fi Protected Access (WPA)

The wireless nature of Wi-Fi is inherently susceptible to security issues. An intruder does not need physical access to the traditional wired network in order to gain access to data communication. As such, data transmission on Wi-Fi must incorporate data encryption techniques and authentication measures that can provide privacy comparable to that offered by the traditional wired network. WEP was an initial attempt to secure Wi-Fi communications. However, the shared encryption key accessible to all users for authentication with the WAP is a weak point in terms of security. The response to this is WPA, which provides a strong measure for authentication and centralized key management. WPA is part of the 802.11i standard, which supports the most advanced encryption available, the Advanced Encryption Standard (AES). WPA uses a fast re-keying algorithm; to enhance the security it implements per data packet dynamic keys that are updated and changed every 10,000 data packets. With increased security, WPA has replaced WEP. The WPA has been further enhanced and expanded into WPA2, which is the current technology providing both data integrity and key management, and combines the data integrity and confidentiality functions of WPA into a single protocol. For healthcare institutions, wireless network security should be checked frequently, because new wireless devices could be easily added by individuals without a proper security setup, which may become easy access points for intruders.

4.7.5 Wireless Transport Layer Security (WTLS)

To ensure secure network connection and data communications, as required by many Web applications, WTLS, which is based on the TLS standard and has special mechanisms for the wireless environment, has been optimized for wireless applications. The WAP gateway automatically manages wireless security for Web applications using standard TLS Internet security techniques. These security measures include data encryption, data decryption, application server and user authentication, and data integrity protection. The cryptographic algorithms and key length used by WTLS is subject to the regulations of the country where it is used.

Different wireless communication technologies can support many different services in compact portable packages. These technologies provide a choice of communication methods, with wired and wireless options available in a single device, which is able to automatically select the most appropriate communication method according to the needs of the user, the type of data to be transferred, and the physical location of the device. It is anticipated that more wireless devices will be adopted for PACS, EHR, and other business operation applications throughout healthcare institutions.

4.8 BACKUP AND ALTERNATIVE NETWORK USING THE INTERNET

Today, access to the Internet is widely available around the world. With the development of SSL and VPN technologies, the Internet is being increasingly utilized to securely interconnect an institution's geographically separated internal networks and users. This provides an opportunity to use an ISP service as an alternative when the primary network connection is down, as a substitute for the primary means to connect remote sites via the Internet. The technologies of ISP are mainly driven by the general consumer market for residential and small business use. Understanding their advantages and limitations can help the user to decide which option meets the healthcare institution's needs for PACS and EHR applications.

4.8.1 Digital Subscriber Line (DSL)

A DSL is a family of technologies providing digital data transmission over wires of the local telephone network. Standard telephone lines consist of a pair of copper wires the phone company installs at residential homes and offices or inside a hospital or clinic. The pair of telephone wires is actually capable of carrying much more information than the phone conversations for which it was originally designed and installed. DSL explores and utilizes this potential capability of carrying digital data for network communication without disturbing the telephone line's function of carrying voice conversation. The concept is to use different frequencies for various tasks. DSL can be used at the same time and on the same telephone line as the regular telephone, as it uses a high frequency, whereas the telephone uses a low frequency. The transmission speed of DSL depends on the specific DSL technology used, the line conditions, and service subscribed to.

4.8.1.1 Asymmetric DSL (ADSL)

ADSL stands for Asymmetric DSL, which is a commonly used DSL technology for many residential homes and small businesses. It is "Asymmetry" because the upload or send speed is much lower than the download or receiving speed, and therefore many DSL services provided by ISPs are actually ASDL. There are a few reasons for ISP providers to market it this way. The primary reason is the assumption that most Internet users surf Web pages or download information much more than they send or upload information. With this assumption, typical users see the benefit most of the time when the connection speed from the Internet is several times faster than the connection speed from the user to the Internet. Though symmetric DSL (SDSL) has the same speed for both downloading and uploading of data, it is much rarer than ADSL.

Since all voice conversations are in the frequency range below 4 kHz, most DSL installations attach low-pass filters to telephone line jacks, which are used for telephone connections not connecting to DSL transceivers, so that high frequencies above 4 kHz for DSL use are all filtered out without interfering with the voice conversation. Most DSL technologies also use discrete multitone (DMT), which divides signals into 247 separate channels, each 4-kHz wide. During operation, each channel is constantly monitored and the signal is separately transmitted and received on the best channels found. Some lower channels are designated as bi directional channels. DMT is complex, but it is flexible for lines with varying quality.

4.8.1.2 Equipment

For a typical DSL setup, there is the DSL transceiver or gateway and a low-pass filter at the customer's end, as shown in Figure 4.16, which shows where data from the user network or computer is connected to the DSL line. For business users such as healthcare institutions and medical clinics, this may be combined with routers, switches, or other network equipment. At the ISP end, there is a DSL access multiplexer connected with many customers and aggregated into a single high-capacity connection to the Internet. ISPs may also provide additional services, such as dynamic IP address assignment and routing.

4.8.1.3 Limitations of DSL

The different versions of DSL technologies have some common features. Usually, they don't require new wiring since they can use the current telephone lines already in place. Unique to DSLs, the connection works better when it is closer to the central office, because DSL is a distance-sensitive technology. The DSL signal quality and connection speed decreases when the connection distance increases. The signal gets weaker when it is farther away from the central office. Also, the connection speed is faster for receiving or downloading data than for sending or uploading data over the Internet. This may be the bottleneck for using DSL to connect remote sites to the central server for PACS or EHR applications, because uploading speed for sending remotely acquired images to the central PACS server may be as important as the ability to download images from the server to a remote PACS workstation.

4.8.2 Cable Internet Access

In addition to DSL, cable television network infrastructure is also capable of carrying a high bandwidth of data communication. Cable Internet is a broadband Internet access method using the cable television infrastructure. It is layered on top of the existing cable television network, just as DSL uses the existing telephone network. At the cable ISP's user end, a cable modem provides two-way data communication in radio frequency channels over a cable television network. Some

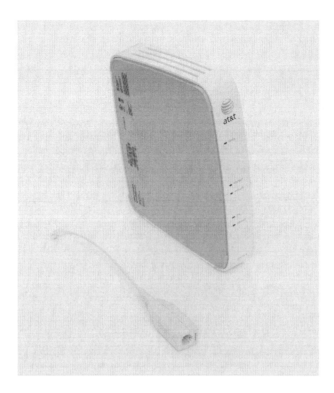

Figure 4.16 A DSL transceiver and a low-pass filter.

cable modems may incorporate router or dynamic IP address assigning functions. At the cable operator's end, there is the cable modem termination system. These two are connected by the coaxial cables originally designed and installed for cable television applications.

4.8.3 DSL and Cable Internet Service Provider Comparison

Cable and DSL are the two dominant ISP methods providing Internet access to residential and small business users. Upload and download speeds vary depending on the service subscribed to; however, the upload speed is usually much lower than the download speed. When selecting either DSL or cable ISP as an alternative way of connecting to the Internet, the healthcare institution as the user should understand the major difference between DSL and cable ISP. Cable ISP users typically share a network loop running through a neighborhood or area. When more users are added, the performance is degraded and connection speed slows down. However, cable ISP is not limited by distance as is DSL, because cable signals can remain strong for long distances. Different from cable ISP, DSL provides a dedicated connection from each user's transceiver or gateway back to the DSL access multiplexer located in the central office, which means that performance is usually not affected when new users are added, until the single high-speed connection to the Internet is saturated and the ISP needs upgrades for their equipment. However, DSL is affected by the distance from the user to the central office. Theoretically, DSL performance is more predictable, because most factors affecting the performance are known, whereas for cable ISP, the number of users in the nearby area who will subscribe to the service or will be online at a particular moment is unknown. When selecting either DSL or cable Internet services, users should discuss with the ISP the intended applications, request and subscribe business service packages with high upload and download speeds available, because depending on the service-level subscribed to, different speed configuration files are downloaded to the user's router when it is in operation.

4.8.4 Satellite Internet Access

Both DSL and cable Internet access require fixed landline infrastructures. Satellite Internet services can be used in locations where telephone landline or cable television network is not available. Satellite Internet service is not only wireless, but also mobile in nature. However, several serious limitations need to be considered when a user chooses satellite ISP as a backup or alternative option. Since satellite Internet access requires bi-directional communication, increasing the size of the satellite dish antenna can improve the signal reception during data download, and generate a higher transmitted signal when sending data, thereby improving the satellite Internet connection reliability (a weak point for satellite ISP).

Compared with land-based wired or wireless Internet access technologies, another issue unique to satellite Internet access is latency, because the signal needs to travel hundreds of miles out into space to reach the satellite, and then back to Earth again. Although the electromagnetic wave signal travels at a lightning speed of approximately 300,000 km/s, the time it takes to travel back and forth is still significantly longer than in ground-based ISPs. Including other network factors, the typical signal delay is 500–700 ms. In addition to the above-mentioned limitations, satellite communication requires a clear line of sight between the dish antenna and the satellite. The signal can be impaired not only by the presence of trees and other vegetation in the direct communication path of the satellite and the dish antenna, but is also susceptible to signal absorption and scattering by moisture and rain. Since most satellite communications operate at very high frequencies, they are sensitive to even minor obstructions such as tree leaves. Other issues include reliability, which is not as good as that of fixed line Internet access, and restricted use of VPN, which is almost a prerequisite for using the Internet in any application where confidential healthcare information is exchanged. Last but not the least, satellite ISPs currently cost a lot more than ground-based ISPs.

4.8.5 Cellular Broadband Internet Access

Another wireless Internet access technology is cellular broadband. Cellular phone technology development and service is also driven by the general consumer market, with the widespread setting up of cellular phone transmission towers. Cellular Internet access uses the cellular communication network infrastructure. Typical cellular devices required for the computer are PC data card, Cardbus, USB modem, phones with data modem, or a router.

Although the most significant advantage of cellular Internet access is its mobile nature and lack of dependence on fixed-line infrastructures, there are issues that need to be considered, mainly the coverage that the cellular phone network can provide. Since the connection speed and performance is dependent on cellular signal strength, in many areas users are not able to achieve the data communication speed indicated by the subscribed service plan. Just like other ISP Internet access technologies, the upload speed for sending out data is much lower than download speed for receiving data. Other issues are:

- Low transmission speed, only occasionally close to claimed maximum speed

- Transmission speed and reliability is very location-specific, and very dependent on cellular signal strength

- Lower download and upload speeds than what is provided by fixedline ISPs

- Requiring detailed discussion with cellular ISP directly for the PACS/EHR usage to avoid future surprises, because many service plans are not designed keeping the transmission of large PACS/EHR data files in mind

New emerging technologies or service plans for cellular Internet access are constantly offered by cellular service vendors. All considerations should be carefully evaluated before relying heavily on cellular Internet access.

In conclusion, the computer network technology is developing at a fast pace, and there are many options available. The healthcare institution should carefully assess its current network environment, the needs of various departments and the entire institution, and the technologies available, paying special attention to network security and reliability. Properly planned, a network thus implemented will meet the needs of the healthcare institution.

BIBLIOGRAPHY

Braswell, B., J. Earhart, and S. Friberg, et al. 2005. *Deploying the IBM Secure Wireless Networking Solution for Cisco Systems*. www.redbooks.ibm.com/redpapers/pdfs/redp3958.pdf. Accessed January 18, 2010.

Cepeda, O., B. Chambers, and J. Mosca, et al. 2000. *Beyond DHCP—Work Your TCP/IP Internetwork with Dynamic IP.* www.redbooks.ibm.com/redbooks/pdfs/sg245280.pdf. Accessed January 18, 2010.

Chernick, C. M., C. Edington III, and M. J. Fanto, et al. 2005. *Guidelines for the Selection and Use of Transport Layer Security (TLS) Implementations.* National Institute of Standards and Technology, U.S. Department of Commerce, NIST Special Publication 800-52.

D'Ambrosia, J., D. Law, and M. Nowell. Ethernet Alliance. 2008. *40 Gigabit Ethernet and 100 Gigabit Ethernet Technology Overview.* www.ethernetalliance.org. Accessed June 5, 2010.

Dreyer, K. J., D. S. Hirschorn, and J. H. Thrall. 2006. *PACS: A Guide to the Digital Revolution.* Springer, New York, NY.

Dutton, H. J. 1998. *Understanding Optical Communications.* www.redbooks.ibm.com/redbooks/pdfs/sg245230.pdf. Accessed January 18, 2010.

Fibre Optics: The Basics of Fibre Optic Cable. www.data-connect.com/Fibre_Tutorial.htm. Accessed January 18, 2010.

Frankel, S., K. Kent, and R. Lewkowski, et al. 2005. *Guide to IPSEC VPNs.* National Institute of Standards and Technology, U.S. Department of Commerce, NIST Special Publication 800-77.

Frankel, S., P. Hoffman, and A. Orebaugh, et al. 2008. *Guide to SSL VPNs.* National Institute of Standards and Technology, U.S. Department of Commerce, NIST Special Publication 800-113.

Gast, M. S. 2005. *802.11 Wireless Networks: The Definitive Guide*, 2nd Edition. O'Reilly Associates, Sebastopol, CA.

HIMSS Healthcare Information Technology Technical Task Force. 2008. Best Practices Defined for Healthcare Wi-Fi, HIMSS 2008. www.himss.org/content/files/BestPracticesWifi.pdf. Accessed June 5, 2010.

HIMSS HIT Technical Task Force IPv6 Work Group. 2008. *IPv6 Features and Benefits: A Technical Update Based on the Federal Government Agency IPv6 Transition Plan Found in the Public Domain.* www.himss.org/content/files/HIMSSIPv6Overview052708.pdf. Accessed August 11, 2010.

Hochstetler, S., H. Tanner, and R. Kulkarni, et al. 2001. Extending Network Management through Firewalls. www.redbooks.ibm.com/redbooks/pdfs/sg246229.pdf. Accessed January 18, 2010.

Huang, H. K. 2004. *PACS and Imaging Informatics: Basic Principles and Applications.* Wiley-Liss, Hoboken, NJ.

McDonald, C. 2010. *Virtual Private Networks: An Overview.* Intranet Journal. www.intranetjournal.com/foundation/vpn-1.shtml. Accessed January 18.

Murhammer, M. W., K.-K. Lee, and P. Motallebi, et al. 1999. *IP Network Design Guide.* www.redbooks.ibm.com/redbooks/pdfs/sg242580.pdf. Accessed January 18, 2010.

Omnitron Systems Technology White Paper. 2008. *Introduction to Media Conversion.* www.omnitronsystems.com/forms/white_paper_request.php. Accessed January 18, 2010.

Parziale, L., D. T. Britt, and C. Davis, et al. 2006. *TCP/IP Tutorial and Technical Overview.* www.redbooks.ibm.com/redbooks/pdfs/gg243376.pdf. Accessed January 18, 2010.

Scarfone, K., M. Souppaya, and M. Sexton. 2007. *Guide to Storage Encryption Technologies for End User Devices.* National Institute of Standards and Technology, U.S. Department of Commerce, NIST Special Publication 800-111.

Tate, J., R. Khattar, and K. W. Lee, et al. 2002. *Introduction to SAN Distance Solutions.* www.redbooks. ibm.com/redbooks/pdfs/sg246408.pdf. Accessed January 18, 2010.

10 Gigabit Ethernet Alliance white paper. 2002. Ethernet—The Next Generation WAN Transport Technology.

Verisign. Secure sockets layer (SSL). How it works. www.verisign.com/ssl/ssl-information-center/how-ssl-security-works/. Accessed January 18, 2010.

VeriSign white paper. How to Offer the Strongest SSL Encryption.

White, B., G. Donny, and H. Howard, et al. 2001. *Fibre Saver (2029) Implementation Guide.* www. redbooks.ibm.com/redbooks/pdfs/sg245608.pdf. Accessed January 18, 2010.

5 Data Storage

Traditionally, patient demographic information, medical history, laboratory test results, and all medical images were recorded on paper or films. However, this is the digital era, and like all other information in digital form, patient medical information too can be stored digitally. With advances in medical imaging, the amount of medical information per patient has increased at an explosive pace.

The smallest digital data unit is the bit, which can be represented by two values, 0 and 1. The next is the byte, which consists of 8 bits. The most frequent units used today for PACS or EHR systems are the kilobyte (kB) for 1000 bytes, megabyte (MB) for 1000 kB, gigabyte (GB) for 1000 MB, and terabyte (TB) for 1000 GB. Before long petabyte (PB), which is 1000 TB, will start to be used frequently.

The safety of digital medical information and its reliability for storage and retrieval all contribute to the success of PACS and EHR. In the following sections, various digital data storage technologies are discussed.

5.1 DATA STORAGE MEDIA TYPES

Unlike the paper and film, which has been used for many years, and from which a patient's medical and image information can be retrieved and reviewed without very special and complicated equipment, the format and standards used to store digital medical information has been evolving constantly at a faster pace, compared to a patient's life span. The entire digital medical information base is composed of hardware and software. The hardware includes storage media—which is constantly changing—such as magnetic spin disks, optical disks, and magnetic tapes. The software controls how the digital medical information is stored and retrieved. Currently, the mainstream short-term and long-term storage medium is the spin disk, though magnetic tape and optical disks are used as back-up storage media or for long-term storage.

5.1.1 Magnetic Tape

Magnetic tape is currently used on some PACS systems as a long-term storage medium, or as a secondary backup for long-term redundant storage. On some legacy PACS, magnetic tape is used as the primary long-term storage medium. The data are stored on and retrieved from a layer of magnetic material coated onto a plastic-based tape. When the tape winds from one reel to the other through a precision-controlled motor, data are stored or retrieved through a fixed tape head. Typically, the maximum data each tape can hold is of the same order as the spin disk of the time: it is approximately 0.5–1.0 TB without any data compression. The advantage of tape for digital data storage is its relative low cost per unit of data stored. However, the disadvantage is that compared to spin disk, the time it takes to store or retrieve data is longer. Magnetic tape is therefore primarily used for backup of PACS/EHR data. With the fast advancement in spin-disk technology, coupled with less progress in tape technology, spin disk is becoming dominant for PACS/EHR data storage.

5.1.1.1 Example of AIT Magnetic Tape–Based Storage System

There are many data storage tape formats that coexist at any time, whether past, present, and very possibly in the future. Vendors continue to release newer generations of tape media and associated drives for higher capacity, higher speed, and better performance.

Take the Advanced Intelligent Tape (AIT) as an example, it is a family of tapes in which a memory chip is built into the tape cartridge case to track the contents of the tape for faster data search and access. The family of tapes includes AIT-1, AIT-2, AIT-3, AIT-4, and AIT-5, offering up to 400 GB native storage capacity in a single 8-mm tape cartridge with a sustained data transfer rate of up to 24 MB/s native.

Shown in Figure 5.1 is an AIT-3 tape. It is an 8-mm dual reel cartridge, and is the third-generation technology implementation with a native capacity of 100 GB, and a native data transfer rate of 12 MB/s. The arrow points to the contact areas where the AIT drive reads the content in the memory chip. AIT tapes are used for computer backups. One feature of the AIT format is that several generations of AIT are both backwards and forwards compatible, so that different generations of tape media can be written and read by different generations of tape drives.

Shown in Figure 5.2 is an internal mechanism of an AIT-3 tape drive. AIT drives use the helical scan method of reading and writing tapes.

Figure 5.1 An AIT-3 tape, which is an 8 mm dual reel cartridge with a native capacity of 100 GB and a data transfer rate of 12 Mbps. The arrow indicates the contact areas from where the AIT tape drive reads content stored on the embedded memory chip.

In Figure 5.3 is a tape library containing 180 AIT-3 tapes with a capacity of 18 TB native (46 TB compressed). When magnetic tape is used as a storage medium for a PACS data archive, it is usually associated with a tape library. A tape library stores tapes in slots and uses a robot arm to load and unload tapes from the tape drives. When AIT-5 tape is used, 72 TB native (187 TB compressed) data can be stored, suitable for a tier-2 long-term PACS storage for stand-alone clinics or small hospitals.

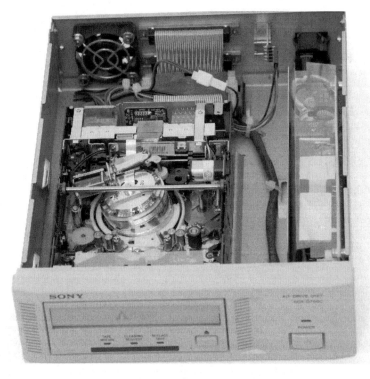

Figure 5.2 The internal mechanism of an AIT-3 tape drive, which uses the helical scan method to read and write a tape.

Figure 5.3 A Rorke Data (Rorke Data, Eden, MN) AIT tape library that contains 180 AIT-3 tapes with a native 18 TB storage space. The tape library uses a robot arm to load and unload tapes from tape drives inside the tape library and stores tapes in slots.

5.1.1.2 IBM3592 Magnetic Tape-Based Storage System

Another example is the IBM3592 tape and drive family developed by IBM for enterprise data center backup. This tape format has a half-inch tape spooled onto a 4 in × 5 in × 1 in data cartridge containing a single reel. A take-up reel is embedded inside the tape drive. Cartridges are available in three lengths and in either re-writable or Write Once Read Many times (WORM) Fibre Channel (FC) formats. The short length cartridge provides rapid data access with lower capacity; the standard length cartridge provides medium data access speed and capacity, while the extended length cartridge provides the highest capacity, up to 700 GB native. The native data transfer rate is up to 100 Mbps, with a 4-Gbps FC interface to attach to a switched fabric environment or a host system. Figure 5.4 shows the IBM System Storage TS3500 tape library using IBM 3592 tapes at a healthcare enterprise data center. It provides tape backup for all enterprise data, including various HIS, RIS, other department information systems, radiology PACS, cardiology PACS, all e-mail systems, and so on. The tape library can hold up to 450 TB of data.

5.1.2 High-Density Optical Storage Media

Optical media, such as compact disc (CD), digital video/versatile disk (DVD), and Blu-ray disk are storage media that can hold digital data that are recorded and read by laser light. These media including their variations and libraries, sometimes also termed jukebox, autochanger or auto-loader, have many advantages and are widely used for storing medical images in healthcare institutions.

5.1.2.1 CD-R and CD-RW

A recordable compact disc (CD-R) is used to store digital medical images, not for archival, but for transferring images between healthcare institutions where direct network transfer is not available.

Figure 5.4 The IBM System Storage tape library using IBM3592 tape cartridges at the Aurora Healthcare enterprise data center, which provides tape backup for all enterprise data including various HIS, RIS, radiology PACS systems, cardiology PACS systems, and other department information systems. It holds approximately 0.5 PB of data.

A standard recordable CD is a 1.2-mm thick disk made of polycarbonate with a diameter of 120 mm, and it can hold up to 650 or 700 MB of data. The polycarbonate disk has a spiral groove molded in before data are written. The polycarbonate disk is coated on the groove side with a very thin recording layer of organic dye. On top of the dye is a coat of a thin reflective layer of silver or silver alloy. Finally, a protective lacquer coating is applied on the metal reflector layer. The spiral groove helps the writing laser to stay on track and to write data to the disk at a constant rate, which is essential to ensure proper size and spacing of data points burned into the dye layer during the writing cycle.

During writing, the writing laser is pulsed to heat areas of the organic dye layer, permanently altering the optical properties of the dye and changing the reflectivity of these areas. During the read cycle, a low laser power is used. The reflected light is modulated by changes in regions of heated and unaltered dye. The reflected laser intensity change is converted into electrical signal, and binary digital information of "0" and "1" can be decoded. The same section of a CD-R disk cannot be erased or rewritten after data have already been recorded on it.

Though CD-R recoding is designed to be permanent, like most writable media, a recorded CD-R is susceptible to material degradation. The quality of the disk is not only dependent on the dye used, the physical characteristics of which may degrade over time, causing read errors and data loss beyond the reading device error-correction capability, but is also influenced by the poly-carbonate, the reflective layer, the top layer, and the sealing. In addition, proper power calibration and correct timing for the writing laser pulses, together with stable disk spin speed are all essential to both immediate readability and longevity of the recorded disk. Therefore, for medical image storage on a CD-R, it is important to have not only a high-quality disk, but also a high-quality writer. In fact, a high-quality writer may generate satisfactory results with medium quality disks. However, a high-quality disk cannot compensate for a low-quality writer. Disks written by a low-quality writer do not produce results of adequate quality.

The life of a recoded CD-R depends on the quality of the disk, the quality of the writing drive, and the storage condition. Actual usable life may be significantly less than the theoretical life of more than 20 years. With proper care, a CD-R can be read about a thousand times and has a shelf life of three to five years. Because of the low cost, and the 700 MB capacity of each disk, which is large enough to accommodate most imaging studies, CD-R is an ideal medium to carry medical images among healthcare institutions without direct network connection.

Compared with CD-R, CD-RW has a lower level of compatibility with standard CD readers and uses more expensive media. A CD-RW is rewritable after data are stored. The recording principle is different from that of a CD-R. A CD-RW disk contains an alloy recording layer composed of phase-change material. During recording an infrared laser beam is used to selectively heat and melt, at 750°F (400°C), the crystallized recording layer into an amorphous state or to anneal it at a lower temperature back to its crystalline state. During the reading cycle under a lower power laser, the different reflection of the resulting areas on the recording layer generates a decoded digital signal in the binary form of "0" and "1."

To users, a CD-RW is very similar to a CD-R; usually CD-RW recorders can also read recorded CD-R disks, and the capacity of a CD-RW is the same as that of a CD-R. However, though in theory a CD-RW can be written and erased approximately a thousand times, in reality this number is much less. Another important difference between the CD-R and CD-RW media is that the re-crystallized alloy of a CD-RW may gradually de-crystallize over time, causing a shorter life expectancy. Because of the higher unit price, compatibility issues, and lower recording and reading speed, CD-RW is not as widely used as CD-R for recording medical images.

5.1.2.2 *DVD + R/DVD – R and DVD + RW/DVD – RW*

In addition to magnetic tape-based storage systems, a recordable DVD is currently also used as a secondary backup for long-term redundant storage, and on certain legacy PACS as the primary long-term storage media.

A DVD is storage with an optical disk medium format. A single-side, single-layer, recordable DVD is composed of two 0.6 mm polycarbonate disks bonded together with adhesive. One disk has a laser guidance groove, coated with a recording dye layer and a silver alloy reflector, the same as the CD-R. The other disk is a flat un-grooved disk, protecting the data recording layer from scratches. The physical dimension of the disk is the same as that of a CD-R disk. There are four types of DVDs currently in wide use: DVD + R, DVD – R, DVD + RW, and DVD – RW. DVD + R and DVD – R are recordable, but the data cannot be erased once recorded. DVD + RW and DVD – RW are recordable and rewritable, which means data can be erased after it is recorded, and the disk used again.

The larger storage capacity of 4.7 GB in a single DVD is achieved by smaller laser spots and smaller track pitches of the spiral groove that guides the laser beam. A DVD uses 650-nm wavelength diode laser with a higher numerical aperture lens as opposed to 780 nm used by a CD-R. DVD – R and DVD + R use dyes different from that used in a CD-R to absorb the shorter wavelength laser properly. The shorter wavelength allows a smaller data record spot on the media surface, approximately 0.74 μm for a DVD versus 1.6 μm for a CD. Consequently, more data record spots can be written on a disk of the same size, enabling a higher storage capacity. The base writing speed for a DVD is approximately ten times greater than that of a CD.

A DVD also allows DVD – R/DVD + R disks to store data in dual layers, increasing the capacity from 4.7 GB to 8.5 GB for a single disk. The drive with the dual-layer capability accesses the second physical recording layer by projecting the laser through the first semitransparent recording layer. Limited use of the dual-layer recordable DVD is mainly because of its significantly higher blank media cost, and because its data writing speed is lower than that of a single layer.

Similar to CD-RW, the recording layer in a DVD – RW and DVD + RW disk is a phase change metal alloy. Depending on the writing laser power, the state of the metal alloy can be switched. Since the crystalline phase and the amorphous phase of the metal alloy layer have a different reflection to the reading laser light, data can be recorded, read, erased, and rewritten.

Just like the CD-R and CD-RW, the durability of a recorded DVD is affected by the reflective layer, the organic dye layer, the sealing method, how it is made, and the storage condition. Since PACS/EHR data need to be maintained for a long period of time, there is no benefit in using DVD + RW or DVD – RW disks, which cost several times more than the DVD + R and DVD – R disks; besides, the recording layer lasts a shorter time and it takes longer to record data.

5.1.2.3 *Blu-Ray*

Recently, with new developments in high-density optical media for the home entertainment industry, a new generation of media and formats are now available. The Blu-ray, also termed Blu-ray Disk (BD), is the most recent optical disk format developed jointly by the Blu-ray Disk Association and a group of manufacturers in consumer electronics, personal computing, and media. The format is for recording of high-definition video and large amounts of data storage. The BD is an improvement on the previous generations of optical disk technologies such as CD and DVD, which utilize red laser to read and write data. This format uses a blue-violet laser.

A BD is an optical medium with the same physical dimensions as a CD and DVD disk. The data recording spot size a laser light can be focussed on is determined by diffraction, the wavelength of

Table 5.1: Data Writing Speeds of Various Optical Disks

	CD-R/CD-RW	DVD + R/RW DVD − R/RW	BD-R/RE
1×	0.15 MB/s	1.35 MB/s	4.5 MB/s
2×		2.7 MB/s	9 MB/s
4×	0.6 MB/s	5.4 MB/s	18 MB/s
6×		8.1 MB/s	27 MB/s
8×	1.2 MB/s	10.8 MB/s	36 MB/s
12×	1.8 MB/s	16.2 MB/s	54 MB/s
16×		21.6 MB/s	
18×		24.3 MB/s	
20×		27 MB/s	
32×	4.8 MB/s		
52×	7.8 MB/s		

the laser, and the numerical aperture of the lens used for focusing. By increasing the numerical aperture to 0.85, making the disk cover thinner, thereby eliminating undesired optical effects, and reducing the laser wavelength to 405 nm versus 650 nm for DVD and 780 nm for CD, the blue-violet laser can be focussed to a smaller 580 nm data spot with greater precision. This enables a data recording spot more tightly packed and stored in less space, making it possible to store more data on the same physical 3.5″ CD/DVD sized disk.

The recordable BD includes BD-R, which can be written once, and BD-RE, which can be written and erased many times. A single-layer BD can hold 25 GB data, and a double layer BD can hold 50 GB data. Feasibility studies from various manufacturers demonstrate that when 4, 6, 10, or even 20 multirecoding layer techniques are used, a much greater data capacity, up to 500 GB, can be achieved. Because of the much larger data size each disk can hold, and because the base writing speed is much faster, 3 times faster than that of a DVD, it is anticipated that with the BD disk technology advancement, Blu-ray will replace the regular DVD for PACS/EHR secondary back-up long-term storage. The data writing speed for a recordable CD, DVD, and BD are listed in Table 5.1.

5.1.3 Spin Disk

Spin disks/hard disks/hard drives store digital data on fast rotating platters covered with magnetic material. Usually a spin disk is sealed in a fixed enclosure to keep dust out. It is used widely in today's PACS/EHR systems.

All magnetic storage media store digital data as 1 or 0, two distinctive digital states. This is achieved when ferromagnetic material, which is the cobalt-based alloy on the platter surface, is magnetized. A platter is a circular disk. When data are retrieved, the magnetization of the ferromagnetic material is detected. As illustrated in Figure 5.5, a spin disk is composed of several platters made of nonmagnetic material such as aluminum or glass. The platters are covered with a very thin layer of ferromagnetic material. These platters are held on a spindle, which is driven by an electronically controlled motor. Each spindle disk has its own electronics to control the movement of the spindle arm/actuator, for read or write tasks. For spin disks used in PACS/EHR systems, a typical spindle spins at a speed of 7200–10,000 rotations per minute (rpm). The read/write head located on the actuator passes through various parts of the platter's surface when it moves; the magnetization on the platter's surface is thus detected during data read or modified during data write. For each platter inside a spin disk there is one read/write head. In a modern read/write head, the read element is separated from the write element.

With the ever-increasing demand for larger storage space and the decreasing cost per unit of storage space, the use of spin disks is increasing. Currently, the largest single disk has a size of 1TB. Multiple disks can be combined into one virtual disk and the total size is summed together, so the PACS/EHR can use it as a much larger disk.

5.1.3.1 Reliability of Spin Disks

Because of the way it operates, with its delicate moving parts, spin disks are susceptible to failures. The probability of a spin disk being replaced during a service call, which may or may not be caused by a malfunctioning hard disk, is much higher than the Mean Time to Failure (MTTF)

Figure 5.5 The key internal component of a spin disk/hard disk includes several platters made of nonmagnetic materials such as aluminum or glass, which are covered with a very thin layer of ferromagnetic material. These platters are held on a spindle that is driven by an electronically controlled motor. Each spindle disk has its own electronics to control the movement of its actuator head for read and write tasks.

published by hard disk drive vendors. No matter whether the hard disk has actually failed or not, it is considered failed if it is replaced in a service call, because the consequence is the same for the PACS/EHR system using that disk. So far, there is no good monitoring tool to effectively predict when a hard drive is going to fail. To maintain the integrity of the PACS/EHR system, data redundancy is absolutely essential, which is discussed later in the chapter.

5.1.4 Solid-State Drive

One recent development in hard disk technology is the Solid-State Drive (SSD) storage device. SSD stores data using solid state memory, either random access memory (RAM) or flash memory. The function of SSD is just like a hard disk drive, to store large amounts of digital data. There are three types of SSD devices.

5.1.4.1 RAM-Based SSD

The earliest SSD is RAM-based SSD, which combines a large RAM and a large hard disk. Typically, Dynamic RAM (DRAM) is used in a RAM-based SSD. Since RAM is volatile memory, it needs power to be switched on to maintain the stored data. When the power is switched off, the data stored is not preserved. In a DRAM-based hybrid SSD, usually there is a built-in battery and backup spin disk to keep the stored data. When the power is switched off, the battery will be used to maintain the data on DRAM, which is copied to the backup spin disk. When the power is switched on, the data stored on the spin disk is copied back to DRAM, and the SSD is fully functional again and ready to be used. The access speed for RAM SSD is extremely fast compared to hard disk drives, especially when used with high-speed interfaces.

5.1.4.2 Flash Memory–Based SSD

Flash memory uses nonvolatile flash memory to store data. Flash memory was first developed in 1984. It is widely used in consumer electronics like digital cameras, flash USB drives, MP3 players, and so on. Though the data access speed of flash memory SSD is much slower compared to RAM SSD, it is still much faster compared to hard disk drives because there are no moving mechanical parts. It is also very reliable compared to hard disk drives, because there is no spinning movement of platters. Since flash memory is nonvolatile, there is no battery needed, data stored on it are preserved during a sudden power outage. Currently, most SSDs available on the market are flash memory-based SSDs.

5.1.4.3 *Hybrid Flash Drive*

A hybrid drive actually combines a large amount of memory with a conventional hard disk drive. Hard disk drives equipped with RAM have been available for a long time. Most modern hard disks have some RAM as a buffer/cache. A hard disk with a large amount of RAM is called a hybrid RAM drive. Hybrid flash SSD uses large amount of flash memory as a buffer to cache data during normal operations. The spin disk part of a hybrid drive does not need to be constantly spun as a conventional hard disk drive. Most of the time, the spin disk platters are at rest. The major benefits of this type of hybrid drives are fast speed and improved reliability.

The major concerns with regard to the SSD are price and capacity. The price of SSD is around 15 times higher than that of hard disk drives for the same amount of storage capacity, and the SSD capacity is also much lower than that of hard disk drives. Another limitation of SSD is that the write cycles of flash memory SSD are limited; it wears out after 300,000–500,000 writes; however, RAM-based SSD does not suffer from such a limitation.

5.1.4.4 *Application in PACS/EHR Workstations*

The flash SSD continues to drop in price and increase in capacity, and it has therefore started to be used in some consumer-grade laptop computers. Because there is no mechanical disk spin and read/write head, the benefits of such storage devices include that it is fast, has random access, and has a very low latency for read/write tasks.

For a PACS workstation which is mainly to review and analyze images, speed is critical, the image data are not permanently stored, and since there is no need for a very large-capacity SSD, and the flash SSD rewrite cycle limit may not be a concern. Therefore SSD is a good candidate to replace the hard disk drive, especially when the maximum RAM allowed in the CPU/operating system is already being used. Also, since there is no mechanical failure in SSD, it is highly reliable in a severe environment. Thus, flash SSD is very useful for PACS/EHR applications where mobile laptop computers or personal digital devices are used in healthcare institutions.

5.1.4.5 *Application for Short-Term and Long-Term PACS Storage*

In PACS, medical image data are constantly sent to storage after images are captured, and, once stored, these data are usually not changed in the database, but it may be read several times, occasionally very often. Because the image data associated with a particular study can be very large, the read speed is critical. However, there may not be many writing cycles except for the first write into the database. In this case, the SSD is suitable for storing actual PACS/EHR data files. For long-term PACS storage, the SSD may not be used in the foreseeable future, mainly because of the cost associated with large data storage size. The high speed may not be fully realized because of network speed limits. The prefetch function on a PACS can preload image data to the PACS workstation.

5.2 DIGITAL STORAGE DEVICE INTERFACE

There are several interfaces used for storage devices to be connected with servers and workstations. The frequently used types include ATA, SATA, SCSI, iSCSI, and FC.

5.2.1 ATA

Advanced Technology Attachment (ATA) is a legacy hard disk parallel interface. It is also called Integrated Drive Electronics (IDE) or Parallel ATA (PATA). ATA has several standard versions. ATA-1 through ATA-3, which are now considered obsolete, suffer from the low storage size of each disk, and the slow data rate. However, new developments in ATA, which include parallel ATA-4 through ATA-8, and serial ATA (SATA), are emerging. ATA disks are generally used less in PACS/EHR servers because of the relative low data access rate compared to SATA and Small Computer System Interface (SCSI) disks. However, it may be found in some PACS/EHR workstations.

5.2.2 SATA

The Serial ATA (SATA) replaces older ATA with a high-speed serial cable, which has only 8 pins versus 40 pins for ATA. The benefits of SATA compared to ATA include faster and more efficient

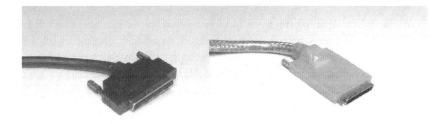

Figure 5.6 Two types of SCSI cables with 68-pin connectors.

data transfer, and the ability to remove or add devices without turning off the host system, which is termed "hot swapping." The maximum bandwidth of current SATA (sometimes termed SATA II or SATA 2) is 3 Gbps, which is comparable to SCSI's maximum bandwidth of 3.2 Gbps. However, the sustained data rate of SCSI is higher than that of SATA drives. Advances in SATA technology will potentially bring SATA speed to 6 Gbps, which will meet the high data speed needs of SSD drives.

5.2.3 SCSI

SCSI is a standard with several different interfaces to connect computer servers and hard disks, tape drives, or optical disk drives. Typically, SCSI has a physical SCSI connector composed of 50, 68, or 80 pins to connect the storage device to a server, or to connect to each other. Figure 5.6 shows two types of SCSI cables with connectors, and Figure 5.7 the 68-pin SCSI interface of an AIT-3 tape drive. SCSI also includes a complex set of command protocols. With a traditional parallel SCSI bus interface, up to 16 storage devices, such as hard disks or optical/tape drives, can be connected together. Each device on the same SCSI bus is assigned a unique SCSI identification number, which is called the Logical Unit Number (LUN). SCSI generally costs more than SATA but offers a faster data transfer rate. Many PACS/EHR use SCSI-interfaced disk arrays in their servers.

5.2.4 Serial Attached SCSI

Serial Attached SCSI, also called SAS, is a more recent development of parallel SCSI. It is used mostly for mission-critical servers, such as PACS/EHR. The benefits of SAS include faster data access speed than most parallel SCSI and hot swapping, which means a faulty storage device can be replaced without completely shutting down the computer server as required by traditional parallel SCSI. This character is very important for fault tolerance in disk arrays, and will be discussed later in this chapter.

5.2.5 iSCSI

iSCSI is a promising interface. In addition to utilizing traditional SCSI standards, Ethernet and TCP/IP are also used to remotely attach hard disks together. To servers, iSCSI disks appear locally connected. A Storage Area Network (SAN) using iSCSI can use the existing switching and network infrastructure without dedicated cabling. This is going to compete with FC-based SAN,

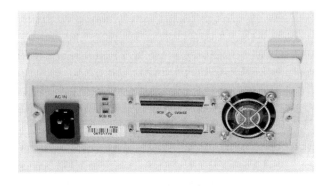

Figure 5.7 The back of an AIT-3 tape drive with a 68-pin SCSI interface and SCSI ID setup.

which has the same function but requires more expensive dedicated FC Fabric networks. In PACS/EHR, iSCSI can be used to access remotely located storage arrays, and various storage arrays can be consolidated.

5.2.6 Fibre Channel

Between two processors, or between a processor and a peripheral device, there are two types of data transmissions: network and channel. A channel is a closed direct data transmission structure among relatively few devices. An S-ka-zi is an example of traditional channel technology.

In today's PACS/EHR implementation, the explosive increase in data that require transmission is an unprecedented challenge. FC is an easy and reliable solution to the information storage and retrieval challenge. Shown in Figure 5.8 is a short-term PACS storage RAID which is interfaced with FC. Currently, FC is a gigabit-speed network technology primarily used for a SAN in enterprise storage.

Fibre Channel Protocol (FCP) is a computer communication protocol that can meet the demand for large amounts of data transfer. FCP transports SCSI commands over the FC. Though it is called an FC, the topology it uses is neither a channel nor network architecture.

FC provides a direct or switched point-to-point connection between two of such devices as servers, tape libraries, disk arrays, and so on. Once the channel is built, it is a hardware-intensive environment and can transfer data at a high speed without much software involvement. Despite its name, FC can run on both twisted-pair copper wire and fibre optic cables. All that an FC port does is to manage a simple point-to-point connection between the port and the fabric. In contrast, though networks include and support a large number of the same types of devices such as servers, workstations, and peripheral devices, these devices may use their own protocols to communicate with each other. These communication processes are, therefore, software-intensive, for routing data successfully from one point to another, and inherently slower compared to channels. An example is the Ethernet network, which was discussed in Chapter 4.

5.2.6.1 FC Layers

Unlike the International Standards Organization open systems interconnect a seven-layer model, which includes the physical, data link, network, transport, session, presentation, and application layers, FC uses a five-layer model, as shown in Figure 5.9. These are:

1. *FC-0*: The FC-0 is the physical layer including fibre optics, cables, hardware connectors, and so on. This is the lowest layer defining the physical media used for linking two FC ports, their maximum distance, and the noise limit. Both copper and optical cables are supported, with copper cables used mainly for connecting short-distance storage devices, and optical cables for connecting long-distance devices because of its low-noise character. Optical cables can use both single-mode and multimode for data transmission, and copper cables mainly use the shielded twisted-pair cable. Thus, either optical or copper cables link the two FC ports, the transmitting port and the receiving port.

2. *FC-1*: FC-1 is the data link layer for data encoding and decoding implementation during transmission. It also defines the transmitting and receiving ports and their operations.

3. *FC-2*: FC-2 is the network layer consisting of the FC core and protocol definitions, providing FC classes of service information with various connection types and managements, and defining the physical models for the FC components, including FC topologies, node definitions, ports, and so on. FC-2 defines the transport mechanism for the channel.

4. *FC-3*: FC-3 is the common service layer used for advanced service definitions, such as using multiple FC ports and links to transmit one information unit so as to multiply the bandwidth, or using multiple receiving ports to receive data so as to reduce the chance of a busy responding port.

5. *FC-4*: FC-4 is the protocol mapping layer, translating other well-used high-layer protocols for use on FC. It is the highest level in the FC, defining application interfaces. FC allows both network and channel protocols to run over the same physical interface and media. Currently, the most used network and upper layer protocols are the SCSI and IP.

Overall, FC-0 through FC-2 function as the physical layer, while FC-3 and FC-4 define interfaces with other applications and some specific protocol information.

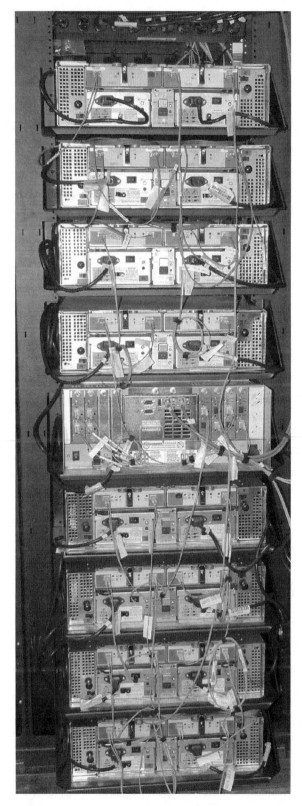

Figure 5.8 A rack-mounted short-term PACS storage RAID is linked by an FC with fibre optical cables.

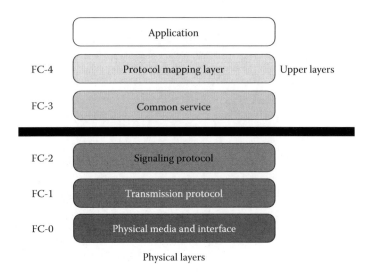

FC-4   Protocol mapping layer   Upper layers

FC-3   Common service

FC-2   Signaling protocol

FC-1   Transmission protocol

FC-0   Physical media and interface

Physical layers

Figure 5.9   A five-layer model FC0-4 for FC.

5.2.6.2 FC Topologies

There are three frequently used FC topologies that connect FC ports. An FC port is not necessarily a physical hardware port; it is an entity communicating actively on the FC. Usually, a storage disk array or an FC switch is equipped with a port. As shown in Figure 5.10, the three FC topologies are:

- *Point-to-point.* This is the simplest limited topology connecting two nodes; a node is a device such as a computer server or a storage device. The link between any two nodes is exclusively used by them, not allowing the full bandwidth to be shared with any other node.

- *Arbitrated loop.* The Arbitrated Loop topology for the FC is used more than point-to-point topology in SAN storage applications. After a connection is established between two nodes on the loop, data traffic follows one direction around the loop. Any disconnection of a device interrupts the loop. Once the data flowing through the loop reach the destination end, the loop is again available for arbitration and for establishing the next connection. Although, theoretically, 126 nodes can be set up in a single loop without any switches or hubs, the number of nodes is typically limited to no more than 15, and the total bandwidth is shared among all these nodes. Both point-to-point and arbitrated loop topology can support a distance up to 6 miles or 10 km, which is in the range of a typical LAN.

- *Switched fabric.* Similar to Ethernet implementations, all devices or loops of devices are connected to FC switches. An FC switch is a network switch also compatible with the FC data-communication protocols. One or more switches are used in this topology, with each node possessing the full bandwidth as opposed to the shared bandwidth in the arbitrated loop topology. This topology is very similar to the meshed network topology discussed in Chapter 4. It provides optimized interconnections for FC devices, simultaneous communication with multiple pairs of ports, and no disruption to the FC if a port fails. The switched fabric is an essential component of SAN, providing a high-performance environment for all enterprise applications related to data storage. The fabric architecture also offers fault-tolerance, ensuring connectivity among servers and storage devices when isolated failures occur, and preventing the failures from spreading out to disrupt the system's operation.
  All these three basic FC topologies can be standalone, or combined to form a fabric.

5.2.6.3 FC Bandwidth

FC devices typically have data speeds of 1 Gbps, 2 Gbps, 4 Gbps, and 8 Gbps. Currently, 1 and 2 Gbps devices are widely used. However, since higher-speed devices are cost effective and usually back-compatible with lower-speed devices, 4 and 8 Gbps devices are used for disk and tape storage systems. Because of the significant difference at the FC1 level, devices with 10 Gbps and 20

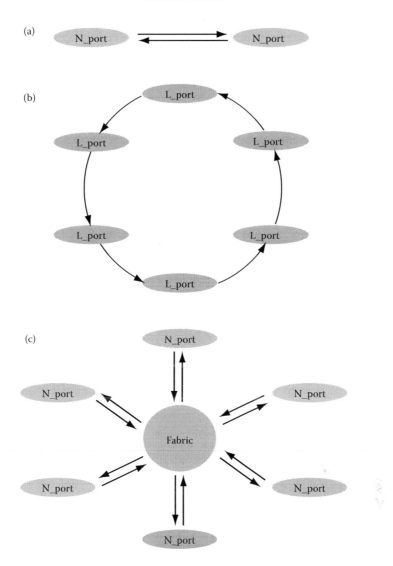

Figure 5.10   Three primary FC topologies: (a) point-to-point; (b) Arbitrated Loop; (c) Switched Fabric. L_port is a loop-capable node or switch port. N_port is not loop-capable, but is used to connect an equipment port to the fabric.

Gbps data speeds are mainly used for linking multiple switches on a stack or within a data center, and they are not compatible with lower-speed devices.

In summary, an FC combines the best features of network and fabric communications into a new type of interface to meet the need for massive data storage and retrieval. It is specifically designed to use hardware for highly efficient operations with little transmission overheads. Compared with traditional SCSI, the FC maps the SCSI command sets and brings faster speed, accommodates a far greater number of connected devices, and allows longer distances between devices. The application of FC is expanding as the preferred storage interface technology for PACS/EHR.

Listed in Table 5.2 are data speed comparisons for various interfaces. Spin disks with SCSI and FC are usually more expensive. They are mostly used in servers and disk arrays. Spin disks using ATA or SATA interfaces usually cost less, and thus are widely used in personal computers or workstations. However, the reliability of a particular spin disk is dependent on the quality of the spin disk itself rather than on the interface it uses. The selection of hard disks used for PACS/EHR storage array also depends on the quality of the disk rather than the interface it has.

Table 5.2: Data Speeds of Various Interfaces for Storage Systems

Interface	Data Speed (MB/s)	Cable Connection
Ultra SCSI	20	50 pin
Ultra Wide SCSI	40	68 pin
Ultra2 wide SCSI	80	68 pin; 80 pin
Ultra3 SCSI	160	68 pin, 80 pin
Ultra 320 SCSI/Ultra 4 SCSI	320	68 pin, 80 pin
Ultra 640 SCSI/Ultra 5 SCSI	640	68 pin, 80 pin
SAS	300	SAS cable
FC-AL 4Gb	400	Fibre Channel cable
FC-AL 2Gb	200	Fibre Channel cable

5.3 DATA STORAGE ATTACHING CONFIGURATIONS

Healthcare institutions must retain massive volumes of critical fixed-content patient data for long periods. This presents tough storage management challenges such as storing/retrieval protected data, disaster recovery, hardware/software upgrades and expanding capacity, centralized management simplification, and automated proactive warnings. Currently, there are several data storage attaching configurations used for PACS/EHR to meet these needs, each with their own advantages and limitations.

5.3.1 Direct Attached Storage

Direct Attached Storage (DAS) refers to a range of storage devices directly connected to a computer or server via a cable a few meters away, or contained inside the computer server case. The term DAS can also be used to refer to more common storage devices having a one-to-one connection with a computer.

5.3.1.1 DAS History

DAS has the longest history as data storage for a computer or server. Typical DAS systems include disk array enclosures, spin disks, optical/tape drives or libraries, and so on. It is designed for optimum efficiency and performance as a storage device for the server. A DAS is interfaced to a server through any one of ATA, SATA, SCSI, and point-to-point connection FC, which are all standardized cables/connectors and protocols for DAS, providing rules for data exchange between storage device and server. Figure 5.11a explains such a DAS approach, where each server or client owns and only sees DAS. The original design of DAS does not take into consideration sharing data on the storage device among various servers and clients systems.

In a legacy PACS system, short-term spin disks and long-term tape, or optical storage library archive are usually directly connected to the PACS server without a storage network between the server and storage. Though the resource or data stored on DAS cannot be shared by multiple servers, and it lacks the flexibility of expanding the capacity when needed, DAS may still be suitable for small PACS used in independent imaging centers or clinics.

5.3.1.2 Advantages of DAS

Storage devices that use DAS to connect with servers benefit in the areas of:

1. *Easy installation*: the linking cable between a DAS and the server is either contained inside the server enclosure case, or is connected across a very short distance, in the order of a few meters, requiring minimum technical skills.

2. *Low cost*: Except for a point-to-point FC connection, which costs more because of the extended length of the fibre optical cables and specialized hardware, other DAS cables are relatively inexpensive and widely available, of minimum maintenance and operational cost.

3. *High data access rate*: DAS is designed for data communication between storage devices and servers, so data access is at the block level and hardware intensive, and the software overhead is minimal.

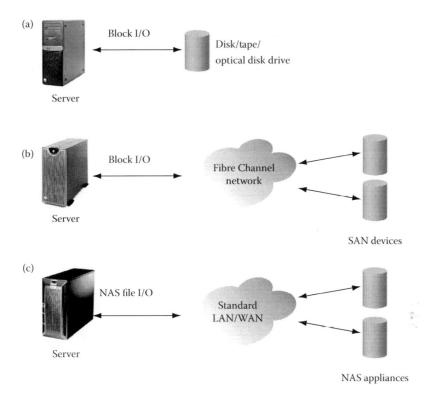

Figure 5.11 Data storage device attaching configurations: (a) DAS: devices are directly attached to workstations/servers with data access at the block level; (b) SAN: devices linked to each other and to workstations/servers through dedicated FC network with data access at the block level; (c) NAS: devices are linked to each other and to workstations/servers through standard LAN/WAN network with data access at the file level.

5.3.1.3 Limitations of DAS

The reason that there are other storage device configurations on the market is that DAS has its limitations in the following areas:

1. *Scalability limitation*: the DAS cable connection has very limited length between the server and the storage device, and the number of storage devices that can be concurrently connected to the connection cable is also very limited. This constrains the total capacity of the linked storage devices.

2. *Inefficient management*: although relatively low cost, DAS cannot share data with other servers. Network Attached Storage (NAS) and SAN architectures discussed in the following sections address this issue; however, they introduce their own issues as well.

5.3.2 Storage Area Network

Today's enterprise data centers for business, including healthcare PACS/EHR, demand centralized management, data storing and retrieval speed, capability of sharing data with remote servers and devices, and high scalability to expand the storage infrastructure capacity with ease. All these require storage devices being able to be linked by networks.

5.3.2.1 Defining SAN

As we already know, traditional data storage devices are either inside or directly connected to a server with a dedicated interface channel, and servers communicate through a LAN or WAN.

Because servers use different operating systems and the amount of disk storage capacity attached to servers has increased exponentially, moving large amounts of data among various servers and storage devices through LAN and WAN not only causes significant performance degradation, but also exceeds the bandwidth limit of the channel that attach storage devices to a server. It is obvious that a new way of connecting storage devices and servers needs to be adopted.

One of the emerging technologies addressing the challenge is SAN. A SAN is a data storage system used for data transfer among computer systems and storage devices. It has switches to connect storage devices and various servers with a dedicated network, so there is direct connectivity among storage devices, and a single storage device can be accessed by multiple servers.

As illustrated in Figure 5.11b, a SAN is a dedicated high-speed high-performance fibre optic or copper cable network connecting servers and data storage devices only for data communication, not for any other generalized applications. Using SAN makes it possible to share heterogeneous storage devices such as disk arrays, optical storage libraries, and tape libraries with many servers running on different platforms. It replaces the traditional local connection between a server and storage devices as in DAS, so that storage devices can be shared across different geographic locations.

5.3.2.2 SAN Fabric and Topology

One of the important components of a SAN infrastructure is its FC topology, offering significant improvement in speed, with data transfer speed up to 20 Gbps. FC networks support three types of topologies, as discussed earlier in this chapter. The SAN network uses FC as the architecture, a SAN fabric is a connection of multiple FC switches linking FC devices together for communication, fault-tolerance, eliminating single point of failure, and increasing the maximum linking distance between interconnected devices.

An FC-based network has many similarities to the Ethernet LAN, but also differs from it considerably. Since FC is not as widely used as Ethernet, the high cost associated with it prevents it from being used widely. However, data centers of a large healthcare enterprise or institution may be good candidates for SAN. SAN can provide faster and more reliable data access than NAS because it uses block-level data access and has a dedicated FC. Unlike conventional LAN/WAN, there is no client/server network traffic on SAN other than the data exchange among servers and storage devices. Among the various FC topologies, switched fabric is the most frequently used topology in SAN today.

5.3.2.3 Fibre Channel for SAN

In contrast to DAS where storage devices are directly connected to a server via its bus, SAN attaches storage devices to the FC network, where servers are also attached. Though most FCs used in SAN utilize fibre optical cabling, FC itself is not limited to optical cabling, and twisted pair copper cabling can also be used in lower bandwidth implementations.

SAN is a dedicated storage network carrying data traffic only among servers and storage devices. There is no application traffic in this SAN network, eliminating traffic jams caused by various applications. A SAN includes physical connections as infrastructure for data communication, storage devices, computer servers, and a management software control layer for the organization of data connections. Though physically SAN can be in a remote location, to servers SAN appears locally connected. The protocol used the most for communication among servers and storage devices in SAN is a modified SCSI, which does not include traditional SCSI interface cables.

5.3.2.4 SAN Servers and Storage Devices

SAN infrastructure allows heterogeneous servers to be used in a particular SAN without any problem as long as vendors adopt the same SAN standards. Vendors usually test their product's interoperability with those from other vendors in performance and reliability before certifying their support for the combinations.

Storage devices in a SAN can be a hard disk array, an optical disk array, or a tape library. One major difference between SAN and NAS is that SAN provides data access at the block level, whereas NAS provides data access at file level. Though the size of a data block can be changed, a typical data block has 512 bytes of data, and a medical image data file from PACS storage contains many blocks of data.

5.3.2.5 Advantages of Using SAN

In today's enterprise environment, SAN offers solutions to many problems faced by healthcare institutions, especially in the following areas.

- *High Speed Performance.* PACS application benefits from efficient FC data exchange, which is implemented with much hardware. The data exchange rate is several times higher than typical SCSI, and many times more than LAN, even a Gigabit Ethernet.

- *LAN Relief.* A conventional data backup using LAN increases traffic significantly, which affects the performance of normal applications relying on LAN. Therefore data backup is usually scheduled at a time when LAN is used the least. However, with the ever-increasing data size, there may not be a large enough backup window. Using SAN to backup data frees LAN from such high volume traffic.

- *Storage Selection Optimization.* Since storage devices are no longer directly tied with specific servers as DAS, users are free to choose the best storage device meeting their unique function, cost, and performance needs, independently of servers.

- *Storage Capacity Scalability.* Unlike DAS, which limits the number of disk storages connected to a single server, and in which the operation of the disk storage is interrupted every time new disk storage is added to the server, SAN allows disk storage to be independently scaled without limit and without interruption of the server's operation. In addition, the physical location of storage devices can be much farther away from servers.

- *Storage Consolidation.* Since SAN disconnects storage devices physically attached to particular servers, and allows storage devices to be connected to servers across a much longer distance using FC, the capacity of storage devices such as disk arrays can be consolidated, which improves enterprise storage management.

- *Data Migration.* Since the storage devices are all attached to SAN instead of to servers, when new storage technologies are available and data need to be migrated from old storage devices, data can be transported directly from old to new storage devices without using servers.

5.3.2.6 Considerations for Planning SAN

Although SAN has many advantages, it does not mean it is a perfect choice for everyone. There are several issues that need to be considered before a decision is made to adopt SAN for a healthcare enterprise.

- *Cost.* SAN requires a dedicated FC network infrastructure. Currently, the costs of FC fabric, switches, and technical staff with special expertise are significantly higher than that of its Ethernet counterpart, because SAN requirements are very specialized and the size of market is much smaller.

- *Distance Limit.* Limited by the FC architecture, currently FC SAN covers distances in the order similar to LAN, which is approximately 10 kilometers or 6 miles. Connecting multiple SAN located at farther distances requires other network transport technologies, and the cost is high.

- *Interoperability.* The FC used for SAN is still relatively new compared to Ethernet LAN, which has been adopted and used for many years by the IT industry, and by most healthcare institutions. Therefore, there is still room to improve the interoperability of FC hardware and software products from various vendors. This may cause interoperability issues when implementing multiple vendors' heterogeneous SAN solutions.

After the above issues are taken into account, at the planning stage of a SAN implementation, users need to know the answers in the following areas too:

1. The current number of nodes and the numbers projected in a few years

2. Geographic locations of storage devices and servers

3. The distance among all servers and storage devices

4. Server and storage device redundancy requirement, and fabric topology design for fault-tolerance

5. The heterogeneous/homogeneous nature of the fabric (multivendors, multiplatforms, or not)

No matter whether the size of the SAN fabric is small or large, good planning is crucial for successful SAN implementation.

In summary, SAN provides block-level access, while NAS provides file-level access. SAN can include hard disk arrays, optical disk libraries, or tape libraries. Although physically SAN can be in a remote location, to servers it appears locally connected.

5.3.3 Network Attached Storage

Network Attached Storage (NAS) is itself a storage device located on a conventional network. NAS also refers to any storage device, sometimes termed storage appliance, linked to a network. The major hardware component of a NAS device is similar to a computer server. It is essentially a LAN-based file server, except that the NAS software and its functions are tailored only for data storage, and for providing access to the data to multiple clients, which can be PACS servers or workstations connected to the LAN. A NAS device with its own server and with a limited operating system is targeted to provide file-level data management and access services, which is shared by other application servers and clients over the network.

5.3.3.1 Introducing NAS

A NAS device has its own mother board, CPU, RAM, interface functions, and limited operating system capabilities, though the computational power is not as strong as a typical server. A NAS device can be accessed and managed by the storage administrator remotely. It provides file-level data access, which is a high-level data access: it specifies the file to be accessed without directly addressing the storage device. The server or workstation to which NAS provides data storage services still handles all data processing. Unlike DAS, where the storage devices such as disk arrays or tape/optical libraries are directly attached or are part of the server, a NAS device, as illustrated in Figure 5.11c, is not directly attached or located inside the server; it can reside anywhere in the LAN network. Clients and application servers import data files stored on remote NAS devices/appliances across the LAN as if the files were residing on their own local disk. Multiple NAS devices can provide data file storage services to a single server using conventional Ethernet with appropriate file-based communication protocols. This is different from SAN, in which an FC is used to connect storage devices to provide data storage services at data block level. The performance of a NAS device relies heavily on the performance of the network it resides on, and the CPU power and the RAM it is equipped with. It allows multiple PACS server/workstations to access the same file system over the network and synchronizes their access. The performance of NAS is limited if there are too many users, too many simultaneous data accesses, and the computational power of the CPU reaches its limit. When in need, more NAS devices can be added to a network to increase data storage space for the server without shutting the server down. Another typical feature of a NAS device is that it is almost a plug-and-play device; its implementation does not require much configuration and effort before integrating into the network, which is especially attractive for those that lack technical resources.

5.3.3.2 Benefits of Using NAS

NAS overcomes many of the limitations of DAS and the complexities of SAN implementation mainly in the areas of:

1. *Full utilization of existing matured IP technologies and network infrastructure*: Since standard IP technologies and LAN are used to connect NAS devices with servers, the widely available equipment, infrastructure, technical skills, and expertise make NAS implementation and also the subsequent operation and administration easy and less costly.

2. *Improving scalability and data management with consolidation*: NAS allows storage capacities to be consolidated and shared on the network at much longer physical distance from servers and clients. Consolidated data file storage can be centrally managed to ease data storage administration. NAS can also be easily scaled up for increased capacity and performance by adding more NAS devices to the network LAN.

3. *More purchase options*: Because data storage is independent from the server, users may have more servers and storage options to choose from to meet specific institutional needs.

5.3.3.3 Considerations for Using NAS

Like other storage technologies, NAS too has its own unique issues that need to be considered during a NAS implementation:

1. *LAN bandwidth constraint*: This is important, because Ethernet LAN is not designed for continuous large data transmission, and NAS may cause significant transmission overheads and low transmission efficiency. The consequence is a congested network with reduced network performance, impacting other computers and applications sharing the same network. Therefore, the network supporting NAS must have enough bandwidth or capacity. The scalability of NAS is limited, and therefore NAS may not be a good choice where the applications used, demand high network bandwidth.

2. *Software overheads*: Since TCP/IP on Ethernet is designed for messaging applications, maintaining data integrity during transmission is not the top concern. Data transmission reliability, therefore, is achieved by using software-intensive network design with extensive processing overheads, which consumes much processor resource.

NAS storage provides acceptable performance and security, and it is often less expensive and easy for servers to implement. Therefore, a NAS device is a good solution for users to access file storage through the Ethernet TCP/IP based network already in place, while block level data storage SAN with FC implementation is too expensive or impractical. It is expected that NAS and SAN will be implemented as complementary solutions, together with DAS and iSCSI devices, as illustrated in Figure 5.12. Healthcare institutions may choose to implement a combination of these storage technologies to meet their institution's specific application and budget needs.

5.3.4 Grid-Oriented Storage

Grid-oriented storage (GOS) is a general term for any approach that uses multiple interconnected storage nodes for data storage, so that any two nodes can exchange data without going through a

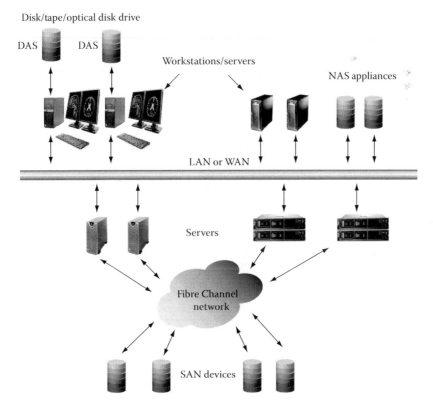

Figure 5.12 DAS, SAN, and NAS can be implemented as complementary solutions.

central switch. Instead of using a central switch with many ports, multiple switches are deployed, each with a few ports to connect a few nodes together as a cluster.

5.3.4.1 What Is Grid Storage?

The concept of "grid" comes from the electrical power grid, telephone grid, and grid computing, which is a decentralized large network with extensive interconnection and unified coordinated management. In grid computing, to significantly improve performance, many distributed computers are connected and coordinated together in grid fashion. Borrowed from the grid computing concept, one of the developments in PACS storage is grid storage. Basically, it integrates smaller storage building blocks (which are physically distributed, self-contained storage units with their own storage medium, microprocessor, indexing capability, and management layer) into a virtual storage unit seen by servers and workstations on the grid. Typically, several grid storage building blocks may share a common switch forming a storage cluster, but each is also connected to at least one other storage cluster. In a meshed grid network, there is no large centralized control switch for routing, and therefore there is no restriction imposed by using ever larger central switches when the network scales up. Generally speaking, grid storage is the natural expansion and advancement of NAS systems, where software on each of the storage nodes manages its own limited storage resources, and they are interconnected over IP networks of differing bandwidths. The interconnections among all nodes are in a mesh network. In a grid storage system, a single storage node has its own storage medium, microprocessor control unit, and associated indexing and management capability, so it is a self-contained storage unit. The key elements are that the capacity of NAS storage be scaled up and that a single file system is managed and synchronized. The most important benefit for the grid storage is that even though physically separated, grid storage on the network is accessed by PACS clients/users as a single storage device.

5.3.4.2 Grid Architecture

To illustrate grid architecture, a simple example of nine storage nodes can be used, with the requirement that all nodes be able to access each other. Figure 5.13 is an example of a star configuration with two large switches, with the second switch for fault-tolerance. In this structure, each switch is connected to each of the nine storage nodes, and each storage node is connected to both switches. This results in 18 wired links and two switches each with nine nodes. Between any two storage nodes there are two possible linking paths. When the network is expanded by adding more storage nodes, switches need to be replaced with ones having more ports. For illustration purposes, there are only nine storage nodes in the example. In a real-world situation, where there are a lot more storage nodes, central switches with a number of ports get expensive, and they need to be replaced when no more ports are available and additional storage nodes are added to the network.

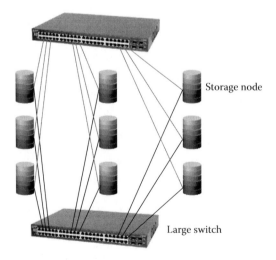

Storage node

Large switch

Figure 5.13 Example of a conventional 9-node storage configuration with two large switches. Each switch is linked to each of the nine storage nodes, and every storage node is connected to both switches for fault-tolerance.

In a grid arrangement, storage nodes can be arranged in three groups of 3; each storage node is connected to a simple switch with at least 3-ports. There are a total of six such switches needed, as shown in Figure 5.14, and 18 wired links as before. Each storage node is also connected to two switches. In case of any switch malfunction or link interruption, there are many alternative paths to link any two storage nodes. The end result for this simplified example is that six switches each having at least three ports are used instead of two switches each with at least nine ports. This grid configuration is more resilient and less susceptible to catastrophic failure compared to conventional star architecture. When the grid is in need of expansion, more storage nodes and inexpensive small switches can be added to the grid without significant costly change to the entire structure.

5.3.4.3 Grid Storage Characteristics and Benefits

Most storage networks are constructed in star topology, which means storage devices and servers are all connected to a central switch. However, for grid storage networks, they are set up with many smaller interconnected switches, as just illustrated. The centerpiece of grid storage is a methodology for synchronized file sharing based on clustered NAS or other interconnected distributed topologies, so that multiple, heterogeneous, distributed storage elements are unified into a logical entity without a single point-of-failure (POF). A typical grid storage system is a structure that links and governs independent storage nodes with common proprietary managing software. At the control layer there is a single control interface to treat all independent nodes as if they were a single data storage pool, to assign different functional roles to nodes such as archival or online storage, to access data at either block or file level, and to provide fault-tolerance among all storage nodes.

Usually GOS systems have the following common characteristics: (1) Devices and servers of the system can be SAN with block access storage arrays, or file access NAS servers and gateways; (2) Data are replicated several times across nodes in the grid, providing redundant data access, fault-tolerance, and disaster recovery; (3) Management is simplified across all nodes for data security, data migration, capacity expansion, and new node auto detection. The entire storage grid appears as a single logical pool of storage resources available to users. Bringing down one node for maintenance or repair, or adding nodes to expand storage capacity, does not affect the operation of the remaining nodes on the grid.

Because of the above mentioned unique characteristics, the benefits offered by grid storage systems can be categorized as:

1. *Easy expansion*: It is easy to use relatively inexpensive switches with a small number of ports as building blocks to expand the grid for improved capacity, bandwidth, and performance. When a new node is added to the grid, the rest of the grid can recognize it automatically without

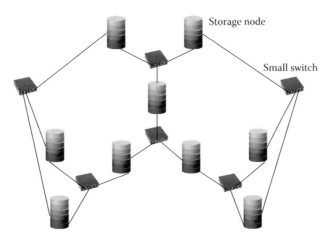

9-node grid storage configuration

Figure 5.14 In a grid storage configuration, for the same nine storage nodes of Figure 5.13, there are smaller switches. Each storage node is also linked to two switches, and there are many alternative paths to link any two storage nodes in case there is a link interruption or switch malfunction.

much hardware upgrade investment and downtime, so additional storage capacity and servers can be easily added to the grid.

2. *Improved redundancy for fault-tolerance*: Grid storage can provide multiple paths between any two nodes on the grid. If a single link between any two nodes is interrupted, the grid network can re-route access through a different path or to a redundant node. It is easy to replace faulty components for repair and service without much impact on the operation of the entire grid, practically eliminating downtime.

3. *Functional enhancement*: Since multiple paths are provided to any two nodes on the network, load-balancing can be realized to ensure that the grid storage maintains optimum performance under fluctuating load conditions; the entire grid system can perform consistently, eliminating possible bottlenecks associated with conventional use of centralized switches and many ports.

Grid storage provides scalability, because it is not restricted by having to change to larger central switches when the network expands. Therefore, it is easier for a grid network to expand in a way that makes it resilient and reliable.

5.3.4.4 Considerations for Grid Storage Implementation

Many factors need to be considered during procurement and implementation of a grid storage system for PACS/EHR:

- Current status of PACS/EHR and data retention policy
- Current PACS/EHR data capacity needs and projected rate of growth
- The network infrastructure to be used for the grid
- The storage protocol, applications, and services
- The grid storage structure management and configuration for grid changes, such as adding new storage devices or removing existing storage devices
- Data migration path from current system to grid storage
- Data migration path from grid storage to future storage systems
- The real benefits grid storage can bring to the users in a particular environment
- The cost associated with vendor proprietary hardware/software, and human labor

A healthcare institution must consider upfront costs, convenience for storage reallocation when needed, data backup and restoration improvement, and block- or file-level data access that grid storage architecture can provide. Ideally, when additional storage capacity is needed and a new node is added to the grid, the grid automatically reconfigures the system making the new node a part of the entire pool of storage, thus reducing storage management costs. In the real world, various vendors sometimes market grid storage capabilities using terms such as "storage clusters" or "virtualized storages." Storage cluster means that independent storages are interconnected by networks, while virtualized storage means different physical storage pools are managed as a single logical virtual storage pool.

In summary, grid storage has two key components: it has a grid topology to scale the capacity of NAS to meet application needs, and a technology capable of managing a single file system for increasing storage capacity. To implement grid storage, the WAN and the bandwidth connecting various grid storage components need to be carefully assessed. Enhanced WAN and bandwidth in this case may be necessary.

5.4 REDUNDANT ARRAYS OF INDEPENDENT DISKS

For many businesses including healthcare, data availability is very important in addition to routine backup for recovery. RAID stands for Redundant Arrays of Independent Disks, an important architecture technology designed for business-critical system implementation to preserve data in case of disk failures. RAID uses multiple hard disks in conjunction with data striping and mirroring for improved data redundancy that can lead to reliability, improved performance, increased total hard disk storage space, or all of these combined with various hardware/software configurations. RAID combines hard disks either using hardware or software solutions. To the attached system, RAID presents as a single logical hard disk drive. Various RAID configurations, together with their benefits and limitations are discussed in the following sections.

5.4.1 Data Striping and Data Mirroring

In RAID "mirroring" and "striping" are key concepts. Data mirroring is a mechanism in which all data are duplicated physically on multiple hard drives automatically. Data striping is the process of breaking down data into small pieces, and storing them in several disks. These two techniques are used in various RAID levels either separately or combined together, depending on how the system is designed. The purpose of RAID is to improve both reliability and performance, allowing greater fault tolerance and data access speed.

RAID multiple hard drives can be located nearby or remotely. When the mirror disk is placed nearby, it is usually placed so for improvement of disk reliability. However, if the mirror disk is placed remotely, it is usually part of a comprehensive disaster recovery plan. Sometimes the data mirroring is done both nearby and remotely, to meet the data redundancy requirement. When a disk failure occurs, the integrity of the entire database and access to the database is still maintained, so the system operation is not interrupted. Disk mirroring not only provides data redundancy in case of hard drive failure, it also improves data reading because multiple disks can be accessed separately for load balance. Since data loss can be contributed by several factors not just limited to hard drive failure, data mirroring cannot replace regular backups.

5.4.2 RAID-0

RAID-0 combines at least two hard disks. When data come in for storage, they are striped into even portions among disks in the array. Data striping means that a single data file is segmented into several parts and each part is stored on different physical disks. The benefit of RAID-0 is that the server can read or write data simultaneously to multiple disks in the array, so data access is faster and throughput is improved, significantly increasing the bandwidth for read and write operations. The trade off is the greater risk of data loss without fault-tolerance. The total storage size of a RAID-0 array can be expressed as:

$$N \times S,$$

where N is the number of disks in the array, and S is the minimum size of the N disks. In contrast to data mirroring, there is no data redundancy for RAID-0, if one disk fails, the entire database is lost, because every file is segmented and saved separately on each disk in the array; none of the files is complete when one disk fails. The reliability of RAID-0 is less than its individual disks. If N disks in the array are all identical, the failure rate of the array of N disks, which means at least one disk would fail, is:

$$Fn = 1 - (1 - F)^N,$$

where F is the failure rate of one disk in the array. Overall, RAID-0 has the best performance among all RAID configurations, but does not have redundancy.

5.4.3 RAID-1

RAID-1 is also called disk mirroring. Different from RAID-0, RAID-1 stores identical data on multiple disks, with data mirroring. Each disk on the array contains the same complete database. When data are duplicated on each disk, data redundancy is preserved. When a single disk fails, the same data are also kept on other disks on the same array, so the array can still be functional, providing fault-tolerance from disk malfunction of all but one of the physical disks, which means the array continues to function as long as at least one disk is functional. The RAID-1 array fails only when all disks in the array fail. If there are N disks in the RAID-1 array, the failure rate Fn of the RAID-1 is:

$$Fn = (F)^N.$$

The reliability of the disk array is increased when the failure rate of the RAID-1 array Fn is reduced compared to F. RAID-1 has the benefit of improved reading performance too, because multiple disks can be read simultaneously. However, during writing, the performance is the same as that of a single disk, because the same data needs to be written on each disk. The simplest RAID-1 has two disks, where the second disk is a copy of the first disk. Failure of either disk does not affect the functionality of the system to which the RAID-1 is connected. A replacement disk can be rebuilt by the RAID-1 controller with data from the remaining disk, realizing fault-tolerance and restoring redundancy. The benefits of using RAID-1 are improved fault-tolerance and faster data read, with the trade off of slower data write and higher cost.

5.4.4 RAID-3 and RAID-4

RAID-3 uses a striped data set with dedicated byte-level parity. A parity bit is a binary digit for error detection when digital data are transmitted or accessed from storage disk media. This array carries the same benefits as RAID-5, providing improved fault-tolerance and performance, except with a dedicated parity disk instead of having parity data striped and distributed across the disks. Although the dedicated parity disk failure does not affect the performance of RAID-3, the single parity disk is a bottleneck for writing operations, because the parity data are updated every time data are written.

RAID-4 uses a separate disk to store parity information, the same as RAID-3, except at block level. A block of data typically has 512 bytes, although the length can be variable. In RAID-4, parity information is calculated when each block of data is accessed. It is similar to RAID-5 in that parity information is checked. However, in RAID-4 the parity information, which is usually calculated through hardware implementation, is stored on a single disk, whereas in RAID-5 it is stored across all disks in the array.

5.4.5 RAID-5

RAID-5 is the most used RAID mechanism that builds redundancy efficiently for PACS. It is similar to RAID-4 in that the data are striped at block level. It differs from RAID-4 in that instead of parity information being stored on a dedicated parity disk, it is interleaved and stored with striped data across all disks in the array, as illustrated in Figure 5.15. If there is a disk failure, data stored on the failed disk is rebuilt from the remaining disks. Compared with RAID-1, the number of disks required for data redundancy is less.

Using RAID-5, the available total storage of all disks in the array is optimized, performance improved, and redundancy achieved. Thus, it is the most widely used RAID configuration through hardware or software control in today's PACS. The total storage size of RAID- 5 can be calculated as:

$$D = (N-1)d,$$

where D is the array's total usable storage size, N is the number of disks in the array, and d is the minimum size of the disk in the array. If all disks have the same size, then d is the size of a single disk. The minimum number of disks needed for a RAID-5 build is 3. When a single disk fails, the data can be reconstructed mathematically from the data stored on the rest of the disks in the array. In RAID-5, if a second disk failure occurs before the first failed disk is detected, replaced, and the data on it rebuilt, unrecoverable data loss is inevitable. Therefore, a hot spare disk, which is powered on but does not participate in normal array operation, is typically built into the system. When disk failure is detected, this hot spare disk can immediately start the rebuilding process. When the rebuild is done, it replaces the failed disk. When the number of disks in RAID-5 is increased, the time it takes to rebuild the first failed disk also increases, because information from all disks in the array is needed to rebuild the failed disk. Hence, the probability of a second disk failure even before the first one is rebuilt is high. Therefore, although theoretically there is no upper limit for the number of disks in RAID-5, the actual number is limited. Using disks from different vendors or different manufacturing batches to build RAID-5 can potentially reduce the

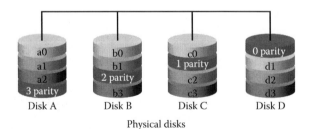

Figure 5.15 In the RAID-5 configuration, parity data are interleaved and stored with striped data across all physical disks in the array.

risk of this type of unrecoverable data loss, because disks from the same vendor or the same batch tend to fail at a similar time frame during their entire life expectancy.

5.4.6 RAID-6

RAID-6 is similar to RAID-5 in many aspects. In RAID-5, if a second disk fails before the first failed disk is detected, replaced, and rebuilt, which takes time especially when there are many disks in the array, unrecoverable data loss occurs. In this case, RAID-6 is the solution. Because of the nature of the importance of PACS/EHR data, the probability of a second disk failure can and should be minimized. This is especially important for today's PACS/EHR where the storage has increased dramatically compared to legacy systems. Many and larger capacity disks are built in the array, which translates into longer rebuild time if a disk is faulty. In RAID-6, there are 2 parity blocks. All data blocks and parity blocks are distributed across all disks in the array. Data can be fully recovered when up to 2 disks in the array are detected, replaced, and reconstructed. This is very useful since the number of disks in an array is getting larger. The total size of RAID-6 can be calculated as:

$$D = (N - 2)d,$$

where D is the array's total usable size for storage, N is the number of disks in the array, and d is the minimum size of the disk in the array. If all disks have the same size, then d is the size of a single disk.

In RAID-5 and RAID-6, usually at least one spare disk is installed, so that when a disk failure is detected, the system will automatically start the disk-rebuild process. When rebuilding data after a hard-drive failure in RAID-5 and RAID-6, the overall performance of the PACS/EHR system is affected because some of the server resources are allocated for the database rebuild process.

Although RAID implementations typically use disks with SCSI interfaces, RAID configurations are independent from the specific disk and interface technologies used. Special RAID controllers are used to connect disk arrays and to present them as a single virtual disk to the processor. The primary purpose of using RAID is to prevent system failure caused by faulty disks during business-critical operations. For RAID implementations there are various basic configurations, and also spanned array configurations, which means a disk in the array is not a single physical array, but another RAID array of disks. The number of the RAID levels does not necessarily associate with ranking of performance or cost, and the most-used RAID configurations are RAID-0, RAID-1, and RAID-5, especially in PACS implementations.

5.4.7 RAID-Spanned Arrays

In basic RAID configurations, when the storage size of the logical drive is desired to be very high, more physical disks need to be configured. This increases the possibility of one disk being failed from the entire RAID. RAID-spanned arrays, which define the array of disk arrays, solve this issue. Each element of the spanned disk array is itself composed of a lower-level disk arrays. Data are striped over multiple lower-level arrays rather than on individual disks with RAID-0 configuration. Spanned arrays have the benefit of allowing more disks to be grouped as a single logical disk. RAID-10 and RAID-50 are examples of the concept of array of arrays.

5.4.7.1 RAID-10

RAID-10 includes RAID-0 striping over lower-level RAID-1 arrays; therefore, disks are mirrored first, then striped together as a large volume logic disk, providing improved redundancy and performance of the increased disk array size for the system it is attached to. The drawback for this RAID configuration is increased cost. The logical disk size is only half of the total physical disks used in the RAID-10 array, assuming all disks have the same capacity.

5.4.7.2 RAID-50

RAID-50 is a configuration that has RAID-5 as the lower-level arrays and RAID-0 striping over these arrays. In this spanned RAID configuration, up to one drive in each lower-level array may fail without data loss, and the higher-level RAID-0 striping enables more disks to be included into a virtual logic disk.

5.4.8 Hot Spare Disk and Rebuilding

Many RAID implementations include hot spare disks, which means that spare disks are installed in the array as standby. Immediately after a disk fails, the system replaces the failed disk with the standby spare disk automatically, and starts the rebuilding process on the spare disk. Failed disks

Figure 5.16 A spin disk is put on a mount as part of a RAID or hot-spare disk, so that it can be easily replaced when it fails.

need to be promptly replaced (shown in Figure 5.16 is a spin disk for RAID applications which has been mounted for easy replacement when faulty), because all disks of one array have the same amount of usage; they may tend to fail within a similar time frame. Before the last-failed disk rebuild is complete, if more disks than the RAID level allows fail, data loss is inevitable. Users need to realize that disk durability varies. Disks designed for enterprise data storage use can last long for an array with heavy use. However, disks designed for the general consumer market may be less durable and not be able to sustain the heavy use in a RAID configuration for long.

RAID is a matured and proven technology to provide safe data storage against potential hard disk failure. The benefit of spanned disk arrays is a larger logical or virtual disk drive size with improved performance and reliability. Implementation of a RAID using the hardware approach requires a special RAID controller. The controller manages the drives and performs calculations for any parity required by the specific level of the RAID. Disks with a variety of interfaces can be used, typically SATA, SCSI, or FC. Disks and controllers may reside in a standalone separate case, or inside the computer or server. The RAID case can be directly attached to the server as DAS, or over SAN. RAID with hardware implementation adds no overhead operations to the system processor, because the RAID controller appears as a single logical disk to the operation systems of the server to which it attaches.

5.5 IMAGE DATA COMPRESSION

Modern medical imaging modalities generate images with large files to be archived for long periods of time. This becomes a big challenge for both image-storage systems and for image transmission through a network. If digital medical images can be compressed without compromising their quality, the resources and cost associated with image storage and transmission is reduced. For medical imaging, DICOM PS3.5 (2008) supports various image compression techniques including Run Length Encoding (RLE), Joint Photographic Experts Group (JPEG), JPEG-LS, JPEG2000, and so on.

5.5.1 JPEG

Created by the Joint Photographic Experts Group (JPEG) committee in the year 1992, the JPEG standard is used for digital compression and coding for continuous tone still gray-scale and color images. It is a widely used method for image compression. JPEG defines requirements and implementation guidelines for coded representation of compressed image data. JPEG images can be interchanged by a broad range of applications and computers. Both lossy/irreversible and lossless/reversible compression schemes are supported by JPEG. The lossless JPEG uses differential pulse code modulation, while lossy JPEG is based on discrete cosine transform. For lossless compression, the compression ratio is typically around 2 ~ 3:1. For lossy compression, the amount of compression is adjustable, enabling a variable selection for tradeoff between image quality and image file size. For JPEG, the common file name extension used is .jpg or .jpeg.

5.5.2 JPEG2000

Created by the JPEG committee in the year 2000, JPEG2000 is an image format based on wavelet image compression. The purpose of JPEG2000 is to improve performance and replace the original JPEG image format, which is based on discrete cosine-based image compression, because the original JPEG was designed to be processed by the much less powerful computers available at that time. JPEG2000 sets standard requirements and implementation guidelines for image data compression coding. The JPEG2000 coding process is intended to be generic for both gray-scale and color still images, so that images coded with JPEG2000 can be interchanged among a broad range of applications and stored and communicated on various computer systems. Up to 16 bits of bit depth is supported by JPEG2000. Compared with JPEG, JPEG2000 has many advantages: it shows superior compression performance using wavelet encoding and has a very flexible and efficient code stream organization, but it requires more complex algorithms and extensive processing power. Similar to the JPEG, the JPEG2000 standard also provides both lossless/reversible and lossy/irreversible compression in a single image compression architecture. Lossless JPEG2000 compression is achieved by using a reversible integer wavelet transform, with a typical compression ratio of around 3:1.

JPEG2000 standard has several parts. Part 1 specifies the core of JPEG2000 and defines the JPEG2000 code stream syntax, the necessary steps for encoding and decoding of JPEG2000 images, and a basic file format called jp2. The later parts of the standard are not essential to a basic JPEG2000 implementation, but are various extensions and file format definitions using extensible architecture of jp2.

5.5.3 Application of JPEG and JPEG2000 in DICOM

DICOM, discussed in Chapter 6, allows various techniques to be used for medical image compression through the encapsulated format. Both JPEG and JPEG2000 are becoming increasingly used in PACS, and in DICOM messages they use specific DICOM transfer syntaxes. A DICOM transfer syntax is a set of encoding rules for a pair of DICOM application entities that communicate the data set portion of the encoding techniques supported by both application entities. For every transfer syntax, DICOM assigns a unique syntax name.

DICOM implementations supporting lossless JPEG compression should also support the DICOM-specified default lossless JPEG image compression defined by the DICOM transfer syntax. For all DICOM implementations that support lossy JPEG compression, there are two default lossy JPEG compressions, each with a corresponding transfer syntax that these DICOM implementations should support. One default is for 8-bit images and the other for 12-bit images.

As for JPEG2000, since it is significantly different from JPEG, DICOM allows encoded images to be encapsulated with specific DICOM transfer syntaxes. DICOM has two transfer syntaxes for JPEG2000 Part1. One transfer syntax specifies the use of only lossless reversible wavelet transformation and reversible color component transformation of JPEG2000 Part1. Another transfer syntax specifies either the lossless reversible mode of JPEG2000 Part1, or the lossy/irreversible mode of JPEG2000 Part1, which uses irreversible wavelet transformation and irreversible color component transformation, or uses a truncated code stream for the lossless/reversible mode of JPEG2000 Part1. The DICOM object sender determines whether the reversible/lossless or irreversible/lossy mode is being used. The purpose of defining two transfer syntaxes is to enable lossless image transfer whenever possible by applications.

Although JPEG2000 has several parts, only the core features of JPEG2000 defined in Part1 are allowed by these two DICOM transfer syntaxes. For any other extensions and features specified in the other parts of the JPEG2000 standard, unless they can be decoded or ignored by all JPEG2000 Part 1—compliant implementations without fidelity loss, they cannot be used with these two DICOM transfer syntaxes. The optional JPEG2000 file format header cannot be included, because its role is already played by the nonpixel data attributes in the DICOM data.

For JPEG2000 Part2, there are two additional transfer syntaxes for multicomponent image compression with multicomponent transformation extensions. Again, one is for lossless only, and one is for both lossless and lossy compression. Any other additional features and extensions defined in JPEG2000 Part2 should not be included in the compressed bit stream with these two transfer syntaxes.

5.6 CONTENT-ADDRESSED STORAGE

Content-Addressed Storage (CAS) is a way to store data, which is significantly different from the conventional file systems used in most storage systems. Most local or networked storage systems except CAS use the location-addressed approach. Location-addressed storage uses the location of the physical medium where data are stored for later data retrieval. These locations are organized as a directory tree. When data retrieval is requested, the data can be found on the physical storage media according to the location.

5.6.1 What Is CAS?

CAS is an object-oriented storage approach. When data are stored into a CAS system, a content address or ID, which uniquely identifies each piece of stored data, is generated to permanently link it with content information, ensuring that a single piece of data is stored as an object only once without any duplicates. When the data object is requested for retrieval, the CAS system determines the address of the actual physical media where the data are stored and retrieves it. Data retrieval is based on the content of the data object, not the location where it is stored. Since the data content address or ID is based on the content itself, any change of the data object causes changes to the content address associated with the data object. CAS is thus efficient for fixed-content data storage. The design of CAS allows fast content search, and ensures that the retrieved document is identical to the one stored originally, otherwise its content address would be different because of its different content.

5.6.2 Application of CAS

A CAS device rarely allows alteration of data objects after they have been stored on CAS, and the institutional policy determines if a data object can be removed from CAS once stored. CAS is efficient for fixed-content data storage that does not change, is not accessed frequently, and is required by retention policy to be kept for long periods of time. Fixed content is a term for data that are stored and never changed, such as legal documents or medical images stored on PACS. CAS ensures that one piece of data as an object is saved only once without any duplicates. CAS is typically used on high-speed storage devices using disk rather than removable media, such as tape-based storage hardware. It requires significant processing and is suitable for large systems. The potential superior searchability of medical images stored on PACS may be its unique advantage for applications.

5.7 SHORT-TERM AND LONG-TERM STORAGE FOR PACS/EHR

Short-term online storage is the storage of PACS images on spin disk hard drives that can be retrieved quickly by the PACS workstations through the network. Usually, the network speed determines how long it takes for the workstation to retrieve the requested images. With the falling cost of spin disk per unit of storage, PACS images on short-term online storage should be retained for as long as possible, usually 12 months. This provides physicians using the PACS much convenience.

In PACS, medical image data are constantly sent to storage after capture, and these data are usually not changed once stored, except for the database. They may, however, be retrieved several times, occasionally more. Because the image data associated with a particular study can be very large, the read speed is critical. Except for the database, there is no need for many writing cycles. In this case, SSD may be suitable for storing actual PACS/EHR data files in the short term. For a long-term PACS storage array, SSD may not be used in the foreseeable future, mainly because of the cost associated with large data storage size. The high-speed benefit may also not be fully realized because of network speed limitations.

The data storage needs of an enterprise with multiple sites are tremendous. If every department deploys its own long-term storage from different vendors, and each system requires dedicated personnel, it is a nightmare for support and service staff, and for the system integration of all referral clinicians using PACS. Using an enterprise-wide, single, replicated data storage for all images from all departments not only improves workflow, but also brings cost savings for the enterprise, because the unit cost is less when more data storage is purchased, and less dedicated maintenance personnel are needed. The enterprise can thus achieve significant cost savings and patient care benefits.

Long-term storage is an ongoing project. Today's newest technology may be obsolete in five years. Planning for long-term storage requires strategic insight; there are many options available in the market. Cost reduction, streamlined storage management, reliability of the technology, improved user satisfaction, and patient care should all be considered to make a sound decision for both enterprise and small clinic long-term storage purchase.

5.7.1 Tiered Storage

Tiered storage is not a specific data storage technology, but rather a data storage strategy consisting of two or more ways of storing data. Although, ideally, PACS data can be stored and accessed on high performance storage all the time, in the long-term with redundancy, it is very expensive for any healthcare institution to store PACS data that way. Instead, long-term and backup PACS data are stored on slower storage, and only recent data are also stored on faster disk arrays.

5.7.1.1 Defining Tiered Storage

Tiered storage for PACS moves data between high-cost and low-cost media. These data storage methods are significantly different in their characteristics in the areas of performance, unit storage price—for example, what is the total cost of storing 1 TB data?—and the total capacity a storage system provides. If two storage systems are very different in at least one of the above-mentioned areas, these two systems are considered separate storage tiers.

5.7.1.2 Tiered Storage for PACS

Usually, time frame is used as the criteria for classifying PACS data for a tiered storage arrangement. In a healthcare institution, patient PACS/EHR data that are recently acquired or retrieved, may be retrieved more frequently, while others may be stored for a long time, only for future use. They pose different requirements on data storage systems. A tiered storage mechanism provides flexibility to meet these needs. In a typical healthcare enterprise PACS with a tiered storage setup, the most recent data files are stored not only on long-term tier-2 storage, but also on short-term, tier-1, expensive, high-quality, high-performance RAID for a period of 12 months or more, determined by the healthcare institution's policy. Later these data files are removed from the tier-1 storage and placed only on the tier-2 long-term storage, SAN. As the tier number is increased, cheaper storage media are used. For disaster recovery, tier-3 storage tapes or optical libraries may duplicate data stored in tier-2, and then be kept in separate geographic locations. In this case, RAID, SAN, and tape/optical libraries are three separate storage tiers significantly different from each other. The advantage is that the total data-storage capacity of the long-term archive is much larger than that of the short-term archive disk arrays, and PACS/EHR data files can be stored for the long-term at low cost, and with disaster recovery capabilities. If the data on tier-2 SAN are reused, they are automatically moved back to the short-term tier-1 RAID, because there is a high possibility that the data are to be used again in the short term. Most users may not feel the slow speed, as only the infrequently used, more-than-12-months-old PACS files are on the slower long-term storage archives.

For smaller healthcare institutions using external storage services, usually a service provider contractor stores client's PACS data on 2-tiered-storage systems, one copy on high-performance disk arrays for fast data access, and one copy on slower tape or optical media library in a different location for disaster recovery.

In an enterprise healthcare institution's data center, data may be stored on 3-tiered storage systems, providing disaster recovery capabilities with added fault-tolerance. However, while purchasing tiered storage systems, users need to know what is needed in each storage tier to meet their specific PACS/EHR needs. This approach avoids paying for unnecessary features and functionalities, because although some functions may provide high availability and performance suitable for a primary PACS/EHR storage tier, they may not be necessary for higher numbered low-storage tiers. Overall, tiered storage provides PACS/EHR data with long-term storage options at better service levels and lower overall costs.

In this chapter, various data-storage technologies and methods for PACS/EHR were discussed. Each has its advantages and disadvantages; there is no perfect solution. For healthcare institutions looking to buy data storage solutions, almost all vendors claim that their products are scalable and can be easily expanded. However, this may not be the case. Some products have better scalability than others. Specifically, buyers need to check if the system uses proprietary software or not, because hardware, such as spin disks, tapes, or optical disks are mostly standardized, though they, especially tape and optical disks, may quickly become obsolete. Reputation, together with the number of years a vendor has been in the data storage business, is also an important factor to be considered. Lack of adequate long-term service support because of company mergers, or any other reason for dropping out of this business, is not unusual in the IT industry in general. Data migration to new storage systems because of a change to a different PACS system several years later may also need to be considered at this time.

In conclusion, data storage for PACS/EHR storage advances as the IT industry does. For large enterprise healthcare institutions, whose constituent hospitals and clinics may be located at geographic locations far away, the data storage needs are significantly different from that of smaller standalone hospitals or clinics. Before making an investment in a particular storage system, buyers should not only understand their unique environment and needs and the available budget, but also look beyond the terms used by vendors, check how the functionality of a system can improve their administration of storage reallocation, data backup and restoration, and data migration path a few years down the road.

BIBLIOGRAPHY

Bycast. 2007. White paper: How bycast storageGRID works. www.bycast.com/resources/files/bycast_resources/product_literature/310-0020_r20_StorageGRID_Common_Topologies.pdf. Accessed January 18, 2010.

CALCE. 2003. Reliability of Hard Disk Drives (HDD). University of Maryland, College Park. www.calce.umd.edu/whats_new/2003/1203.pdf. Accessed January 18, 2010.

Dreyer, K. J., D. S. Hirschorn, and J. H. Thrall. 2006. *PACS: A Guide to the Digital Revolution.* Springer, New York, NY.

Erickson, B. J., K. P. Andriole, and R. G. Gould, et al. 2000. Irreversible compression of medical images. A white paper from SCAR. www.scarnet.org/WorkArea/showcontent.aspx?id=1208. Accessed June 5, 2010.

Erickson, B. J., K. R. Persons, and N. J. Hangiandreou, et al. 2001. Requirements for an enterprise digital image archive. *J. Digit Imaging* 14 (2): 72–82.

Hernandez, R., K. Carmichael, and C. K. Chai, et al. IP *Storage Networking: IBM NAS and iSCSI Solutions.* www.redbooks.ibm.com/redbooks/pdfs/sg246240.pdf. Accessed January 18, 2010.

Huang, H. K. 2004. *PACS and Imaging Informatics: Basic Principles and Applications.* Wiley-Liss, Hoboken, NJ.

Massat, M. B. 2007. Will replacement PACS gear toward grid PACS? www.itnonline.net/node/28379. Accessed January 18, 2010.

Nagy, P. 2004. The ABC's of storage technology. SCAR university, educating healthcare professionals for tomorrow's technology. *Society for Computer Applications in Radiology* 2004: 123–127.

Nagy, P. and J. Farmer. 2004. Demystifying data storage: Archiving options for PACS. *Appl. Radiol.* May, 18–22.

Scarfone, K., M. Souppaya, and M. Sexton. 2007. *Guide to Storage Encryption Technologies for End User Devices.* National Institute of Standards and Technology, U.S. Department of Commerce, Special Publication 800-111.

Scheier, R. L. 2005. "Grid" storage is in the eye of the beholder (and vendor). www.computerworld.com/s/article/102140/_Grid_storage_is_in_the_eye_of_the_beholder_and_vendor_. Accessed January 18, 2010.

Smith, E. M. Storage service provider—The cost effective option? www.himss.org/content/files/EHR/WP_SSP_2006.pdf. Accessed January 18, 2010.

Tate, J., F. Lucchese, and R. Moore. 2006. *Introduction to Storage Area Networks.* www.redbooks.ibm.com/redbooks/pdfs/sg245470.pdf. Accessed January 18, 2010.

Tate, J., B. Cartwright, and J. Cronin, et al. *IBM SAN Survival Guide.* www.redbooks.ibm.com/redbooks/pdfs/sg246143.pdf. Accessed January 18, 2010.

Toigo, J. The grid storage façade. www.networkworld.com/buzz/2004/092704grid.html. Accessed January 18, 2010.

Transforming IT. IBM. University health care system improves patient care with enterprise grid storage system. www-01.ibm.com/software/success/cssdb.nsf/CS/CCLE-6ZNS63?OpenDocument&Site=sttotalstorage&cty=en_us. Accessed January 18, 2010.

Wang, F., S. Wu, N. Helian, et al. 2007. Grid-oriented storage: A single-image, cross-domain, high-bandwidth architecture. *IEEE Transaction on Computers*, 56 (4): 474.

6 Healthcare Information Integration

6.1 HEALTH LEVEL 7

The healthcare industry has very unique settings in terms of data/medical image collection, database, software, payment system, business workflow, and so on. Healthcare institutions usually have many different computer systems to perform various activities ranging from patient registration to tracking of bills, and their interactions with patients' clinical data are all unique. Ideally, all these computer systems should exchange and synchronize information automatically. However, this is not realized for many of the systems today. In today's healthcare environment there are many clinical computer application software programs. The need for clinical data to be shared among these applications is tremendous. There are mainly two driving forces for requiring a common standard among these applications: the dramatic increase of clinical applications of electronic patient data and the establishment of EHR. This hasted to the establishment of a worldwide standard for interchange and representation of clinical data by HL7, an international standard organization.

6.1.1 Introduction

HL7 specifies a number of guidelines and data standards for various healthcare systems. These guidelines and data standards are a collection of rules which allows clinical information to be processed and shared by various healthcare organizations and systems consistently and uniformly.

HL7 generally refers to messaging standards HL7 V2.x, HL7 V3.x, HL7 V3.x RIM, document standards Clinical Document Architecture (CDA), and application standards Clinical Context Object Workgroup (CCOW). Of these, the HL7 V2.x and V3.x are especially important because information exchanged among systems is defined in these. It should be noted that after HL7 V3.x is released, the development of HL7 V2.x is not stopped yet and new versions of HLV2.x are still released, because there is significant difference between HL7 V2.x and HL7 V3.x, and installed base of HL7 2.x is very high.

The name HL7 symbolizes the seven-layer International Standards Organization (ISO) communications model. The ISO defines communications model in seven layers, as discussed in Chapter 4. "Level 7" indicates the highest level of the ISO communication model for open systems connection, which is the application level. The application level determines the exchanged data definition, data interchange timing, and error communication to applications. In addition to the most important data exchange structure, also supported in the application level are security checks, participant identification, data exchange mechanism negotiation, and data exchange structure. HL7 standards are framework of possibilities, not actual implementation guidelines which are developed from standards to meet interface needs.

6.1.1.1 History

HL7 was originally founded in 1987 by an international community of healthcare information technology scientists and professionals. They collaborated to establish the first HL7 standard for exchange and management of healthcare information in electronic forms. In 1994, it was accredited by the American National Standards Institute (ANSI) as a standard developing organization operating in the healthcare industry. For HL7, the original area is hospital administration and clinical data transaction. Today, HL7 is no longer accredited by ANSI but adopted by ISO and several international standards developing organizations. HL7 collaborates with ANSI and ISO in healthcare and information technology to promote the use of HL7 standards, so that appropriate new standards are developed to meet real-world emergent requirements.

HL7 standards were developed by clinical information interface specialists who use software applications to build interfaces. Instead of using extensive customized software coding to build an interface between two software applications, they decided to adopt a better method to reduce the cost associated with every interface development. The importance of HL7 increases when increasing numbers of vendors and healthcare institutions adopt it.

6.1.1.2 HL7 Goal

Before HL7 was adopted, data exchange between two HIS required extensive customization of programming at both the sending and receiving applications. Because each application development is isolated without input and collaboration from other application development teams, it is a great challenge for healthcare institution teams or software vendors to create interface. Many commercial software applications are proprietary in how they are built and so it is difficult to build interfaces for them.

HL7 is a messaging standard that enables clinical applications to exchange data and information freely. Applications used in healthcare institutions that complied with the HL7 messaging standard can exchange information with each other, although applications may be run on different platforms with different operating systems, and written in different languages. There are various levels of interoperability. The lowest level is message exchange, so that clinical messages can be transported from one clinical system to another. The intermediate level is functional interoperability, which is the capability to exchange information reliably without error. The highest level is semantic interoperability. Semantic interoperability ensures the ability to interpret, and therefore, to make effective use of the information thus exchanged.

6.1.1.3 HL7 Organization

HL7 is a nonprofit international organization whose members chiefly are healthcare providers, equipment manufacturers, payers, consultants, and government groups. They collaborate together to create and develop standards for exchange, management, and integration of electronic healthcare information. Their interest is in the development and advancement of healthcare clinical and administrative standards. HL7's domain is healthcare clinical and administrative data, and their exchange. It promotes the use of standards within and among healthcare institutions for improved effective healthcare delivery. It is a global organization, with country- or region-specific organization affiliations.

The HL7 committee compiles message formats and clinical standards to provide a framework for healthcare data exchange, and sets the basis for data exchange. The standard does not provide step-by-step instructions as how to build a clinical application program. Instead, HL7 sets a framework for negotiation in interfacing healthcare equipments to exchange data for electronic healthcare information exchange, integration, sharing, and retrieving. It is widely used across the world by large enterprise healthcare institutions and small clinics to support clinical practice and management.

HL7 as a standard development organization does not develop any software, but develops standard specifications enabling disparate healthcare applications and systems to exchange clinical and administrative data sets. It works closely with other standards developing organizations and various US government agencies. Since 1987, HL7 messaging standard is used internationally and is still being revised to meet the ever-changing healthcare clinical data need for integration.

6.1.2 HL7 Version 2.x

HL7 V2.x messaging standard is an application protocol for electronic data exchange in healthcare environment. The first HL7 V2.x ready for use by healthcare institutions was released in 1990. These versions are known as HL7 V2.x, and they are backward compatible. Because it specifies a series of electronic messages to support clinical workflow, hospital logistics, financial, and hospital administration, HL7 V2.x realizes the interoperability among clinical information systems including patient administration systems, electronic practice management systems, clinical laboratory information systems, dietary, pharmacy, billing systems, EHR systems, and so forth.

6.1.2.1 Introduction

Currently, HL7's V2.x messaging standard is supported by every major medical information system vendor in the United States, and are used by more than 90% of healthcare institutions in the United States. The most recent version of HV7 V2.x is used almost by all medical information systems for enterprise integration. HL7 V2.x is considered by many as the workhorse of data exchange in healthcare, and it is the most widely implemented healthcare information standard internationally.

6.1.2.2 HL7 Version 2.x Data Structure

HL7 V2.x is an implicit model driven by simplicity. It includes a series of message definitions, and its syntax is proprietary. All HL7 V2.x messages follow the same pattern. The building blocks of HL7 V2.x are data items that are combined into logical segments. A HL7 V2.x message is a unit of data exchanged between two systems. A message is composed of a group of segments in a defined sequence. The message type defines its purposes such as trigger, query, or response, and so on. New required elements are added to existing segments, and new functionalities create new segments; while HL7 2.x messages are driven by trigger events, HL7 V2.x messages are strings of data.

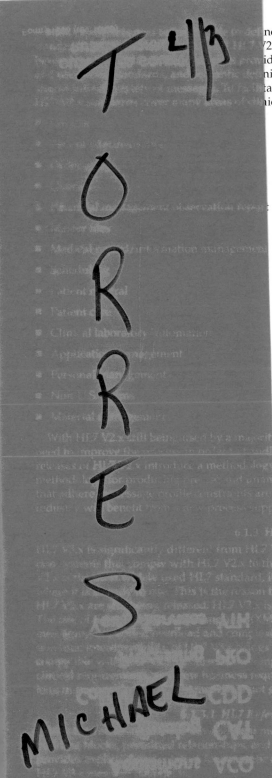

…ne and measure compliance for early HL7 V2.x …V2.x specifications have given little guidance …ided in standards. The lack of universally accepted …nition related to the underlying processes create …ate the concept of conformance, recent releases of …ical care and administration in areas listed below:

…y of healthcare institutions, there is a substantial …e the complexity of challenges faced by HL7. Later …y for defining message profiles. Message profile is a …ambiguous specifications. Later messages of HL7 V2.x …said to be conformant to the profile. The healthcare …orting more rigorous specifications.

HL7 Version 3.x

…V2.x. There is no easy means of converting informa- …ose that are compliant with HL7 V3.x. Because HL7 …is most likely to remain in place for a long time …hat parallel to release of HL7 V3.x, new versions of …Reference Information Model (RIM)-based design. …L) in HL7 V3.x ensures human readability. HL7 V3.x …y compared with HL7 V2.x messaging. The …ocabulary binding. HL7 V3.x is not backward …lexity and added function in privacy, security, …rements. Since HL7 V3.x has not been in use very …ping to replace HL7 2.x in any time soon.

…nce Information Model

…hodology and process of HL7 V3.x. It defines the …ules for composing HL7 V3.x messages. RIM also …d lexical relations in the information contained in

…es and many permitted relationships between them. …are message exchange and reduce implementation costs and efforts. The consensus-based HL7 V3.x standard for healthcare information system interoperability is driven by RIM, vocabulary specifications, and the model-driven process of analysis and design.

6.1.3.2 HL7 V3.x Message Development Framework

HL7 V3.x message development follows the methodology defined in the Message Development Framework (MDF). MDF is a process, which involves defining original requirements to the final implementation for message development. It is a continuously evolving process seeking to develop specifications facilitating interoperability among healthcare information systems. To integrate healthcare information systems, it is not sufficient to translate messages from one application to another. Semantic content of various data need to be preserved for effective interoperation of two applications. The primary document for describing the methodology is the MDF. It is the most current rendition of the HL7 V3.x development methodology. The methodology divides the development process into requirements, analysis, and implementation phases. To ensure consistency of contents created for each phase and between phases, various aspects of the healthcare environment are defined and consistency is checked. The MDF methodology specification addresses updates to messaging specification which is applicable to structured documents and context management. The MDF documents the processes, tools, actors, rules, and artifacts relevant to development of all HL7 V3.x standard specifications, not just messaging.

The HL7 V3.x message development methodology is a systemic process, through which existing software applications are presented as software agents that interact with other software agents across various architectures, systems, platforms, and languages to achieve interoperability at a greater level. Beyond text message, the MDF also crosses all HL7 V3.x standard specifications, including any new standards for EHR architectures and requirements. The following is a list provided by the MDF-defined process:

- Precise functional requirement definition

- Consistent data definition through the entire HL7 V3.x

- Rigorous message content definition

- Use of conformance claims

MDF is a guide for creating new messages for HL7 V3.x and beyond. HL7 RIM plays an important role in this process. The MDF can be fully used only when RIM is created for message development. The MDF presents a series of steps for message development, and the actual process involves iterations depending on the complexity and size of the problem.

6.1.3.3 Clinical Document Architecture

HL7 has developed well beyond the extent for message exchange. The purpose of HL7 CDA is standardization of clinical documents for exchange and EHR. A clinical document is a record of observations and other services with characteristics of:

- *Persistence*: A document that exists for a long time and can be used in many contexts.

- *Stewardship*: A document that needs to be managed and shared by the steward.

- *Potential for authentication*: The document that is intended to be used as medical and legal documentation.

- *Wholeness*: A clinical document that includes its relevant context.

- *Human readability*: This ensures human authentication.

CDA documents are defined with complete information objects including not only text messages, but also images, sound, and other multimedia contents. Typical CDA documents include patient referrals, admission, history and physical, diagnostic report, therapeutic procedure reports, imaging report, pathology report, care record summary, progress note, discharge summary, claims attachments, and so on. CDA is a critical component of EHR. Based on the HL7 RIM, Extensible Markup Language (XML), coded vocabulary, HL7 V3 data types and methodology, the core of the CDA is the clinical statement model which is used across HL7 V3. It is crucial for automated processing which determines semantic interoperability.

6.1.3.4 CDA Structure

The CDA is a part of the HL7 V3 standard. It specifies encoding, structure, and semantics of clinical documents for exchange. Each CDA document consists of a header and a body. The header is composed of metadata for document search, management, retrieval, identification, classification of the document for patient, healthcare provider, and authentication. The body includes text, multimedia content, or coded statements for clinical reports or any content carrying a signature. CDA specifies the body content of each document to include a mandatory text part which can be used for interpretation of human document content and can be reproduced as legally attested clinical content when needed. CDA also has optional machine-readable coded data entries, which are suitable for automated machine process. Because coding clinical statements using a general approach is still a challenge, CDA allows the use of incremental amount of coded data inside the CDA model. CDA has different levels of semantic interoperability; hence, there is a range of complexity users within the specification who can set their own level of compliance. The simplest CDA has only a small number of standard header metadata fields for document type, ID, provider name, date, and so on, and human readable body. The body can be a common portable document format (PDF), word document, or scanned image file. Although the body containing these documents cannot be electronically interpreted by healthcare software applications, the minimal standard metadata set of these CDA documents still enables them to be managed together with more complicated CDA documents, which are encoded with the full power of the HL7 RIM and controlled vocabularies. The different complexities of a CDA indicate various degrees of automated process that a receiver can perform, describing the degree of semantic interoperability of a CDA document.

6.1.3.5 CDA and EHR

CDA provides an exchange model for clinical documents beyond messaging, a framework for full semantics of a clinical document, and is suitable for EHR, which is important for adoption of EHR in the healthcare industry. CDA is considered by many as a primary format for diagnostic imaging reports. It makes documents readable by both machine and human so that these documents can not only be easily parsed and processed by application software, but also easily retrieved and used by people who need them. CDA can also be adapted for Web access because it utilizes XML.

There are many CDA compliant applications developed by vendors for healthcare document generation, management, and viewing. Because CDA is implemented in XML, it can be managed by a clinical repository, which is capable of utilizing XML, a custom database, and can be accessed via e-mail, HL7 messaging, or custom Web services. Currently CDA can be generated or converted as output by some EHR and report dictation and transcription applications, and is used mostly for enterprise healthcare information exchange.

6.1.3.6 Clinical Context Object Workgroup

In addition to the development of a standard for message exchange, HL7 also develops application standard for clinical workflow efficiency improvement. During patient care in a healthcare institution, a clinical user may use several independent applications from various information systems when physically interacting with a single computer. To achieve optimum workflow efficiency, it is necessary for all these applications to virtually act together like a single application, so that the clinical user can access and manage different types of healthcare information related to the same patient. This capability is essential for clinical users and can be realized by using the technique of context management.

CCOW stands for "Clinical Context Object Workgroup," which is a committee within the HL7 group that develops the CCOW standard. CCOW is an end-user-focused standard facilitating applications integration at the point of use, and is the primary standard protocol in the healthcare environment to facilitate context management. It is a process using user, patient, clinical encounter, and so on, to virtually link disparate software applications. Architectures, component interfaces, and data definitions specified by CCOW are all technology neutral. The intent of CCOW is not only to use broadly accepted industry standard computing technologies for CCOW standard implementations, but also to provide details of CCOW specifications which are needed for consistent and robust implementations of CCOW-compliant applications and system services.

The context management is often combined with single sign on applications. It is a process that enables the user to access disparate applications securely through a single identifier and also aids password authentification. The context management standard defines a protocol to ensure various healthcare applications are linked and synchronized to the same context. The context is comprised of subjects. Each subject identifies an entity such as patient, account numbers, and encounters. These subjects become a set of common user interactions with healthcare applications. Elements of the context can either be provided by users when they interact with an application directly, or by other applications automatically. The context management then enables coordination and

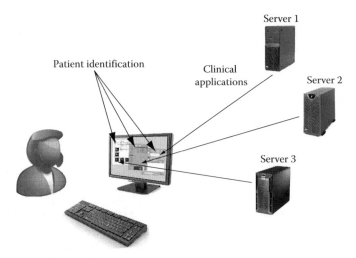

Figure 6.1 CCOW allows patient's context coordination such as patient's identification at the point of use, so that disparate clinical applications from different information systems (e.g., servers 1, 2, 3, etc.) can be synchronized in real-time at the end user interface level.

synchronization of all disparate applications when the user identifies subjects such as patient, ID, or date once. When the subject is updated once in one application, the same subject in all other applications is also updated automatically and seamlessly. This gives the user an aggregated view of all patient information across disparate applications, as illustrated by Figure 6.1.

CCOW is designed to enable disparate applications at the end user interface level to synchronize in real-time, to allow visual integration of healthcare applications, and to enhance the user's ability to incorporate information technology in the process of care delivery. With CCOW's implementation, healthcare institutions have the flexibility to choose the best of the breed applications they want, and CCOW coordinates and synchronizes applications at its point of use. This capability provides uniform, simplified access and authentication for clinical staff, so they can have improved desktop and context management, and native applications can be used seamlessly.

6.1.3.7 Arden Syntax

One of the future areas for HL7 standards development and application is semantic interoperability. The Arden Syntax is a medical knowledge encoding language for semantic interoperability. It is meant for healthcare institutions, information systems, and professionals to share computerized healthcare knowledge. This healthcare knowledge is limited to those capable of being represented as a set of discrete modules. Each module is defined as Medical Logic Module (MLM) in the healthcare environment. Each MLM contains sufficient knowledge and information to make a single medical decision, such as treatment protocol, data interpretation, alert, diagnosis score, patient care management suggestion, quality assurance function, administrative support, and so forth. An MLM has management information for maintaining knowledge base of other MLMs and linking to other knowledge sources. MLMs can be created by healthcare professionals. When an MLM is run on a computer with the appropriate software installed and utilized by information systems conforming to the specification, it can provide advice whenever and wherever needed. HL7 adopts the Arden Syntax standard for semantic interoperability among information systems.

6.1.3.8 HL7 Conformance and What Users Need to Know

One important benefit of HL7 V3.x is that its conformance criteria are clearly stated in terms of healthcare information system interface performance, so that they can be used as the basis for contractual agreements in purchase contract and conformance testing on systems. Healthcare institutions as users and manufacturers as vendors of systems can communicate compliance to HL7 V3.x using conformance statements. Because of the broad scope of HL7 V3.x, few systems conform to all application roles defined in HL7 V3.x. For buyers or users of information systems to know precisely what parts of HL7 V3.x the system conforms to, HL7 V3.x specifications describe entities for individual statement of conformance criteria. By combining a list of claimed individual conformance statements, a vendor creates a conformance claim set which includes a list of identifiers for specific HL7 V3.x conformance criteria statements. There are functional and technical statements of conformance criteria. A functional conformance criteria statement is a commitment to fulfill responsibilities of a HL7 V3.x application role.

During purchase contract negotiation, it is not enough for a vendor to claim "the system conforms to HL7"; the vendor should clarify the part of HL7 to which the system conforms to. Using the conformance claim set before signing a contract, the user understands the system's HL7 capabilities and how it would interact with other systems. The user can expect the system fully conforms to the conformance claim set with all statements. The conformance claim set also serves as a basis for resolution of issues during system implementation.

Because there are many information systems that use HL7 V2.x, and in the foreseeable future it will continue to be used, in a contract negotiation process involving HL7 V2.x, there are a few areas that both the user and the vendor should pay attention to:

- Because there are many versions of HL7 V2.x and they may not be compatible, the version of HL7 V2.x and its message definitions need to be agreed upon, as well as provisions to migrate to newer visions in the future, so that the system stays current.

- In addition to stating compliance to HL7 by the vendor, an agreement needs to be made clearly stating the final HL7 interface will be in accordance with the user specifications. HL7 V2.x is susceptible to variations.

- It is also important to clarify content of data elements with the vendor, so that there is no misunderstanding at a later time when certain data fields are not accepted because of database incompatibility.

- Determine who will be responsible to support HL7 interfaces after it is implemented.

- Establish a mutually agreed testing plan for functionality test according to the user specifications before the implementation, using user-provided test data.

- Determine responsibility for any customization program files that need to be created and maintained for the HL7 interface.

In the future, ultimately it is under the IHE umbrella that information systems should conform to either HL7 V2.x or V3.x, so that various information systems can be integrated seamlessly without much customization work.

6.1.4 Future of HL7

Although HL7 is widely used in healthcare institutions for message exchange, the future of HL7 standards is in the areas of semantic, functional, business, privacy, and security interoperability. Applications of standards to semantic interoperability occur on HL7 V3.x only. Actually, HL7 standards are frameworks that define capabilities and boundaries of interoperability, whereas standard implementation guides select specific elements of a group of standards that give a detailed description of what, when, why, and how data are being exchanged. Not only standards themselves, but also the implementation guides based on standards affect interoperability.

HL7's wide adoption enables easy critical information exchange, which improves every aspect of healthcare workflow for reduced error and efficient operation. With the emergence of IHE and EHR, and more healthcare institutions adopting them for their workflow, HL7 is increasingly becoming a critical component in the healthcare industry.

6.2 DICOM

The HL7 standard is mainly for clinical textual message exchange, whereas the DICOM is a standard for medical image. Since the 1980s, there has been a tremendous growing use of digital imaging modality systems, display viewing workstations, image storage systems, and hospital and radiology information systems in clinical healthcare environment. Although it is vital for all these equipments to connect and be interoperable, because manufacturers of equipment used proprietary designs, it was almost impossible to interface these equipments without extensive involvement of vendors. To improve and simplify equipment interface for connectivity, medical professionals and equipment manufacturers formed a joint force to develop DICOM with international collaborations. The purpose is that medical equipments from multiple manufacturers with built-in DICOM can be directly connected without customized interfaces. DICOM is a standard with very detailed specifications for manufacturers to use in their equipment designs, so that medical images and information can easily be exchanged between medical imaging modalities, workstations, image storage devices, and printers. It is a global standard and is used in virtually all hospitals worldwide. DICOM is an evolving medical imaging standard for transmission, storing, and printing, so that digital medical images and associated information can be transferred freely among scanners, printers, storage media, workstations, servers, and networks made by various vendors.

6.2.1 History

In 1983, American College of Radiology (ACR) and National Electrical Manufacturers Association (NEMA) formed a standard committee, and the first version of ACR/NEMA standard was released in 1985. In 1992 the ACR/NEMA name was changed to DICOM when the third version of the ACR/NEMA standard was released. Since then, the DICOM standard is widely used. Although the version is still DICOM 3.0, it is constantly updated to add new service classes and modalities.

The complete DICOM 3.0 standard consists of 18 parts:

PS 3.1: Introduction and Overview

PS 3.2: Conformance

PS 3.3: Information Object Definitions

PS 3.4: Service Class Specifications

PS 3.5: Data Structure and Encoding

PS 3.6: Data Dictionary

PS 3.7: Message Exchange

PS 3.8: Network Communication Support for Message Exchange

PS 3.9: Retired

PS 3.10: Media Storage and File Format for Data Interchange

PS 3.11: Media Storage Application Profiles

PS 3.12: Storage Functions and Media Formats for Data Interchange

PS 3.13: Retired

PS 3.14: Grayscale Standard Display Function (GSDF)

PS 3.15: Security and System Management Profiles

PS 3.16: Content Mapping Resource

PS 3.17: Explanatory Information

PS 3.18: Web Access to DICOM Persistent Objects (WADO)

The standard has a total of more than 3000 pages and is available from the NEMA Web site. DICOM does not cover issues specific to implementation, such as computer hardware, operating systems, database structure, programming language, or how to process data, and there is no license involved for a medical imaging device to use the DICOM standard.

DICOM consists of many different services, most of which involve data transmission over a network. These services are categorized into application areas addressed by DICOM. These areas are image transfer on network, imaging study management, network image print management, and open media exchange. Some of the most-used and important DICOM service classes and standards are discussed in the following sections.

6.2.2 Various DICOM Service Classes

In the DICOM standard, Information Objects Definition (IOD) defines information objects such as images or reports, and service class defines what types of services are provided to information objects. Each service class consists of a Service Class Provider (SCP) and a Service Class User (SCU). SCP is the system that provides the service, whereas SCU is the system that uses the service. A fundamental unit of DICOM is a Service Object Pair (SOP), which is formed by combining a service class and an information object class. An SOP class is a DICOM function, such as print management service class, storage service class, and query/retrieve service class. In an example to store MRI images from an MRI scanner to the PACS archive, SCU is the user (the MRI scanner), SCP is the provider (PACS archive), storage is the service to store, and IOD is information object (MRI images). DICOM is a family of SOP classes; therefore, DICOM's conformance is to SOP classes and not to a "version" of the standard.

6.2.2.1 Query/Retrieve

The DICOM query/retrieve service provides the capability for two devices to communicate, so that imaging devices can send images, query remote imaging devices, and retrieve images. This service is frequently performed among imaging modality scanners, workstations, and image storage archives. It enables a local workstation to find images, and send a command to get images from a remote image archive server or workstation.

The query/retrieve service class defines an application-level class of service facilitating composite object management. Two DICOM application entities (AE) implement a SOP class of the query/retrieve service class with one serving in the SCU role and another serving in the SCP role. This service is not intended to provide a comprehensive general database query, which is complex. The query/retrieve service instead is for a basic composite object (such as DICOM images). It enables a DICOM AE (e.g., a PACS workstation application) to retrieve DICOM images from a remote DICOM AE (e.g., a DICOM storage archive), or request the remote DICOM AE to start a transfer of images to another DICOM AE.

6.2.2.2 Storage

The storage service class is defined in DICOM for network-based storage of SOP instances. The DICOM storage service is an application-level class of service to send images or other persistent objects (structured reports, etc.) to a remote image display workstation or image archive device. It allows transmission of images, waveforms, reports, and so forth, from one DICOM AE to another. At least patient, study, and series information entities are sent in the storage service class as IOD for SOP instances. Before two AEs perform a DICOM storage transaction, they first determine the SOP class (e.g., MRI image storage) to use the SCU (such as a MRI scanner), the SCP (such as the PACS storage archive), and the transfer syntax (lossless JPEG, etc.).

The storage commitment service class is created to facilitate the commitment to storage. It allows an AE to act as an SCU, to request another AE as an SCP making commitment for maintaining the SOP instances. The SCP determines how the storage commitment is provided. Therefore, SOP instances such as DICOM images and reports can be stored in some medium, and be accessible for some time. For example, a PACS storage archive may be committed to store SOP instances (images and reports, etc.) permanently, whereas some SCP may be committed to store just for a short period of time. In the conformance statement the SCP's commitment to storage is clearly documented. The SCU can decide whether or not to delete its copies of the SOP instances after the SCP committed to store the SOP instances. The DICOM storage commitment is a service used to confirm that an image is stored successfully on a storage media either locally or remotely. A workstation or imaging modality as the SCU uses this SCP's confirmation from the archive to determine whether it is safe to remove images from a local workstation or imaging modality. This is important for a modality scanner, because original images should not be deleted from the modality scanner unless they are successfully archived on the storage device.

Storage SOP classes are used in various situations, such as image transfer from an image modality scanner to PACS workstations or storage archives, from PACS storage archives to workstations or back to imaging modalities, or postprocessed images from a workstation to PACS storage archives. Contrary to what many assume, DICOM does not define or have an archive standard as how images are archived for PACS. Some equipment manufacturers choose to preserve images in a DICOM data format, whereas some use standard or proprietary compression method to compress images before archive. Therefore, it is challenging when PACS vendors are changed, or even upgrade with the same vendor, it is common that both PACS databases and image archives need to be migrated.

6.2.2.3 Print

Among one of the early primary DICOM services, the print management service class is an application-level class of service used for printing of images and image-related data mostly in 11" × 17" size (most-used x-ray film size) on hard copy medium, as shown in Figure 6.2. A DICOM printer is the destination to which the images are sent. The print management service class covers general cases of medical image printing in standardized layouts, and provides the capability to print images on a network printer. Some applications can have more flexible layouts and format with annotations by directly manipulating the pixel matrix used in the DICOM print management,

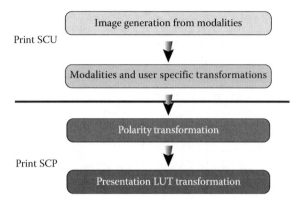

Print SCU

Print SCP

Figure 6.2 DICOM print management data flow.

or use the page description written in postscript or PDF format. The page description is transferred to the hard-copy printer with protocols commonly available. Multiple imaging modality scanners or workstations can share a single printer for image printing, and there is a standard DICOM calibration to ensure consistency for printout from various DICOM printers.

The print management service class defines two grayscale and spatial transformations which transform original images into printed ones. To achieve consistent print image quality, the sequence of these two grayscale transformations is important. However, the relationships of the grayscale transformations with spatial transformation such as enlargement of annotation with images, and the sequence of spatial transformations are implementation-dependent and not specified by DICOM.

For example, the modality LUT first converts vendor-dependent image pixel values into vendor-independent pixel values that are meaningful for the modality. These modality pixel values are transformed into those meaningful values for the application or user. P-values are presentation values related to human eye perception response. They are tailored for particular modalities, applications, user preferences, and are used for consistent display for both softcopy display devices and hardcopy printers. The presentation LUT converts polarity pixel values into P-values; the hardcopy printer then transforms P-values into optical density for printing, which is defined in the presentation LUT SOP class.

6.2.2.4 Modality Worklist

Today's application of DICOM goes well beyond medical image transmission and storage; it is used for workflow efficiency improvement. A typical example is the Modality Worklist (MWL). A worklist is the structure to present information related to a particular set of tasks. It consists of worklist items related to tasks. A patient's imaging procedure is a task. The MWL SOP class is an application-level class of service. MWL provides imaging modality devices with network capability to integrate with various information systems such as Radiology Information System (RIS) and archives. MWL communicates the scheduled procedure steps to the imaging modality and other entities related to the scheduled procedure steps. Some of the information communicated is to be used by the imaging modality scanner itself, whereas much of the rest is to be presented to the operator of the imaging modality scanner. Before the patient's imaging procedure, the MWL enables a modality scanner to get the patient's demographic, procedure, and scheduled examination information automatically and electronically from the RIS, thus reducing redundant patient data entry by the operator. In this case, the imaging modality scanner is the SCU, and the RIS is the SCP. This not only eliminates modality operator input redundant data avoiding possible error caused by manual tasks, but also facilitates image data consistency, which associates images to the correct patient or procedures seamlessly. Right after the modality scanner starts the procedure, the modality scanner sends a service request to the information system, which is the RIS, to inform that the procedure step has been started. This allows the MWL to be updated. After the imaging procedure is successfully completed, the worklist on a PACS workstation indicates the completed status of the imaging procedure ready for physician report. The structure of the MWL is based on scheduled procedure steps, each of which is a unit of service in the context of a requested imaging procedure.

The essential purpose and function of the MWL is to improve the efficiency of the imaging modality operation and workflow. Although patient procedure information is transmitted from RIS to the imaging modality scanner, the MWL SOP is not designed to provide physicians or imaging modality operators access to all other information services that the RIS provides.

6.2.2.5 Modality Performed Procedure Step and Management States

As a complementary service to MWL, this modality performed procedure step service allows the image modality to send a report about a finished patient examination. The report about a performed examination includes information of images acquired, duration of the study, dose delivered, beginning time and end time of the examination, and so forth. This report gives the radiology department a more precise handle on resource use. This service also allows an image modality scanner to coordinate better with image storage servers.

The modality performed procedure step state information is specified by its IOD. The performed procedure step object is only the actual performed part of the scheduled procedure. Each scan event of the performed part is initiated by the imaging modality scanner. Because the scan event cannot be finished by the modality scanner once the performance of a procedure step is started, the event is marked as discontinued if it is terminated or cancelled. Then the modality scanner passes this information to the information system, which is the RIS in the radiology department; hence, the RIS can reschedule or cancel this procedure step. If the procedure is completed successfully, the completed state information is passed to the RIS to indicate that the acquisition of composite SOP instances has been completed successfully, and the SCU has provided all required attribute values for the performed procedure step.

Using the modality performed procedure step, schedules can be updated immediately for better resource management in the department and the billing information can be gathered promptly. In addition, subsequent image postprocessing and reporting are notified immediately for more efficient workflow.

6.2.2.6 Image Hanging Protocol

In today's softcopy reading environment in an imaging department, hanging protocol is a default display protocol that consists of a set of instructions for the layout of a class of images to be displayed on a PACS display monitor. A hanging protocol determines the default layout such as image orientation, order, window, and level. Because a hanging protocol is not a presentation state, it is not specifically tailored to images from a particular patient. The hanging protocol storage service class is an application-level class of service enabling a DICOM AE to send a hanging protocol SOP to another DICOM AE, specifying medical image viewing preferences of a specific user or group, for a specific type of study such as modality and anatomy. DICOM hanging protocol contains information about the hanging protocol, the creator, the type of study it addresses, the type of image sets to display, the intended display environment, and the intended layout for the screen. A PACS workstation can transfer hanging protocol SOP instances to other PACS workstations or archives, or PACS archives can transfer hanging protocol SOP instances to PACS workstations. A DICOM hanging protocol may be exchanged between connecting devices that claim conformance to the DICOM standard. The hanging protocol storage SOP service class uses the hanging protocol IOD, and it requires both the SCU and SCP supporting hanging protocol, storage of the hanging protocol information on medium, and access to the hanging protocol for some time frame. As an example, when images from a patient's exam are to be displayed on a PACS display monitor, potential applicable hanging protocols are all retrieved from the DICOM image storage archive, and the appropriate one suitable for this exam is selected from the list. Then the hanging protocol is applied to the exam, series, or images for display. These displayed images will be laid out as defined by this hanging protocol.

6.2.2.7 Structured Reporting

After successful implementation of DICOM standard for medical image communication and storage, DICOM was then developed into the area of structured documents. Why is there a need to standardize a report in DICOM if HL7 has message exchange feature in a relatively versatile form? The main reason is that at the time DICOM structured report (SR) is developed, most healthcare institutions use HL7 V2.x whereas HL7 V2.x standard does not really allow for a rigorous structure, and there is a need to develop standards for documents incorporating references to images and associated data such as waveforms. Another reason for DICOM SR is that it is well-suited for data mining and can be used for outcome measurement. The advantage of SR is that it allows very effective linking to other DICOM objects such as images, waveforms, measurements, and parts of images or presentations of images.

DICOM SR is an object, which can be exchanged very similar to an image, except that instead of image pixel data, the message body is an SR. The purpose of DICOM SR is to standardize clinical documents in the imaging environment. In a DICOM SR document, imaging-based diagnostic or therapeutic procedure observation report with image references, image measurements and analysis, computer-aided diagnosis results, and procedures logs can be recorded.

Unique features of DICOM SR are that they use DICOM patient/study/series header just like other DICOM images; they are encoded using DICOM standard data elements and they are capable of DICOM query and storage network services. There are many specific DICOM SR document templates defining content constraints for specific types of documents or reports. In addition, there are object classes for DICOM SR, for example:

- Basic text is for text description referencing to images

- Enhanced and comprehensive text is for numerical measurements and region-of-interest references

- CAD is for computer-aided diagnosis results using automated analysis

- Key object selection is to flag images for nonradiology physicians and text note

- Procedure log is for extended duration procedures like vascular imaging

- Radiation dose report is used to report the dose used in the procedure

Complementary to DICOM SR, PDF, and HL7 CDA can be encapsulated. Crucial information such as patient ID and document titles can be extracted, so that DICOM SRs are archived or queried/retrieved on PACS archive just like other DICOM modality images.

DICOM SR integrates very well in the DICOM structure in the imaging environment. When images are linked closely to reports, DICOM SR is used. In addition to radiology and cardiology, DICOM SR can also be used in other extensive image-related medical specialties, such as endoscopy, pathology, ophthalmology, dentistry, and dermatology.

On comparison with HL7 CDA, DICOM SR emphasizes coded semantic content like relation to images, whereas CDA emphasizes human readable text through XML-styled document sheet suitable for EHR. Both DICOM SR and HL7 CDA allow documents such as images and reports for cross-reference. DICOM SR is capable of being coded and transformed into XML-based HL7 CDA document structure, so that it can be distributed as CDA documents enterprisewide. DICOM SR will continue to see expanded use for evidence documents created in the imaging setting, as included in the IHE Evidence Documents Integration Profile. IHE also has profiled coordination use of both standards for improved workflow. In summary, DICOM SR has its advantage in robust image-related semantic content, whereas HL7 CDA has its advantage in human-readable narrative report.

6.2.3 Media Storage for Interchange

This service defined by DICOM PS3.10 provides the capability to manually exchange images and related information like DICOM images and SRs on a removable storage media, based on the DICOM communication model as shown in Figure 6.3. DICOM uses a standard for common file format and image directory. The DICOM media storage service defines a set of operations with media that facilitates storage to and retrieval from the removable storage media. DICOM has two major components: image data file format and network communication protocol. Unlike other image data formats, a DICOM image data file is composed of the image data itself, and header including the patient's demographic information and image acquisition parameters. Figure 6.4 is the model for storage media interchange communication in DICOM. It includes storage of physical medium, media format, and DICOM application message exchange.

The DICOM media storage model emphasizes areas related to data structures, its rules, and services involved for interoperability through removable media interchange. AEs use this media storage model to exchange data on removable storage media. DICOM AE relies on the DICOM file service to access storage media; it is independent of specific file structures and physical media storage formats.

On a removable storage media, DICOM images of a patient are organized so that each image file holds acquired image data; image files with same acquisition protocol are grouped into a series,

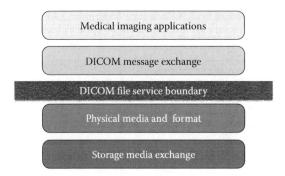

Figure 6.3 General DICOM communication model.

whereas a study groups all series for a given procedure. File names on DICOM storage media is limited to eight alphabetic characters. The extension for a DICOM file is ".dcm". There is a requirement for a media directory file, which is always named DICOMDIR and contains pointers to a list of DICOM files on the removable storage media such as CD-ROM, DVD, MOD, or flash memory. DICOMDIR resides in the root directory of the directory hierarchy. On a single removable storage media, there is only one DICOMDIR file providing detailed information about index and summary for all the DICOM files.

There are several attributes for each DICOM object, including an image attribute containing the image pixel data and attributes for patient ID, names, and so on. For many DICOM objects, the image attribute contains only one image. However, for modalities like nuclear medicine, or cardiac dynamic cine, the image attribute contains a series of image frames, so that multiple images of a movie loop are packaged in a single DICOM object. This also applies to modalities generating three- or four-dimensional image data set. Typically, the image pixel data can be compressed as JPEG or JPEG 2000 images.

DICOM provides a mechanism for supporting the use of both lossless and lossy JPEG imaging encapsulated format. However, DICOM does not specify selection of specific compression parameters such as compression ratio for JPEG lossy compression. DICOM defines a default for lossless JPEG image compression identified by unique transfer syntax unique identifiers (UIDs); it is

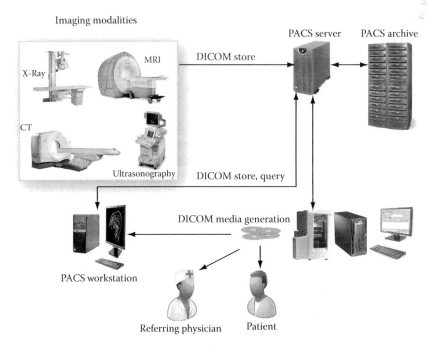

Figure 6.4 DICOM removable storage media interchange communication.

supported by all DICOM implementations that chose to support lossless JPEG compression. DICOM also supports lossy JPEG image compression for 8- or 12-bit images with separate transfer syntax UIDs. JPEG 2000 lossless image compression is also supported by DICOM by a different transfer syntax UID. In addition to JPEG, JPEG Progressive Interactive Protocol (JPIP) is supported for interactive access to large image data files or over slower network links. The main advantage of JPIP is speed. It is not well-suited to transfer complete images with full fidelity.

For images to be stored on or retrieved from a removable storage media, the physical media layer such as CD-ROM or USB flash drive determines physical media form factor, dimension, mechanical characteristics, and recording properties. At the media format layer, digital bit streams from physical media layer are organized into specific data file structures. Data file structures and related hierarchical directory structures allow physical media space management and its efficient access. Typically, this layer is computer operating system dependent. At the DICOM data format layer, elements specified include DICOM media storage SOP classes and associated IOD for medical imaging information exchange, DICOM file format, DICOM media storage directory SOP class, DICOM media storage application profiles, and DICOM security profiles for media storage.

Today, many vendors provide DICOM image viewer software with the CD-R removable media as discussed in Chapter 2, so that when modality scanner images are stored on the media in DICOM format, the viewer software is also stored on the same media. When the CD-R storage is inserted into any non-PACS computer, these DICOM images can be viewed using the built-in viewing software, or can be viewed on a different vendor's PACS workstation capable of viewing DICOM images. Examples of this service use include exchange of images for remote consultation or publication. DICOM CD-R containing patient images are mailed instead of hardcopy film prints for reference or for other specialty physicians requesting these images.

6.2.4 DICOM GSDF Display Function

In Chapter 3, principles of PACS display monitors were discussed in detail. The entire process for a medical image display consists of several steps. First, the image modality scanner makes adjustment to the image when it is acquired. Then a dynamic range with "window" and "level" of the image for display is selected, followed by the LUT outputs, P-values, which become digital driving levels for standardized display systems such as LCD display monitors and dry laser printers/cameras. Finally, the GSDF maps P-values to output the standardized display system, which is an implementation-dependent step.

One of the essential criteria for display monitor quality assurance is compliant to GSDF specified in DICOM PS3.14. A display function defines how grayscale is rendered on display systems such as LCD monitors or dry laser cameras. This is achieved by mapping of the digital driving levels (DDLs) to luminance, which is the luminous intensity per unit area projected in a given direction. DDL is a digital input value to a display system for generating luminance. The set of DDLs of a display system are all possible discrete values capable of generating luminance values on it. The inherent display function of a display system, which maps DDLs to luminance values, is called characteristic curve. It is dependent on operating parameters of the display system and includes the effects of ambient light. The luminance produced by an LCD monitor can be measured with a photo meter as illustrated in Chapter 3.

The DICOM GSDF is based on human observer contrast sensitivity, which characterizes the sensitivity of typical human observers to luminance changes of a standard target. Contrast sensitivity of typical human observers is nonlinear within the luminance range of the DICOM GSDF. The sensitivity of the human eye is less in dark areas of an image than that in bright areas, which results in easier perception of small luminance changes in the bright areas of an image compared with that in the dark areas of an image, according to Barten's model of the human visual system. The purpose of DICOM PS3.14 is to define a mathematical expression of an appropriate GSDF for all image presentation systems, so that P-values are transformed to viewed luminance values by a standardized display system, and the relationship between P-values and typical human observers' perceptual response is approximately linear. This ensures similar grayscale perception and basic appearance for images displayed on display systems with different luminance.

DICOM PS3.14 defines the GSDF of a standardized display system mathematically for grayscale image display. A GSDF is a mathematical definition mapping of input Just Noticeable Difference (JND) indices to luminance values, whereas JND is the luminance difference of an object under specific viewing condition that a typical human viewer can just perceive, and JND index is the value input to the GSDF, so that a single JND step results in a just noticeable luminance difference.

DICOM PS3.14 also provides methods for measurement of display system characteristic curve, which is used for display systems GSDF calibration or conformance. In DICOM, display systems include both softcopy display monitors with associated driving electronics and hardcopy printers generating films reviewed on light box. It should be noted that other than DICOM GSDF for grayscale image display, DICOM does not specify display functions for color image display. However, a color display system such as a color LCD display monitor may be calibrated with the DICOM GSDF for grayscale image display.

The GSDF, including ambient illuminance effects, is defined mathematically with the interpolation of 1023 (10 bits) distinct luminance levels from the Barten's model. The minimum luminance is 0.05 cd/m² corresponding to the lowest practical useful CRT monitor luminance, whereas the maximum luminance is 4000 cd/m² corresponding to possible luminance from the brightest light box for x-ray mammography interpretation. Using this GSDF equation, the luminance value L in the unit of candelas per square meter (cd/m²) can be calculated as a function of the JND index, J, as shown in Equation (6.1), where ln is the natural logarithm, j is the index (from 1 to 1023) of the luminance level Lj of the JNDs, and $a = -1.3011877$, $b = -2.5840191E{-}2$, $c = 8.0242636E{-}2$, $d = -1.0320229E{-}1$, $e = 1.3646699E{-}1$, $f = 2.8745620E{-}2$, $g = -2.5468404E{-}2$, $h = -3.1978977E{-}3$, $k = 1.2992634E{-}4$, $m = 1.3635334E{-}3$.

$$\log_{10} L(j) = \frac{a + c \cdot \ln(j) + e \cdot (\ln(j))^2 + g \cdot (\ln(j))^3 + m \cdot (\ln(j))^4}{1 + b \cdot \ln(j) + d \cdot (\ln(j))^2 + f \cdot (\ln(j))^3 + h \cdot (\ln(j))^4 + k \cdot (\ln(j))^5}. \tag{6.1}$$

The logarithms to the base 10 of the luminance Lj are approximately uniformly interpolated with this function across the full luminance range. A tabular form of the DICOM GSDF is also presented in the DICOM PS3.14. The DICOM GSDF is an essential part for PACS softcopy display monitor luminance calibration/QA discussed in the AAPM-TG-18 report in Chapter 3.

6.2.5 Web Access to DICOM Persistent Objects

DICOM is constantly updating its service classes to meet new clinical needs. WADO defined in DICOM PS3.18 is one of the new developments. There are several functions that DICOM defines in the Web-based environment, which means Internet-related technologies are used for DICOM information exchange. As shown in Figure 6.5, DICOM WADO specifies a Web-based service for accessing and presenting DICOM images and reports. WADO is intended for distribution of images and results to healthcare professionals through the Internet. It provides a simple means for accessing DICOM images and reports from Web pages or XML documents. Data may be retrieved either in a presentation-ready form such as JPEG or GIF as specified by the requester or in a native DICOM format from a DICOM Web server.

A DICOM Web server is a system managing DICOM persistent objects and able to transmit them on request to a Web client system. A Web client is a system using Internet technologies such as Web and e-mail in retrieving DICOM images or reports from a Web-enabled DICOM server. The features supported by WADO is a mechanism that DICOM persistent objects from a DICOM-based server system can be retrieved by Web-based systems, so that images and reports from patients' EHR can be referenced through Web or e-mails, and outside referring physicians can access patients' images

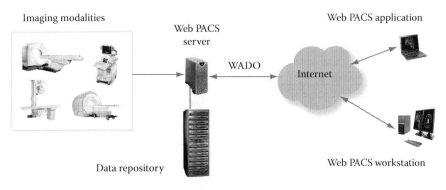

Figure 6.5 Access of DICOM objects through the Web using WADO.

and reports through a hospital Web server. DICOM WADO standard relates only to DICOM images and reports, and does not support or facilities for Web searching of DICOM images.

6.2.6 DICOM Device Setup

Currently, image transfer is the most common connectivity feature supported by DICOM compliant medical devices, so that various image objects such as CT, MRI, ultrasound (US), CR, DR, PET, and nuclear medicine (NM) can be easily transferred from the modality scanner to the PACS or a workstation. DICOM network communication services are used for DICOM AEs such as a PACS workstation or a modality scanner. These services define DICOM data format and the protocol for network communication. Provided by the upper layer protocol for TCP/IP network communications, DICOM network communication services are upper layer services. Using these services, any two network devices are capable of exchanging DICOM files containing images and patient information. Figure 6.6 illustrates TCP/IP stacked protocol that supports DICOM AE communication.

In this layered model for network communication, each layer uses the immediate lower layer's service, while providing a new service to the above layer. The service, which carries results from a protocol operation, plays the role of a mediator between the two layers, whereas a protocol defines communication between two networked devices.

The relationship between a TCP transport connection and an upper layer association is one-to-one based, which means a TCP transport connection supports one and only one upper layer association. The DICOM upper layer protocol is used together with the TCP/IP transport layers. When IPv4 is used for TCP/IP, transmission control program, Internet protocol, Internet control message protocol, and Internet subnet are used. In the case where Ipv6 is used for TCP/IP, IPv6 address allocation management, Internet protocol, and IPv6 specification are used.

6.2.6.1 IP Address

A DICOM network device needs an IP address too. The IP address is used for a network device during communication on the network. The device can be a computer, a switch, a router, a printer, or a server. Some IP addresses are unique in the Internet globally whereas some are only unique locally. IP address was discussed in detail in Chapter 4. Typically DICOM devices or PACS/EHR servers are assigned with static IP addresses. Some clinical PACS (nondiagnostic)/EHR workstations may also employ dynamic IP addresses.

6.2.6.2 Port Number

In a DICOM network setup, a port and AE title need to be set up. A unique TCP port number not only serves as the transport selector when a TCP connection is established; it also identifies a DICOM upper layer entity of a given system in the network. Port numbers of remote DICOM upper layer entities such as imaging devices can be easily configured. For example, "104" is a common port number used by medical imaging devices supporting only one single DICOM upper layer entity.

6.2.6.3 AE Title

Another important parameter for DICOM network set up is the AE title. An AE title uniquely identifies a service or application residing on a specific system. It is independent from where the

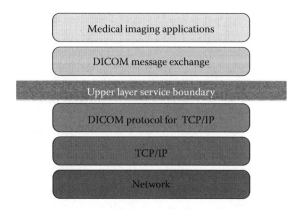

Figure 6.6 TCP/IP stacked protocol for DICOM application entities communication.

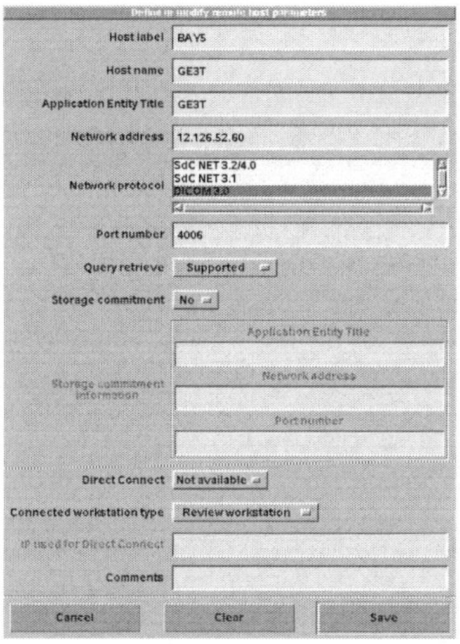

Figure 6.7 DICOM network setup for a GE Advantage Workstation with a GE MRI scanner.

specific system is physically located in the network. AE titles identify AEs in calling or called roles. They are used in the called/calling AE title fields of the upper layer service and in various attributes of the DICOM message data set. A single AE title can be associated with multiple network IP addresses assigned to a single imaging device, or multiple TCP ports with the same or different network addresses. However, a DICOM system in a network can support multiple application processes with various AE titles.

IP address, port number, and AE title are essential parameters for a DICOM device setup. Figure 6.7 illustrates the actual DICOM setup page for a DICOM workstation to communicate with an MRI modality scanner. Typically IP addresses can be obtained from information service department, port numbers are determined by the manufacturer, and AE title is assigned by consulting with the manufacturer. The user responsible for maintaining the PACS needs to document these parameters for all PACS servers, workstations, modality image scanners, and printers.

6.2.7 DICOM Conformance Statements

After introducing some of the commonly used DICOM service classes, the question is how to determine which DICOM services are provided by a specific piece of medical imaging equipment. The answer is to check the DICOM conformance statement. A DICOM conformance statement is a formal document prepared by a medical imaging device manufacturer. It describes how a product from the manufacturer adheres to the DICOM standard. It provides a methodology for implementers, integrators, and end users to assess whether or not two DICOM implementations can interoperate. DICOM conformance ensures that medical imaging devices can work with current or future imaging modalities and related peripheral equipments regardless of vendors. A DICOM conformance statement following a strict format is mandatory for DICOM implementations. Many medical imaging device vendors have their DICOM conformance statements listed on their Web site on the Internet. There are three basic components covered in the conformance statement: network support, physical media support, and IOD descriptions. In the DICOM conformance statement SOPs supported by the product are listed, together with descriptions of the product DICOM implementation details and behaviors.

For example, connectivity between any two pieces of medical imaging devices can be evaluated by the use of their DICOM conformance statements. Buyers from healthcare institutions and vendors are able to determine whether and to what extent two pieces of equipments can connect with each other by comparing their conformance statements. From the DICOM conformance statement one can tell if a new imaging device to be purchased would communicate with existing equipments successfully, and to what extent. It should be noted that DICOM connectivity does not equal application interoperability. DICOM connectivity means that through DICOM message standard for establishing connections, information objects such as MRI, CT, US, or NM images sent from one application is received intact by the receiving application. However, this does not guarantee application interoperability, which means data received is sufficient for receiving application to properly function, such as three-dimensional reformat, surgical planning, radiation therapy (RT) planning, and quantitative analysis. Therefore, DICOM connectivity is necessary, but not sufficient for application interoperability.

It should be noted that the DICOM standard does not provide specific validation or test procedures for evaluation of an imaging device's conformance to DICOM, or whether an implementation matches the device's conformance statement. DICOM conformance is to SOP classes, not to a specific version of the standard. Comparing one product's SCP table with another product's SCU table will reveal if these two products have interoperability in services such as storage, query/retrieve, and print.

During the decision-making process for any new medical imaging device purchase, buyers should fully understand contents of the DICOM conformance statement of the device. If needed, assistance from a third party expert or the vendor should be requested for interpretation of the DICOM conformance statement. Discussing the DICOM conformance statement by customers and vendors together provides understanding for possible problems before purchase contract is signed. For application interoperability, although it might not be possible to verify for advanced applications, it can be determined with the DICOM conformance statement for some simple applications such as three-dimensional reformat.

During system integration, the DICOM conformance statement provides critical information to determine whether applications interoperate. In case any problems occur during system implementation, the DICOM conformance statement is a source of information for potential solutions. The purpose and structure of a DICOM conformance statement provides a framework as how DICOM conformance information should be placed into a conformance statement. In addition, a consistent template for DICOM conformance statements not only benefits those who prepare these conformance statements, but also helps potential buyers of imaging devices understand the meanings of these conformance statements, thus making informed purchasing decisions.

6.2.8 New DICOM Development

With successful DICOM applications in radiology and cardiology since its inception, the list of supported modalities and services in DICOM is constantly expanding. New services are added when there are such needs, and some services are retired when needs no longer exist. Today, other medical fields are increasingly using DICOM for imaging studies in their areas. DICOM is currently being used in many medical specialties such as vascular imaging, NM, ophthalmology, dermatology, dentistry, imaging guided surgery, pathology, veterinary medicine, visible light (endoscopy, photography, microscopy), and RT. For example, RT requires imaging modality images from CT or x-ray to be imported, and other objects such as treatment plans, RT images, and treatment records to be created and stored. DICOM RT is created to meet these demands.

With previously supported imaging modalities, enhanced services of new SOPs are expanding to three-dimensional ultrasound, CAD, x-ray radiation dose report, image registration, and segmentation for image postprocessing. For MRI, new objects are added to accommodate new acquisition techniques such as perfusion, diffusion, cardiac, angiography, spectroscopy, palette color, and raw data. In general, acquisition for imaging modalities is moving toward multidimensional, three-dimensional, or more, and the trend is followed by DICOM. New SOPs also enable easy handling of more than 1000 image studies for multislice CT, multislice/frame/stack MRI, and other modalities by packing all images in one object.

6.2.9 Future of DICOM

DICOM provides connectivity to other information systems such as RIS and HIS as well. The DICOM standard is constantly being updated to keep current with industry innovations. Every year the standard is updated four to five times; on an average, approximately 10 new SOP classes are added to the DICOM. It is republished once a year or two. There are more than 20 active workgroups inside DICOM, and the DICOM standard itself is a product of the DICOM standard committee and many of its own international working groups. DICOM works with other standard development organizations, utilizing relevant parts of other well-established standards. DICOM is an integral part of IHE initiative to help users and equipment providers to develop approaches for integrating medical information and imaging systems. With advances in medical imaging technology, DICOM standard is expected to follow and accommodate new advancements. The latest additions to the DICOM standard can be found at the NEMA Web site.

6.3 INTEGRATING THE HEALTHCARE ENTERPRISE

So far, HL7 and DICOM standards for medical information and image exchange in healthcare environment were discussed separately. In this information age, improving healthcare for patients relies on seamless inter- and intra-enterprise digital healthcare information exchange. It is not uncommon for a healthcare institution to have dozens of clinical information systems, such as admission/discharge/transfer (ADT) system, RIS, cardiovascular information system (CIS), laboratory information system (LIS), PACS, and billing system, each furnishing a specific task. To fully realize the benefit of each information system for patient care improvement, all these systems should be integrated together for more efficient information exchange and communication, so that healthcare professionals and clinicians can make better and informed decisions. Historically, all these systems were designed and manufactured as stand-alone systems to meet specific needs of a specialty area. Various customized interfaces among all these clinical information systems are required if information exchange is desired, which makes it complicated and costly not only to manage these information systems, but also to upgrade or change systems at various time frames. IHE is a concerted effort to bring all HL7, DICOM, IT and any other relevant standards together to improve the entire patient care process. IHE is not a standard, but an initiative producing consensus document.

6.3.1 IHE Purpose

Although healthcare information systems are constantly changing in a fast pace, standards used by the medical information industry have a much longer life span. Fortunately, existing established medical imaging and information standards such as DICOM and HL7 provide a standardized approach for patient clinical information exchange and improved efficient workflow. Using established existing standards as tools, the IHE initiative provides a reliable and convenient methodology to achieve the integration goal. IHE is not a standard, but an initiative producing consensus document. This document is a technical framework for healthcare information systems designed to be interoperable in healthcare environments. IHE realizes the benefits of IT by integrating a patient's clinical data from multiple clinical information systems, improving quality and efficiency for patient care.

6.3.2 IHE History

IHE is an initiative by healthcare professionals and IT industry to improve communications among clinical information systems and equipments. It was formed in 1997 by a group of visionary radiologists and IT professionals. The goal of IHE is to establish a process which will ease integration of various healthcare information systems or equipments, so that information of each system can be freely exchanged. In IHE, actors are defined as information systems or components of information systems that produce, manage, or act on categories of information required by operational activities in the enterprise. An integration profile defines the problem it solves, actors required to build the solution, and transactions they use to interact. It is a representation of a real-world capability. IHE itself is not a data standard, but a process. IHE defines integration profiles utilizing existing data standards, through which various systems and their workflow are integrated. Using established standards such as DICOM and HL7, IHE actors are defined in an integration profile with much detailed transactions they need to perform. Systems from manufacturers supporting IHE integration profiles are easier to implement and communicate better. Using IHE results in optimized use of patient clinical data and enhances quality of care for patient. Currently, IHE is sponsored by the HIMSS, RSNA, the American College of Cardiology (ACC), and there are several hundred products from manufacturers that support one or more IHE profiles.

6.3.3 IHE Domains

As there are many clinical and technical areas in healthcare, currently IHE is organized into several domains. The organization of IHE consists of operational and clinical domains. Each domain is closely coordinated with other domains and has its own so-called technical framework documents. In each domain, clinical and technical experts form a technical and a planning committee. The primary task of the technical committee is to identify healthcare information exchange and integration issues, prioritize them, and address them by developing consensus and standards-based solutions. Each solution is called *IHE Integrating Profile*. Within each domain, there are groups of integration profiles which are specific to that particular domain. The technical and planning committee review and republish technical framework documents on an annual base. Hence, expanding supplements for defining new integration profiles can be added easily. Each new profile is first published for public comment and then revised for trial implementation. Finally, if the testing is successful, the profile is published as final text. The number of operational and clinical domains in IHE is growing, with the current list including:

- Cardiology
- Eye care
- IT infrastructure
- Laboratory
- Pathology
- Patient care coordination
- Patient care devices
- Quality
- Radiation oncology
- Radiology

6.3.4 IHE Process

In a healthcare institution there are many information systems using HL7, DICOM, and other standards. These systems include HIS, RIS, PACS, imaging modality scanners, and many other department information systems. Historically, there is no consensus before IHE on how to use existing standards for information communication in an integrated way. This lack of interoperability hampers automated process, such as physician order entry, patient and examination registration, and report creation and access. The result is inefficient workflow despite information systems deployment. In 1998, RSNA and HIMSS made a joint effort to clearly define the way existing standards should be used for information system integration in radiology. Its primary goal is to provide correct and readily available healthcare information to healthcare professionals beyond radiology whenever and wherever needed. The IHE initiative defines a consensus-based model and framework for seamless healthcare information systems integration. To reach the goal, the IHE process can be divided into three major steps:

1. Under IHE healthcare system users and manufacturers get together to define a real-world task for system integration.

2. IHE defines the technical framework for specific definitions using integration profiles. Each profile is supported by a set of actors that interact through transactions. A profile specifies exactly how existing standards such as HL7 and DICOM are used by healthcare information systems to finish a set of well-defined transactions. These transactions together accomplish one real-world task. The IHE technical framework precisely defines a common information model and human vocabulary, so that HIS can exchange clinical information with this information model, and system users can communicate with manufacturers using the standardized common language.

3. In the annual week-long IHE Connectathon event, manufacturers come together to test interoperability of their products and their compliance to the IHE technical framework. Imaging modality and medical information system manufacturers are among the architects and supporters of the IHE Connectathon. If a product is tested successfully, the manufacturer then prepares and publishes an IHE integration statement to list the IHE profiles supported by its specific release of a particular product.

This is illustrated by Figure 6.8. This whole process is repeated every year, so that constant improvements can be made to improve interoperabilities among various clinical information systems and medical equipments.

Again, IHE is neither a standard nor a certifying authority. Users and manufacturers utilize existing standards (such as HL7 and DICOM) to achieve IHE goals for system integration. The IHE initiative is now extended to meet challenges in medical specialties other than radiology, and standards used by IHE are no longer limited to HL7 and DICOM.

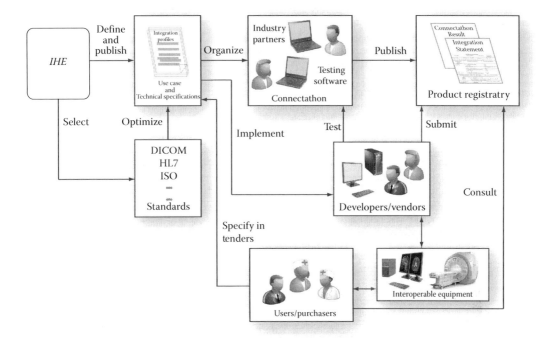

Figure 6.8 The IHE process.

6.3.5 Relation with Standards

IHE is a framework briefing the means of using existing established standards for healthcare systems integration. These systems can be imaging modality scanners, PACS, medical devices, HIS, RIS, or any department information systems. From the early days of IHE, HL7, the DICOM from the ACR and the NEMA are standards employed mostly by healthcare system manufacturers. However, as more and more other healthcare processes can be improved and integrated with the same approach utilized by radiology with IHE, the IHE initiative is expanding into areas well beyond radiology. This inevitably brings more standards under the umbrella of IHE. However, with complex information system integration demands, existing standards such as HL7 and DICOM are also evolving independently to meet new challenges that arise from more complex clinical information systems' integration needs.

6.3.6 Integration Profiles

The IHE is organized into several domains covering various areas in the healthcare environment. Each domain has its own technical framework, which defines profiles for real-world integration solutions. IHE profiles can be divided into four major categories based on their functions in the entire scope of information system integration: (1) infrastructure profiles; (2) content profiles; (3) workflow profiles; and (4) presentation profiles. In the IHE, the technical framework of each domain may include a combination of integration profiles from different categories. IHE profiles are constantly updated, some frequently implemented, and the important ones are described in the later sections of this chapter.

6.3.7 IHE Connectathon

IHE has a process for system manufacturers to test their IHE actors and integration profile implementations. This process is called IHE Connectathon, which allows manufacturers to obtain basic assessment and receive valuable feedback for their products' implementation conformance. IHE has been organizing the Connectathon for more than a decade. It is now an annual week-long event held in several places around the world for testing interoperability of healthcare information systems.

At the IHE Connectathon, trained technical experts supervise testing of manufacturers' products in a neutral environment. Advanced testing software developed by IHE and its partner organizations are utilized to test the conformance of IHE technical framework claimed by manufacturers for their products. During the event, healthcare system manufacturers test their IHE integration profiles, integrated products with other vendors, and products equipped with same profiles. A product from a vendor fully tests the roles that it selects to play, verify all IHE integration profiles it implements, and exchanges information and performs transactions with multiple systems from other vendors. Manufacturers can also fine tune their products based on interactions and feedbacks received during the IHE Connectathon, to achieve interoperability of their products if not yet achieved. However, the IHE Connectathon is not for independent evaluation and for ensuring products' compliance to IHE technical frameworks.

6.3.8 IHE Integration Statements

After a vendor successfully tests the systems' interoperability with other vendors' products at the annual IHE Connectathon, the vendor prepares and publishes IHE integration statements to list the IHE profiles supported by the specific release of a particular product. The product is confirmed for conformance with the IHE technical framework. Specific IHE capabilities of the product designed to support are identified in the statement. A statement describes which IHE integration profiles are supported by the product, and which IHE actor roles does the system play in these profiles. Integration statements are to be used together with conformance statements to specific standards such as HL7 and DICOM. Standards conformance is a prerequisite for vendors adopting IHE integration profiles.

Users from healthcare institutions or system integrators can study and understand the conformance statement. This helps to determine the level of integrations that vendors put in their products, and workflow improvements associated with these integrations. Vendors publishing statements are responsible for the accuracy and validity of their specific IHE integration statements. IHE and its sponsoring organizations are not the authority for testing, evaluating, or certifying these products.

II IE itself is not a standard, but a framework for clearly defined use of established existing standards. Whenever necessary, IHE refers to the relevant standards. Therefore, it is not appropriate to claim conformance to IHE. However, conformance need to be made with reference to specific standards, stating that the product is implemented in accordance with the IHE technical framework or in compliance with the IHE technical framework, and specifying which actors' roles the product plays.

The IHE initiative and integration profiles should be known to any buyer who makes healthcare information system purchasing decisions at a healthcare institution. In addition to specifying conformance to standards such as HL7 and DICOM, the buyer can request compliance with the IHE technical framework. This significantly reduces effort and cost during system integration, and realizes workflow efficiency improvement. Compliance with HL7, DICOM, and IHE technical framework should be an integral part of the system for purchase, not an option to be purchased.

6.4 IHE TECHNICAL FRAMEWORKS

The previous section described the entire IHE process. This section emphasizes on some of the important details of IHE. IHE profiles are compiled into IHE technical frameworks, which are detailed technical documents that serve as implementation guides and are available online. IHE profiles put real integration issues and their solutions into models. Established existing standards such as HL7 and DICOM are used to support IHE profiles. In IHE, an actor is defined as a subsystem or the system itself, which acts, manipulates, or manages healthcare data; whereas a transaction is defined as an interaction among actors for healthcare information exchange and communication. IHE technical frameworks define integration profiles and associated transactions. Systems such as HIS, RIS, PACS, or imaging modalities as actors play different roles in performing transactions. Equipment manufacturers and users can determine the role played by a system in a specific department environment.

IHE technical frameworks of various domains are all process-oriented. They are consensus-based documents of solutions to deal with integration issues using existing established standards. However, a technical framework requires users of a healthcare institution to modify their workflow to match with the one defined in the framework. Otherwise benefits of the technical framework are not fully realized, and users still need customized integration.

It is challenging to integrate various systems and software applications in the healthcare environment. It requires clear communication between healthcare institutions as the user and vendors as the provider. The IHE technical framework provides well-defined set of workflow patterns and appropriate elements as the common language for such communication. If both healthcare institutions and vendors agree on definitions and workflow patterns defined in the framework, most specifications and possible questions and answers for integration are already clearly documented for both parties.

6.4.1 IHE Integration Profiles

IHE integration profiles are the functional units of the IHE initiative providing precise descriptions for communication and security standards implementation to solve clinical issues. These profiles are based on established standards for intra- and interenterprise information exchange. IHE integration profiles provide solutions for interoperability issues in the areas, such as clinical workflow, patient information security, administration, information infrastructure, and information access for healthcare providers and patients. In each profile, a solution for a clinical use case is provided by using appropriate standards and defining actors, transactions, and information content involved. IHE profiles define unambiguous description for both healthcare system buyers and manufacturers to communicate the integration capabilities of information systems. For a manufacturer, an IHE profile provides a consistent clear implementation path and guide for utilizing standards for their equipments in a way carefully documented, reviewed, and tested by other industry partners. For healthcare professionals as the buyer, an IHE profile provides standardized language for requesting integration requirements, and a convenient and reliable methodology of defining a level of compliance to established standards, so that interoperability can be achieved efficiently in acquiring new systems or upgrading current systems with reduced complexity and interfacing cost.

6.4.2 IHE Radiology

Before IHE Radiology was formed in 1998, there was no agreed upon methods for various systems used in radiology, such as HIS, RIS, imaging modalities, PACS, and printers to work together for typical patient care. IHE Radiology was established to address interoperability and information exchange issues impacting patient care in radiology. These radiology profiles are standards-based solutions for radiology-related important, common, and core processes. There are many commercial radiology-related imaging and information systems compliant to these profiles and implemented across the world.

In the radiology department, delivering services to patients requires accomplishing a number of tasks. IHE radiology technical framework defines a methodology of creating, managing, and exchanging elements for successful accomplishment of these tasks. Integration profiles specified in the radiology framework make use of existing HL7 and DICOM standards in a realistic radiology environment. IHE has processes in a fully defined manner. Therefore, different manufacturers of information systems have a consensus-based, mutually agreed, precisely defined common context to implement their products for successful radiology process performance.

Currently, IHE Radiology technical framework defines 16 integration profiles. Each of these profiles consists of a group of actors and transactions to accomplish a typical workflow task in the radiology department. The following is a list of these profiles, and a few core profiles are discussed in following sections to illustrate their processes:

- Radiology Scheduled Workflow (SWF)

- Patient Information Reconciliation (PIR)

- Consistent Presentation of Images (CPI)

- Presentation of Grouped Procedures (PGP)

- Access to Radiology Information (ARI)

- Key Image Note (KIN)

- Simple Image and Numeric Report (SINR)

- Charge Posting (CHG)

- Postprocessing Workflow (PWF)

- Reporting Workflow (RWF)

- Evidence Documents (ED)

- Portable Data for Imaging (PDI)

- NM Image

- Cross-Enterprise Document Sharing for Imaging (XDS-I)

- Mammography Image

- Import Reconciliation Workflow (IRWF)

6.4.2.1 Scheduled Radiology Workflow

The IHE SWF integration profile is fundamental to the IHE model. It is for the radiology department to integrate ordering, scheduling, image acquisition, storage, and viewing for radiology exams. The SWF combines HL7-defined order and ADT messages, together with DICOM-defined scheduling, worklist, status notification, storage commitment, and other services. The profile defines precisely those transactions needed to accomplish typical operations in radiology.

Figure 6.9 illustrates the process flow. The profile starts with the patient registration actor of ADT registering a patient.

The MWL of DICOM is a key component of the profile. It sends patient demographics and procedure information to the imaging modality automatically, and let the imaging modality operator retrieve a list of scheduled procedure steps to be performed. It also puts patient and procedure information in image headers, making it convenient for radiologists to display and compare current and previous studies from a patient. The increased information accuracy compared with manual input at imaging modality scanners results in fewer study repeated and lost, hence improving patient care and operation efficiency. The SWF integration profile uses nine actors and 51 transactions for collaboration among various information systems and devices used in the radiology department. It provides all needed details for performing the typical radiology workflow. Without IHE SWF, interfacing each system is a customized process characterized by its associated complexity and cost. The SWF is the foundation for the IHE initiative. It is essential for achieving significant workflow benefits that IHE integration implementation can promise. The SWF is the first profile developed in 1998, remains to be the fundamental part for the radiology technical framework, and serves as the model for other IHE domains.

6.4.2.2 Postprocessing Workflow

With the advancement in imaging technology, many applications for modality image postprocessing are used in imaging departments. The PWF integration profile has clear description for

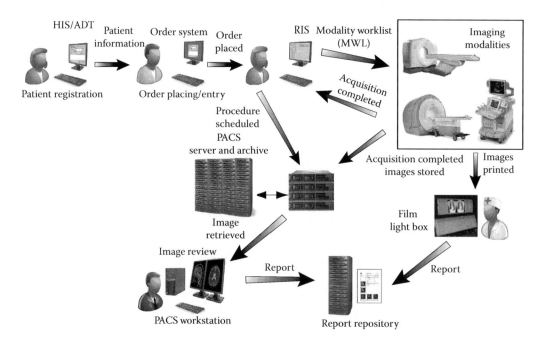

Figure 6.9 Process of radiology scheduled workflow.

mechanisms to automate the distributed image postprocessing. Currently, image postprocessing such as three-dimensional reconstruction, surface rendering, or CAD are not integrated seamlessly into the radiology department workflow. They are often implemented on stand-alone workstations, and data sets need to be manually stored or retrieved on PACS or local workstations. The workflow involves radiologists, technologists, referring physicians, the stand-alone workstation, and it is not efficient. By using PWF, worklist status and result tracking are provided for image postprocessing.

6.4.2.3 Patient Information Reconciliation

In an imaging department, it is not unusual to be in a situation that the patient is unknown to the healthcare institution and yet the imaging procedure needs to proceed. The PIR integration profile supports this type of clinical situations. The PIR profile complements the SWF profile by dealing with clinical situations in which information system is not available for SWF process, or patient information input is incorrect. Without IHE, these errors are hard to be caught, and need significant manual work to detect and correct.

6.4.2.4 Consistent Presentation of Images

For healthcare providers it is crucial to have images presented consistently. It has great clinical significance for image characters to be presented and perceived as equally as possible under very different presentation environments. The CPI integration profile ensures that images are displayed and perceived as closely as possible on different display devices both in the soft copy reading environment and on films displayed using light box.

As indicated by DICOM part 3.14 which is the core for this profile, consistent presentation for image is a complicated task. Before the DICOM grayscale softcopy presentation state and IHE CPI integration profile is available, there is no method to record and exchange image presentation information among various display devices. The CPI profile defines precisely how to use the DICOM GSDF and the DICOM Grayscale Softcopy Presentation State. It ensures that images are displayed and perceived by human viewers as similar as possible on various display devices such as LCD monitors and films from laser printer. The CPI profile impacts the radiology department workflow significantly with improved efficiency. It also plays a key role in the following discussed presentation of the PGP profile.

6.4.2.5 Presentation of Grouped Procedures

Many a time, a patient comes to the radiology department with several imaging procedures of a single imaging modality. To improve the patient's comfort, operational efficiency of the workflow, reducing patient dose and acquisition time, and multiple orders for a single patient can be grouped into one acquisition. However, when the single acquisition is completed, it is necessary to be able to distinguish them as separate ordered procedures for reporting and billing purposes. This PGP integration profile is used for multiple imaging procedure orders completed in a single acquisition. Using this integration profile, the imaging modality operator can perform several procedures and acquire them in a single exam, and store images to the image archive just as any other stored DICOM images. Then the operator or technologist manipulates acquired images according to separately ordered procedure needs, generating one or more grayscale soft copy presentation states associated with subsets of images. Each subset of images are thus properly grouped together and presented depending on the requested procedure, ready for reading and reporting. The PGP profile provides presentation states pointing to corresponding images with correct window and level adjustment for display, saving radiologists' time during reading and reporting. Because the profile provides the modality with separate accession number and report for each individual grouped procedure, lost revenue is avoided.

6.4.2.6 Access to Radiology Information

In today's healthcare institutions, although RIS are used by many radiology departments for efficient workflow, there are still many departments without an RIS in place. This IHE access to ARI integration profile provides radiology information to users without an RIS by making use of the DICOM SR. The profile provides access to radiology images, reports, and presentations. As discussed earlier in this chapter, DICOM SR is a relatively complex part of the DICOM standard, addressing most radiology reporting needs. This profile uses subsets of DICOM SR. The ARI profile defines report creator, report reader, report manager, report repository, and external report repository access actors. The report creator actor sends a radiology report in the form of DICOM

SR to the report manager who maintains the report state. The report manager sends the final report to the report repository, which permanently stores the DICOM SR reports. When query/retrieval request is received, the report manager provides the report. The external report repository access actor functions as a gateway, translating external reports from other information systems for query/retrieval requests with DICOM SR. The ARI also defines details for images query/retrieval by image display actors.

6.4.2.7 Key Image Notes

The KIN integration profile is for radiologists or technologists to flag images as important and attach notes to them when they dictate reports or perform procedures. The profile uses the key object selection document from DICOM. A user of the image acquisition modality can generate flagged images as an actor and send them to an image archive such as PACS for storage as part of the procedure. A workstation as an image display actor can later retrieve them to be displayed. A KIN can be related to more than one image, and several image notes can be associated with one image. Using this profile, clinically significant findings from a large amount of original modality images such as from a multislice CT scan can be conveniently communicated by radiologists to other specialty clinicians. Clinical, technical, or QA- and QC-related incidents during a radiology procedure can also be easily documented with KINs.

6.4.2.8 Simple Image and Numeric Reports

In a radiology department, the reporting process is complex. To address most of the needs from a typical radiology reporting process, a SINR integration profile is created. The IHE defines this profile to create, exchange, store, and display reports based on DICOM SR and content mapping resource templates, which specify the structure of the documents to be used. The basic image diagnostic report template used in IHE requires that a simple image and numeric report must include a title and a section or sections, with each section containing a title and report text, referenced images, measurement data, and coded entries. In addition, further image references and measurements can also be documented in the report text and coded entries.

6.4.2.9 Charge Posting

The CHG integration profile provides timely procedure detailed information from modalities where procedures are done to billing systems. The profile uses standardized messages and real-world assumption model adopted from HL7, so that financial information pertaining to the patient is stored as accounts properties. The assumption used by CHG is that at any time each patient may have more than one active account, and an account may have financial data related to more than one patient visit, but one patient visit cannot associate to several accounts. Following with local rules and regulations, each individual healthcare institution may decide when to get each type of charges posted. The requested procedure is for radiologists' billable work, and the technical component may be posted at a different phase of the entire workflow process.

The CPG integration profile provides the healthcare institution's billing system with details of the radiology workflow information directly and precisely from the department information system. The profile allows accurate and timely department charges to be posted, while not exposing the complex department workflow processes to the billing system.

6.4.3 Relevance of IHE to the User

After much discussion about the IHE, a user may ask how IHE is relevant to the user's work. The user can read through the IHE User's Handbook and the IHE Integration Statement. They provide practical guidance on purchasing or upgrading new equipment.

6.4.3.1 IHE User's Handbooks

For healthcare institution information system buyers or implementers, IHE handbooks are valuable tools for selecting, specifying, purchasing, and deploying healthcare information systems with advanced integration capabilities. Currently the *IHE Radiology User's Handbook* and the *IHE Radiology: Mammography User's Handbook* are available for download from IHE's Web sites. Specific topics, such as practical planning before purchase of a modality including digital mammography, highly integrated PACS, upgrading RIS, issues during their implementation, and verifying their functionality with tests after systems installations are all covered. Other IHE domains will be addressed by future editions.

6.4.3.2 How to Understand IHE Integration Statement

IHE Integration statements are documents prepared and published by vendors to describe intended conformance of their products with the IHE Technical Framework. In these integration statements-specific IHE capabilities of a product are identified in terms of Actors and Integration Profiles, which are key concepts of IHE. Users can use vendor's Integration Statement to determine the level of integration the vendor's claims for their product, and the clinical and operational benefits such integration might provide. Conformance to specific standards such as HL7 and DICOM should always be part of an Integration Statement.

IHE provides a specific template for vendors to use for their integration statement. Table 6.1 is a typical IHE Integration Statement in a format suggested by IHE requirements, where in the table, according to IHE:

1. Vendor name

2. Product name

3. Product release version

4. The IHE Integration Statement publication date

5. The IHE required standard statement

6. A list of all supported specific IHE Integration Profiles for this particular product

7. A list of all Actors supported in each Profile listed in (6). All Profiles, Actors, and Options use names defined by IHE

8. The Internet address where this particular IHE Integration Statement is listed

9. The Internet address where the product's conformance statements to specific standards such as HL7 or DICOM are listed and

10. A list of Internet addresses for general IHE information

Among all items listed above, items (6) and (7) are core components. By comparing these two items of two Integration Statements from two products intended for integration, if the same profile(s) and actor(s) are listed in both IHE Integration Statements, IHE capabilities

Table 6.1: IHE Integration Statement

1. Vendor	2. Product name	3. Version	4. Date
ABC Medical System	Integrate ABC Medical	V3.0	10 Feb 2010
5. This product implements all transactions required in the IHE Technical Framework to support the IHE Integration Profiles, Actors and Options listed below:			
6. Integration profiles implemented	**7. Actors implemented**	**Options implemented**	
Scheduled workflow	Acquisition modality	Patient based worklist query	
		Broad worklist query	
		Assisted acquisition protocol setting	
Patient information reconciliation	Acquisition modality	none	
8. Internet address for ABC Med IHE information: www. ABCmedsysco.com/ihe			
Links to standards conformance statements for the implementation			
9. HL7	Not applicable		
DICOM	www. ABCmedsysco.com/dicom/ Integrate ABC Med Info.pdf		
10. Links to general information on IHE			
In North America: www.ihe.net, www.rsna.org/ihe, www.himss.org/ihe	In Europe: www.ihe-europe.org	In Japan: www.jira-net.or.jp/ihe-j	

supported by the profile(s) are confirmed for both products and they can be integrated without customization.

IHE is relatively new and is under constant update, therefore, there is still a long way to go for the complete integration of all healthcare information systems.

BIBLIOGRAPHY

Alschuler, L. 2008. The HL7 clinical document architecture. Healthcare Information and Management Systems Society 2008 *Annual Meeting*. Orlando, FL.

Beeler, G. W., S. Huff, and W. Rishel. 1999. Health Level Seven. Message Development Framework.

Clinical Context Object Workgroup. 1998. CCOW white paper.

Clunie, D. 2009. *IHE Radiology Technical Framework Supplement: Basic Image Review (BIR)*. www.ihe. net/Technical_Framework/upload/IHE-RAD_TF_Suppl_Basic_Image_Review_2009-06-21.pdf. Accessed June 5, 2010.

Department of Health & Human Services, Office of the National Coordinator for Health Information Technology. 2008. *The ONC-Coordinated Federal Health Information Technology Strategic Plan: 2008–2012*.

DICOM Standard. ftp://medical.nema.org/medical/dicom/2009/. Accessed January 20, 2010.

DICOM Standards Committee. 2008. Working Group 20. *DICOM Supplement 135: SR Diagnostic Imaging Report Transformation Guide*.

DICOM Standards Committee. 2009. Working Group 27. *Web Technologies for DICOM. DICOM Supplement 148: Web Access to DICOM Persistent Objects by Means of Web Services*. Extension of the Retrieve Service (WADO Web Service).

Digital Imaging and Communications in Medicine. 2009. *DICOM Strategic Document, Version 9.5*. medical.nema.org/dicom/geninfo/Strategy.pdf. Accessed January 17, 2010.

Dreyer, K. J., D. S. Hirschorn, and J. H. Thrall. 2006. *PACS: A Guide to the Digital Revolution*. Springer, New York, NY.

EMR Experts. 2009. *Electronic Medical Record eBook*. www.emrexperts.com/emr-ebook/index.php. Accessed January 17, 2010.

Frisse, M. *Lessons Learned from State and RHIOs: Organizational, Technical and Financial Aspects*. www. himss.org/content/files/LessonsLearnedfromStateandRHIOs.pdf. Accessed January 17, 2010.

Healthcare Information and Management Systems Society (HIMSS). 2009. *HIMSS Healthcare Information Exchange National/International Technology Guide White Paper*.

Health Level Seven. Beeler, G. W. *HL7's Version 3 Standard: The Essence of Model-Driven Standards*. www.hl7.org/events/himss/presententions.cfm. Accessed June 5, 2010.

Health Level Seven CCOW Technical Committee. 2006. *HL7 Context Management "CCOW" Standard: Best Practices and Common Mistakes, Version 1.0*. www.hl7.org/implement/standards/ReferenceMaterials/hl7_ccow_best_practices_1_0.pdf. Accessed June 5, 2010.

Health Level Seven. 1998. *Health Level Seven Implementation Support Guide, Version 2.3*.

Health Level Seven. 2004. *HL7 Development Framework: HDF*. healthinfo.med.dal.ca/HL7intro/HDF. html. Accessed June 5, 2010.

Health Level Seven. Weaver, E. 2006. *An Overview of the HL7 Context Management Standard ("CCOW")*.

HL7 message examples: Version 2 and version 3. www.ringholm.de/en/whitepapers.htm. Accessed June 5, 2010.

Horii, S. C. *A Nontechnical Introduction to DICOM*. www.rsna.org/technology/dicom/intro/index.cfm. Accessed January 17, 2010.

Huang, H. K. 2004. *PACS and Imaging Informatics: Basic Principles and Applications*. Wiley-Liss, Hoboken, NJ.

IHE-Radiology Technical Committee. 2004. IHE radiology technical framework 2004–2005: Department workflow. www.ihe.net/Technical_Framework/upload/ihe_tf_whitepaper_DWF_TI_draft_2004-04-15.pdf.

Integrating the Healthcare Enterprise. 2005. *IHE Radiology User's Handbook*. www.ihe.net/Resources/upload/ihe_radiology_users_handbook_2005edition.pdf

Integrating the Healthcare Enterprise. 2007. *IHE Technical Framework Vol. I: Integration Profiles*. www.ihe.net/Technical_Framework/upload/ihe_tf_rev8.pdf. Accessed June 5, 2010.

Integrating the Healthcare Enterprise. 2007. IHE NA 2008 connectathon fact sheet.

Integrating the Healthcare Enterprise. 2009. IHE quality, Research and public health White paper 2008–2009—Quality measurements data element structured for EHR extraction. www.ihe.net/Technical_Framework/upload/IHE_QRPH_White_Paper_Performance_Data_Element_Structured_for_EHR_Extraction_2008-06-10.pdf.

Joint DICOM Standards Committee/ISO TC215 Ad Hoc Working Group on WADO. 2004. DICOM *Supplement 85: Web Access to DICOM Persistent Objects (WADO)*.

Langer, S. 2008. *The Society for Imaging Informatics in Medicine 2008 Annual Meeting*. PACS/RIS/DICOM (The Traditional Informatics Tools), Seattle, WA.

Leavitt, M., A. Ray. 2010. EHR certification town hall. *HIMSS 2010 Annual Meeting*. Atlanta, GA.

Marshall, G. F. 2008. IHE: A proven process for HL7 Standards implementation. *Healthcare Information and Management Systems Society 2008 Annual Meeting*. Orlando, FL.

Myer, C. and J. Quinn. 2008. HL7: An overview. *Healthcare Information and Management Systems Society 2008 Annual Meeting*. Orlando, FL.

Painter, J. and A. Kirnak. 2009. IHE IT infrastructure white paper: A service-oriented architecture (SOA) view of IHE profiles. www.ihe.net/Technical_Framework/upload/IHE_ITI_TF_WhitePaper_A-Service-Oriented-Architecture_SOA_2009-09-28.pdf. Accessed June 5, 2010.

Siegel, E. L. and D. Channin. 2001. Integrating the healthcare enterprise: A primer. *Radiographics* 21: 1343–1350.

Simon, D. 2005. Basic DICOM concepts and healthcare workflow. *DICOM Conference*. Budapest, Hungary.

Solomon, H. 2008. Electronic reports: HL7 CDA (clinical document architecture) and DICOM SR (structured reporting) for advanced reporting. *RSNA 2008 Annual Meeting*. Chicago, IL.

Weddle, S. 2007. Cardiology workflow considerations for non-cardiology PACS professionals—SIIM news summer 2007. www.scarnet.org/index.cfm?id=3686. Accessed January 17, 2010.

Weisfeiler, L. 2006. DICOM supplement 106: JPEG 2000 interactive protocol. *SPIE Med. Imaging*. San Diego, CA.

7 Teleradiology

Teleradiology, one of the special services of telemedicine, is the electronic transmission of radiological images from one location to another for interpretation and/or consultation. A teleradiology service usually consists of an image-acquisition party and an image-interpretation/diagnosis party connected by an image communication system. A PACS, which allows storage and archiving as well as transmission of digital images within a healthcare enterprise, is often attached at the interpretation side or at both sides. Teleradiology is a method of providing radiology services at places that are geographically apart from where the radiologists are located. It can provide access to specialists' expertise that may not be available where the patient is located, which in some sense brings the experts to the patients. Today and in the future, teleradiology and telemedicine in general will attract patients from around the world toward expertise without geographic constraints.

7.1 TELERADIOLOGY OVERVIEW

By strict definition, people had been doing teleradiology locally for years even before the broad implementation of PACS and the Internet, although most of those "teleradiology" services were done at different buildings or rooms within a healthcare institution. For instance, within the boundary of a healthcare institution, imaging studies may be performed at multiple locations that are geographically apart, and multiple reading rooms may be scattered around the institution. Radiologists routinely use their institution's RIS to access and review images acquired at different locations, make diagnoses, and generate reports into it. Essentially, they are performing teleradiology services within the boundaries of a single healthcare enterprise.

In the past few years, the transmission of images between healthcare institutions has begun to integrate with PACS, which has traditionally been handling image acquisition, management, and transmission within a single enterprise. Today, the lines between healthcare institutions are blurred as these institutions begin to establish links to their affiliated hospitals and clinics.

In this chapter, teleradiology refers to applications and services rendered at different healthcare institutions that are geographically apart from each other. Teleradiology service providers are generally those groups or institutions receiving images and providing image interpretation and diagnosis. Teleradiology service customers are those who send images to the providers for image interpretation and diagnosis. Usually, the teleradiology providers are individuals or institutions possessing the expertise and proper credentials, whereas the customers are the ones who choose to send some of their workloads to the providers for various reasons. Teleradiology providers and customers may or may not belong to the same healthcare enterprise.

7.1.1 History of Telemedicine and Teleradiology

The history of telemedicine would not be complete without mentioning the comprehensive reviews assessing telemedicine by the Institute of Medicine Committee on Evaluating Clinical Applications of Telemedicine (IMC) in 1996. Since the 1960s, clinicians and researchers in healthcare have explored the use of advanced technologies to improve healthcare delivery and efficiency. Telemedicine from its beginning has combined technical advances in electronic information and communication technologies with healthcare delivery. These technologies provide support for healthcare when participants are separated geographically. One of the oldest, longest-lasting, but often overlooked telemedicine applications is the 911 emergency call center, which uses telephones to direct emergency healthcare.

As noted above, radiologists had been performing teleradiology locally for years before the broad implementation of PACS and Internet, although most of those services were restricted to within a healthcare institution. An American College of Radiology (ACR) survey in 1990 of about 1000 practices showed that nearly 71% had some teleradiology systems in place, although only a small fraction of their image interpretation was done via teleradiology; at the same time, 30% of the solo practices had used some sort of remote or off-site reading. Over the past two decades, the explosive expansion and broad adoption of Internet and computer network technologies in every aspect of society has accelerated the adoption of teleradiology. The fast growth of teleradiology, particularly medical imaging, has been fueled by advances such as the transition from analog film to digital imaging, the standardization of medical imaging, and the broad adoption of computer networks. Widespread use of the World Wide Web by the general public has made transmitting image data over the network affordable. Nowadays, teleradiology and remote reading have become part of the day-to-day routine for most radiology practices nationwide.

7.1.1.1 Early Development

Telemedicine in its earliest form began with the adoption of the telephone when some patient consultations were done via telephone calls instead of face-to-face house calls.

In the late 1920s, in what may be the earliest teleradiology attempt on record, there was an attempt to send dental x-rays via telegraph to a different location. In late 1950s and early 1960s, telefluoroscopic examinations were first transmitted using coaxial cables in Canada, and direct microwave video links were established between Massachusetts General Hospital and the outpatient clinic at Boston's Logan International Airport. These probably are the earliest modern teleradiology applications on record. Other teleradiology projects from the 1960s to the 1980s were done mostly in the United States and primarily used broadcast or closed-circuit television technologies for image transmission and communication. Most of these projects were sponsored by government grants, and because of the limited spatial and contrast resolution of television technologies, and because the economic model of such projects was not sustainable, most of the projects stopped when government funding ran out. However, closed-circuit television technologies continue to be used for education and conferencing purposes today.

As computer technologies developed rapidly in the late 1970s and 1980s, teleradiology and telemedicine efforts in general turned their attention to computer-based (thus digital) technologies. Images and other medical data could be captured and stored digitally and then sent forward for interpretation or further processing. Unlike television communication, these "store-and-send" systems do not require all parties involved in telemedicine and teleradiology to be present simultaneously. This change has minimized the need for real-time face-to-face interaction and has led to more efficient use of specialists' time and effort.

The National Aeronautics and Space Administration (NASA) and the U.S. military have been pioneers in the use of advanced technologies to improve telemedicine. Several spin-off projects derived from lessons learned from the space program were tested in the 1980s to provide telemedicine care to underserved areas such as Indian Reservations. These government-backed projects pushed forward the technologies needed for telemedicine and teleradiology, including the technology needed for PACS.

Even though there were some commercial ventures in telemedicine and teleradiology in the 1980s, none was sustainable due to many technical and economic reasons. On the technical side, the digital technology at the time was cumbersome to use and had limited spatial resolution. This plus the high cost of high-performance computers and high-speed networks hindered the acceptance of teleradiology and telemedicine in the marketplace. For instance, because most medical imaging modalities used at the time were not digital, medical images often had to be first captured on film and then digitized before being sent over a computer network. These steps reduced the image quality and were difficult to use. On the economic side, these early teleradiology efforts were too expensive and the demand for remote diagnosis was not high enough to justify their use.

The early 1990s was a period of rapid growth in teleradiology efforts. Many factors contributed to this growth, but two of the most important driving forces came from outside the healthcare field. First, rapid advances in digital communication technologies were taking place. Information transmission services, such as telephone calls, telegrams, image and document transfer, and television programming all began using a digital format. As a result, digital communication became faster and costs dropped dramatically. Telemedicine, because of its high demand for speed and bandwidth, attracted venture capital and organizations hoping to tap into this potentially huge market for data transmission. Second, as the world grew "smaller," there was an increasing demand worldwide for better access to high-quality medical care and expertise, and telemedicine had the promise of improving medical care in rural and underserved areas.

Since managed care had created strong competition among healthcare providers, telemedicine services could potentially be used to increase revenue by winning more healthcare contracts, reducing economic and medical risks, and providing lower-cost specialty services.

7.1.1.2 Financial Model of Early Telemedicine Projects

Historically, improving access to specialists' expertise has always been the primary driving force behind telemedicine applications. Nearly all the early telemedicine projects were funded by government research grants and most were focused on providing specialists' expertise to scattered populations in large, remote areas where access to specialists was rare. Unfortunately, since these projects were funded by grants from the U.S. government, they failed mainly for economic reasons after the grants ran out. In those early days, the costs of communication and equipment were too high, and the business model of telemedicine projects was not economically viable. Those teleradiology projects could not justify themselves on a cost-benefit basis, and could not be sustained long term without government subsidy. Additionally, limited acceptance by physicians resulting from poor image quality and difficulty of use may also have played a role.

7.1.1.3 Technical Challenges in Early Days of Teleradiology

In addition to financial issues, all telemedicine projects in the early days had to overcome many technical obstacles including a lack of standards, which caused incompatibilities among medical imaging equipment. This in turn made exchanging of information very difficult and costly because customized solutions had to be implemented. The inflexibility of early telemedicine systems made them cumbersome to use, further reducing their acceptance among healthcare providers. Some major challenges of early telemedicine initiatives were:

- The rapid advancement of information and telecommunication technologies often resulted in the quick obsolescence of key hardware and software components for telemedicine applications, causing a telemedicine system under development to move from state-of-the-art to outdated.

- Complex and often unwieldy communication infrastructure made the transmission of electronic medical data very expensive and economically unviable without government subsidy.

- Lack of standardization for medical imaging and electronic records in general resulted in a dizzying array of telemedicine technologies, making integration between different vendors' equipment impossible or cost-prohibitive.

- Because of incompatibilities and lack of standardization, telemedicine demanded an unusual level of cooperation from users, which made independent institutions and individuals reluctant to participate in telemedicine initiatives. The cost of telemedicine hardware, long distance communication, and software were prohibitively high.

Since the late 1990s, three major developments in technology as well as regulation have occurred: the successful and broad adoption of standards such as DICOM and HL7 discussed in Chapter 6 in medicine (especially medical imaging), the dramatic cost reduction of data transfer via the Internet, and HIPAA (discussed in Chapter 12). These changes have introduced another wave of teleradiology applications.

7.1.2 Scope of Teleradiology

Teleradiology usually involves the transmission of radiological images (x-ray, MRI, CT, etc.) of patients via electronic means for the purpose of interpretation, diagnosis, and consultation by radiologist experts at locations distant from where the images are acquired. For instance, radiological imaging studies are acquired at an outpatient imaging center, a physician's office, or a small regional hospital, but there is no radiologist on site. Therefore, patients may have to carry the images or have them sent to a large hospital where specialists are available for the final diagnosis and consultation. It may take days, if not weeks, before the final diagnosis becomes available. This not only causes anxiety for the patient, but also may delay treatment and potentially exacerbate the medical condition. With teleradiology, images can be sent electronically in minutes and the diagnosis may be available in hours if not faster. Depending on the technology used, patients and their primary physicians at local hospitals may have face-to-face consultations with specialists located far away, greatly enhancing the efficiency of patient care and reducing the cost at the same time.

7.1.2.1 Driving Forces for Teleradiology Development

Several fundamental technical and socioeconomic changes that occurred in the past two decades have laid the foundation for the recent rapid growth in teleradiology and telemedicine. The first is the separation of image acquisition and image archiving that has been enabled by advances in the digitization of medical imaging and other medical records. The second is the affordable computer network propelled by the broad adoption of the World Wide Web, making it economically viable to transmit large amounts of medical data over distances. Furthermore, the rapid growth in the use of imaging services by referring physicians and the increasing demand for the services of imaging specialists have fueled growth in teleradiology and telemedicine in general. Other motivations for teleradiology applications include the following:

- Improved use of expertise can be achieved by moving the image interpretation tasks to the radiologists or other specialists instead of having the specialists travel to specific locations.

- The associated changes of increasing globalization, world travel, and blurred national boundaries have started the trend of demanding the best services money can buy. Therefore, demand for access to the best specialists' knowledge has increased.

- Economic and competitive pressures force the consolidation and expansion of radiology services beyond their traditional geographic regions and institutions.

- The consolidation of radiology services has caused commoditization of radiology diagnostic services.

- Healthcare providers seek improved lifestyles, balanced workloads, and after-hour call coverage.

There are several successful major teleradiology service providers in the United States, which may be an early indication of what is coming in the area of telemedicine. The need for teleradiology and the growth in this field are going to continue in the next decade.

7.1.2.2 Economic Aspects of Teleradiology

One of the major economic factors enabling the deployment of teleradiology was the dramatic reduction of communication costs through the broad use of the Internet. As with any economic activity, the sustainability of telemedicine must come down to the balance of cost and benefit, and ultimately profit, for both the service provider and the service customer. There is no question about the value of telemedicine. The after-hour and other lifestyle-related demands that drive teleradiology services are well established, and the advantages and benefits of teleradiology in increasing patient access to specialists' services and expertise are clear. However, the fundamental questions still remain the same: What are the costs of teleradiology applications and services, and who is paying for those costs? In the United States today, Medicare and Medicaid laws require that radiologists interpreting imaging studies be on U.S. soil in order to qualify for reimbursement for final diagnostic reports. The business model is not quite clear at this point, and it is going to take some government involvement and policy changes (whether we like it or not) for the broad adoption of teleradiology services to remote or underserved areas where access to specialists is limited or nonexistent.

At present the successful teleradiology service model without direct funding from the government is the consolidation of radiology diagnostic services. The primary customers of these services are small hospitals lacking after-hour or reliable radiology services. It mostly involves expansion of image interpretation services at the expense of solo practice radiologists and small radiology practices, which cannot afford round-the-clock staff. Because they have economies of scale, these teleradiology service providers are making headway into other geographic areas, starting in rural and other areas which lack specialists. In order to have a successful and sustainable teleradiology service, it is critical to provide supporting services beyond image interpretation and diagnosis, such as coding and billing, which are also critical for the sustainability of teleradiology services, and the economics of the teleradiology application is ultimately the driving force that determines the viability, the ultimate success or failure of the service.

7.1.2.3 Enabling Technological Advancements for Teleradiology over the Past Decades

The fundamental change—the separation of image acquisition and image archive media (i.e., from film to digital for nearly all medical imaging)—has led to the separation of the point of image data acquisition and the point of diagnosis. Teleradiology, as part of and the leading specialty of telemedicine, has become technically possible because of the standardization of medical imaging (e.g., DICOM and HL7) and the broad use of computer networks and the Internet. These technologies also enable simultaneous access to image data (or a patient's medical record) by multiple healthcare providers involved in the patient's care. This has changed the traditional linear pattern of patient care in which a patient had to be seen by one specialist and then another, and the patient's medical history was sent separately to each specialist. Nowadays, the paradigm change to a parallel model of patient care can increase the efficiency of healthcare if workflow and patient care are arranged logically and carefully designed.

7.1.3 Applications and Benefits of Teleradiology

At present the major applications of teleradiology and telemedicine can be placed in the following categories: (1) after-hour imaging diagnostic service coverage, (2) access to specialists' expertise and second opinions, and (3) consolidation of imaging interpretation and diagnostic services.

Although specific business models and practices of existing teleradiology services have evolved over the past decade, the common models of service and target markets for these teleradiology services are:

- Providing image interpretation and diagnosis after hours and for emergency rooms

- Furnishing subspecialty image interpretation for hospitals and primary care clinics and for domestic and international clients

- Making radiological image consultations available to medical facilities that do not have on-site radiology expertise

- Giving a second opinion on imaging studies performed elsewhere and access to subspecialty expertise

- Providing timely radiology image interpretation for emergent and nonemergent clinical care facilities in remote and underserved areas

- Reducing on-site interpretation calls for radiologists

- Managing workloads and equipment use via enterprise-wide teleradiology and virtual PACS

- Providing access to advanced image processing

- Promoting efficiency and quality improvement

- Enhancing educational opportunities for practicing radiologists

- Supporting research and multicenter trials

- Facilitating quality assurance and peer review

- Enabling a physician's remote/home reading, on-site call reduction, and other lifestyle improvement needs

The teleradiology service can be modeled as shown in Figure 7.1.

The overarching advantages and benefits of teleradiology and telemedicine include the promise of providing quality imaging interpretation anytime and anywhere. The technical advancement in worldwide connectivity via the Internet has provided a powerful platform that offers many possibilities and opportunities as well as challenges to the existing model of radiology services. The outsourcing of image interpretation and other internationally competitive services may not necessarily be a completely bad thing for radiologists in developed nations because there are demands from the developing nations for access to specialists' expertise in imaging.

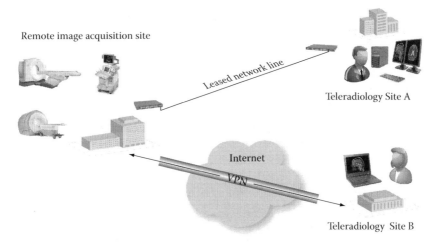

Figure 7.1 In a teleradiology setup, images can be acquired by modality scanners located in a remote site, and then transmitted to teleradiology providers in different locations (Teleradiology Sites A and B) through leased communication network lines, or through the Internet with security technologies such as VPN.

7.1.3.1 Reduced Turnaround Time

Fast turnaround time is crucial for the success of a sustainable teleradiology service. In the past, obtaining a second opinion or expert consultation might require a patient to travel to the specialist or the images to be printed on film and sent to the radiologists. Today, nearly all medical records can be digitized, stored, and transmitted electronically. The broad adoption of computer network technology has enabled the fast and inexpensive transmission of medical images around the world, making it possible to provide value-added teleradiology services with fast turnaround times, which are especially critical for emergency room coverage.

Fast turnaround time and quality of interpretation may be dependent upon several other factors, some of which are technical and some not. On the technical side, the network speed affects access to the images by the interpreting radiologists, and the network integration and preset channel of communication to send a report back to the image-sending party are critical as well. This involves integration issues of allowing the teleradiology provider access to the RIS systems of remote sites and the establishment of proper procedures to inform the remote sites about urgent findings on the image studies. Unfortunately, there is no standard way of doing this for the time being.

7.1.3.2 Subspecialty Utilization

In teleradiology, there is a trend moving toward subspecialty interpretation, which signifies the expansion and maturation of teleradiology services. However, one of the challenges for subspecialty teleradiology is to efficiently match the qualified reader with the right imaging studies. This has been the major challenge for most teleradiology service providers.

The demand for subspecialty teleradiology is mostly driven by the ever-growing number of outpatient imaging facilities, increased imaging volumes because of ongoing advancements in imaging technology, competition among radiology services, and other economic considerations. The shortage of subspecialty-fellowship-trained radiologists has been exacerbated further by the dramatic increase in image volume (e.g., the use of multislice CT and the ever-expanding use of MRI). With the increasing complexity of imaging studies, it has become impossible for a single radiologist to be an expert in all of them. Teleradiology services address these supply and demand mismatch issues by providing customers with a network of subspecialty experts. Such a change has happened and is happening in other specialties of medicine as well. The trend of further subspecialization in radiology and other specialties is expected to continue for the foreseeable future.

7.1.3.3 Workload Balancing

In a large and geographically scattered radiology group practice, it is common for one site to be much busier than others. Teleradiology provides technical solutions for efficiently balancing the workload among the radiologists in the group practice. In such a scenario or other comparable ones, access to images via teleradiology technology provides virtual reading rooms and virtual worklists for all radiologists in the group, improving efficiency and turnaround time.

7.1.4 Remote Image Access

Teleradiology images can be accessed in two ways. One is that all images are stored in a PACS. The remote teleradiology workstation accesses the PACS server through a secure network, which can be a private, leased line from a communication carrier, or through the Internet with security measures such as VPN and SSL (see Chapter 4). The teleradiology workstation is essentially a remote PACS workstation connected by a WAN or Metropolitan Area Network (MAN). The user interface is identical to that of PACS workstations connected to the server by a LAN in the healthcare institution's central reading room. This type of configuration is suitable for teleradiology services within the same healthcare enterprise for remote sites or to connect to the radiologists' home for a reading. Once logged on to the teleradiology workstation, the user has full access to all functionalities and features. Shown in Figure 7.2 is the teleradiology access to an enterprise PACS. The user interface of worklist in Figure 7.3 and all PACS user tools in Figure 7.4 are identical to that of a PACS diagnostic workstation (see Chapter 2). The only difference is that the teleradiology workstation may not have the same hardware and software configurations, operating system versions, and number of monitors as the on-site PACS workstation.

In another approach for teleradiology image access, images are sent from image providers to a Web server using PACS or other imaging modalities. This approach primarily can be used for

Figure 7.2 Teleradiology remote access to the St. Luke's Medical Center PACS (Emageon, Birmingham, AL) through a Web browser for log-on. This is primarily for use by radiology staff.

image access by referring physicians outside the image provider's healthcare enterprise. The specialist who logs on to the teleradiology Web server using a Web browser, as shown in Figure 7.5, sees a user interface on the screen of the teleradiology workstation (Figures 7.6 and 7.7). This interface is determined by the design of the Web access teleradiology product, which is usually different from that of a diagnostic PACS workstation with limited features and imaging display tools.

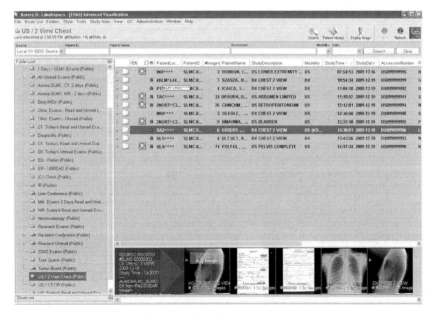

Figure 7.3 Teleradiology remote access to the PACS's worklist.

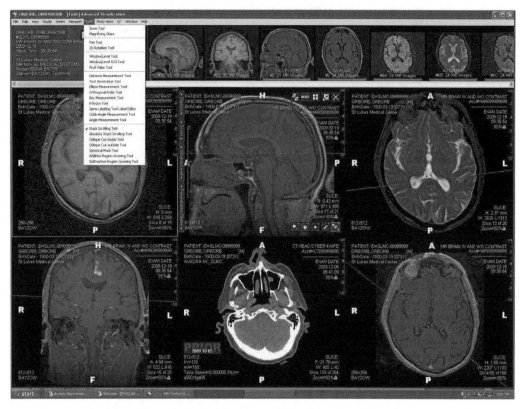

Figure 7.4 Teleradiology remote access to the PACS's tool list. The user interface is the same as those diagnostic workstations located in St. Luke's Medical Center, Aurora Healthcare.

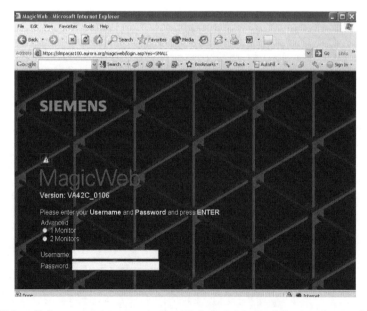

Figure 7.5 Teleradiology remote access to the (St. Luke's Medical Center, Aurora Healthcare) clinical Web-based PACS (MagicWeb, Siemens AG Medical Solutions, Health Services), which is primarily for use by referring medical staff.

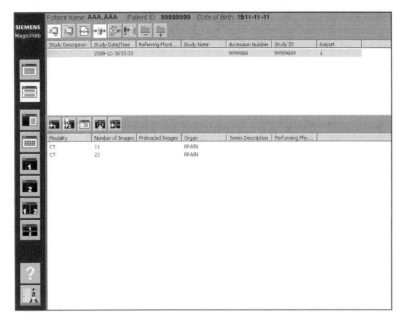

Figure 7.6 Remote access to the Web-based teleradiology MagicWeb (Siemens AG Medical Systems) worklist.

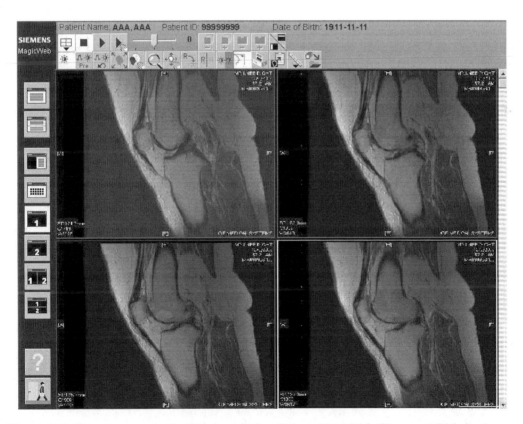

Figure 7.7 Remote access to the Web-based teleradiology MagicWeb (Siemens AG Medical Systems) user interface, which provides essential capabilities to navigate through imaging studies.

The two approaches to teleradiology image access can satisfy different needs. In either case, if the Internet is used as the network infrastructure for teleradiology image transmission, typically, VPN and SSL are used to ensure data security during transmission.

7.2 TECHNICAL CONSIDERATIONS FOR TELERADIOLOGY IMPLEMENTATION

A teleradiology system usually includes the following major components: data acquisition, data transmission, data interpretation, and diagnosis feedback. There are many complex issues involved in the implementation and operation of teleradiology applications. Although it is beyond the scope of this book to describe all the technical details involved in teleradiology, we will address some of the most important technical aspects of teleradiology applications along with some closely related nontechnical aspects.

7.2.1 Image Acquisition and Digitization

At present, all medical imaging modalities in radiology such as MRI, CT, and x-ray are capable of acquiring images in digital form. With the broad adoption of industry standards such as DICOM and HL7, these images can be stored and transmitted via computer networks to anywhere in the world. This was not possible till as recently as the mid-1990s when the lack of standardization made communication impossible among equipment manufactured by different vendors and sometimes even between parts of equipment made by the same vendor.

Since all of today's medical imaging modalities are inherently digital, images acquired on the latest imaging equipment no longer need to be processed by a film digitizer. However, old images stored on films are needed for comparison or other diagnostic purposes. In these cases, a film digitizer converts the images on film into digital images in a standard format (e.g., DICOM), and then these digitized images are sent to PACS or other teleradiology systems for review and diagnosis. Users should be aware that there is information loss or image quality degradation in the process of film digitization. However, some of these losses may not be clinically significant, that is, the loss of information may not cause any significant differences in diagnostic quality or results. The quality of the digitizer should be checked routinely according to the vendor's recommendations.

7.2.2 Data Compression

Before image data are sent over a computer network, image compression is routinely done to reduce the amount of these data, decreasing the time required (thus potentially lowering the cost) for data transmission. There are two main categories of image compressions, lossless and lossy. The lossless compression process does not involve information loss, while the other does. There are various specific techniques to both lossless and lossy compression. The detailed technical specifications of these compression techniques are beyond the scope of this book because data compression itself is a constantly progressing field of research and development. For more information, look for articles about new developments in the IT journals or see Chapter 5 of this book.

7.2.3 Roles of PACS in Teleradiology

In order to have a successful and efficient teleradiology service, it is essential to have a PACS at the image-receiving end. There are various levels of sophistication for teleradiology services; some may involve the full integration of multiple PACS and EHR systems at different institutions, and others may operate with only a film digitizer from which images are first captured and then sent to a centralized facility for interpretation and diagnosis. Most of today's large-scale teleradiology operations are dependent on the direct communication and transmission of images over computer networks via DICOM rather than proprietary systems.

7.2.4 ACR Standard for Teleradiology

ACR published a technical standard for teleradiology in 1996 and updated it in 2002; the updates became effective in 2003. Although some of the technical specifications need updating (e.g., the requirements for image display devices) because the technology has progressed rapidly in the past few years, the ACR standard should serve as a good guideline for the minimum requirements of teleradiology services.

7.2.5 Image Transmission in Teleradiology

The speed of data transmission is critical to the success of teleradiology applications, as it will impact the turnaround time for the diagnostic report. To transmit the image data, the user needs a computer workstation equipped with the necessary software and computer network connections for sending images in a standard format (e.g., DICOM for radiological images). At the receiving end, there must be a gateway workstation that handles the communication issues involved in receiving images. Security measures and patient confidentiality concerns require both the sending and the receiving workstations to be configured properly so that one workstation "knows," or authenticates, the other, as discussed in Chapter 4, and are thus authenticated to send and receive patients' images. These security measures are important for patient privacy protection as required by HIPAA and also to ensure proper billing and financial charges which is an important aspect of teleradiology services.

The speed of data transmission is determined primarily by the computer network's bandwidth. A detailed discussion of the various computer network technologies can be found in Chapter 4. In general, the higher speed of transmission comes with a higher cost. The selection of an appropriate level of network service should consider the following factors:

- The average amount of image data being sent over the network

- The speed with which images need to be sent to the receiving site

- The costs of various levels of network services

After careful planning and evaluation, users should be able to choose the network bandwidth and services that are appropriate and economically sensible.

7.2.6 Integration of Teleradiology with Information Systems and EHR/PHR

After the diagnosis and interpretation of imaging studies, the results must be properly and accurately entered into the patient EHR for future reference. Also, the diagnostic results must be communicated to the parties who requested the image interpretation. It is critical to communicate the results (especially for those findings that require immediate attention) back to the patients' attending physicians and maybe even the patients themselves in life-threatening situations. This may involve interfacing and integration between the EHR systems of the teleradiology service provider and the customer.

There are potential contractual and technical issues that need to be addressed in order to accomplish this integration. Ultimately, it is highly desirable to have a seamless integration of the RIS at both ends—the image-interpretation side and the image-sending side of the teleradiology service—but this is not always possible. Sometimes, a manual communication mechanism or procedure has to be put in place to complete the feedback loop. This feedback mechanism and the integration of the teleradiology service's RIS to the service customer's is probably the weakest link in teleradiology and telemedicine today. This lack of integration among different healthcare providers has already received attention from all levels of the healthcare industry, including government; there is a U.S. government initiative to implement EHR as the first step toward ultimate integration.

One technical solution for integration is to give the teleradiology provider proper access to the EHR or PACS system of the teleradiology customer, if such systems exist. Otherwise, it can be arranged to give the customer proper access to the reports of imaging studies of their patients. Because customers will almost always have Internet access, Web-based access to the images and reports is desirable. Of course, security issues such as patient confidentiality must be addressed when using Web-based access. There is currently no single, complete product that meets all these needs for teleradiology applications.

7.2.7 ACR Technical Guidelines for Data Storage

After the completion of image interpretation, patient image data should be properly stored for future use. Currently, the ACR technical standard requires that teleradiology systems provide image storage capabilities that comply with all facility, state, and federal regulations regarding medical record retention. Images stored at either site of the teleradiology service should meet the jurisdictional requirements of the image-sending site. ACR guidelines do not require the image-interpretation site to store the images if a copy is stored at the sending site. However, if the images are retained at the receiving site, the retention period of that jurisdiction must be met as well. This may have some technical implications for pediatric cases, for which the retention period may be quite long. The agreement between the teleradiology provider and customer regarding storage policy and record retention must be documented.

ACR also states that the examination data files should be accurate and include the patient's name, identification number, examination date, type of examination, and facility at which the examination was performed. ACR recommends that space be available for a brief clinical history and that prior examinations be retrievable from archives within a time frame appropriate for future clinical needs. ACR requires each facility to have written policies and procedures for the archiving and storage of digital image data that are equivalent to the policies for protection of patient records on hard-copy (film) storage media.

7.2.8 Teleradiology Implementation Considerations

Before being put into clinical service, teleradiology equipment should be tested in advance to measure the speed of data (especially image data) transmission and the estimated turnaround time to get the report back to the customer. To accurately gauge the network speed at different times of the day, these tests should be performed during the hours that most images are expected to be transmitted.

At the image-receiving site, teleradiology specialists need to be notified of incoming images to their PACS so they can interpret the images in a timely fashion. A standard operating procedure and mechanism (preferably automatic) should be put in place to inform the on-call specialist about the incoming images and remind the specialist to interpret the images. This is critical for the success of a teleradiology service. Delays in providing timely interpretation could cause potentially serious consequences for patients and reduce the quality of care. Financial losses could also stem from such delays.

It is not only important to transmit images to the interpretation site in a timely manner, but also to make sure the diagnostic reports are communicated back to the sending site in an effective and verifiable way. In teleradiology services, the importance of effective and precise two-way communication cannot be underestimated.

7.3 TELERADIOLOGY QUALITY CONTROL

The quality of image interpretation may be compromised if the available teleradiology system does not provide images of sufficient quality to allow specialists to perform the indicated tasks, and in any teleradiology system, there should not be any clinically significant loss of information in the entire process of image acquisition, image transmission, and image display. The image quality should be sufficient to satisfy the needs of the specific clinical circumstances; otherwise the quality of image interpretation could be compromised. The ACR technical standard for teleradiology indicates the minimum quality that image display systems must meet.

7.3.1 Overall Quality Control Requirements

Although the ACR technical standard does not give specific quality control requirements, it specifies that policies and procedures be implemented to ensure teleradiology services at a level consistent with on-site imaging studies within a facility or institution. These procedures should include internal redundancy systems, data backup, a disaster recovery plan, and other standard quality control measures.

7.3.2 Image and Data Integrity

Mechanisms or procedures must be implemented to ensure information accuracy and integrity both for incoming images and for interpretations and diagnosis sent back to the image-sending sites. Managing different patient identification systems from multiple image-sending sites may be challenging, but it is very important.

7.3.3 Data Security and Patient Privacy

In accordance with HIPAA, the issues of patient confidentiality and data security must be addressed in teleradiology services. Audit trails, electronic signatures, password protection, and data encryption must be implemented in teleradiology applications. Additionally, data archive and security responsibilities should be clearly laid out in the agreement between the teleradiology provider and customer. The teleradiology system should include measures to safeguard data against intentional or unintentional corruption.

Since patient images are transmitted mostly via public computer networks, important issues concerning the security, privacy, authentication, and finally the integrity of these data need to be addressed. Although ensuring patient privacy is at least partially the responsibility of the network provider, which provides security measures such as firewalls and password protection, the involved parties in teleradiology services also need to have mechanisms in place to ensure data security and data integrity. There are several technical alternatives for addressing these issues (see Chapters 4 and 11), and teleradiology administrative personnel need to keep these in mind when implementing their service.

7.3.4 Verification and Validation of Display Quality

The display devices at workstations used for image interpretation need to be routinely checked to ensure optimal display quality. Although the image-interpretation party of a teleradiology service is responsible for the quality of their interpretations and diagnosis, the ultimate responsibility for image quality rests with the facility where patients are cared for. Therefore, the image-sending party should require the image-receiving party to provide evidence and routine reports about the quality of the image display devices used in the teleradiology service. Details about the quality control procedures for display devices can be found in Chapter 3.

7.3.5 Image Quality Considerations for Handheld Devices

Today, many handheld devices are capable of image display; however, users should be cautious about using them for medical image display, because these devices were not designed for this purpose, and the quality of display is highly dependent on the environment and the display hardware. Therefore, images displayed on handheld devices should only be used for quick reference and not for diagnostic purposes.

7.3.6 Peer Review and Quality Control

Finally, the clinical quality of the interpretations provided by radiologists must be ensured. Interpreting radiologists must have proper training and credentials, and routine quality control peer reviews and double reading may be necessary, as in other radiology services.

7.4 LEGAL AND SOCIAL CONSIDERATIONS

Similar to the implementation of PACS, teleradiology applications bring about important changes including workflow redistribution, technical challenges, and social and legal issues. Since there are multiple parties involved in teleradiology services, other practical issues such as local laws regarding licensure, credentialing, and quality assurance/quality control also come to the forefront. In teleradiology services, these practical operational issues need to be properly handled in order to make the service successful.

Different states in the United States and different regions around the world have different laws and requirements defining who can legally make interpretations and diagnosis for imaging studies and get reimbursed. Although case law has yet to catch up with the rapid development of telemedicine applications, legislation is underway in many states to address these issues. As evidenced by activities such as taxing online sales, state governments are gradually catching up to the new paradigm of business transactions in today's world, and healthcare is no exception. As time progresses, the details will become clear concerning which practices are legal and which are not, which services can be reimbursed and which cannot be, and what constitutes providing patient care in the scenario of telemedicine and teleradiology.

Teleradiology plays a multifaceted role in the radiology field by improving efficiency and productivity and providing access to imaging specialists' services in previously underserved areas. However, when image interpretation is outsourced to other countries, many issues, including turf battles and increased competition worldwide, become very contentious. The seemingly inexorable increase in the use of teleradiology will certainly bring about some changes in the daily routines of radiologists and in their lifestyles. Teleradiology will provide economies of scale and more timely access to specialists' expertise, but at the same time it will increase competition and potentially commoditize imaging services, as has been the case for other fields affected by the macro trend of globalization.

7.4.1 Considerations of Remote Reading in Teleradiology

Remote reading as a special type of teleradiology has become commonplace for radiology practices in the United States, most of which manage image interpretation services for multiple small clinics, hospitals, and other specialty practices. Technology advancements in the past decade have dramatically reduced the cost of high-speed data transmission, making remote reading technically and economically possible. Additionally, the widespread availability of low-cost, high-performance workstations, and software packages added the necessary ingredient for this particular form of teleradiology to become popular. However, this trend of convenience for radiologists also has caused concern among some of the leadership in the radiology community, which has always emphasized the importance of clinical contact with patients and other physicians. Additionally, the financial incentives involved in the interpretation of imaging studies have led to fiercer competition among imaging specialists.

7.4.2 Ownership and Responsibility of Data Storage

The issue of where and how to store images after their interpretation and diagnosis by teleradiology services needs to be considered. There are legal and regulatory requirements regarding the duration of storage for patients' images (especially for pediatric patients). Also, HIPAA and other recent regulations require that these images be available to patients and their designated healthcare providers for future diagnosis and treatment. Should the teleradiology customer (the party that sent the images) be responsible for the long-term storage of the images, or should the image-receiving party be responsible for keeping those records? These are some of the nontechnical issues that must be determined through negotiations between the teleradiology service customer and provider. As cross-enterprise image-sharing initiatives become more prominent in the near future, making those image studies available and accessible online will add further financial and technical burdens for the parties involved in teleradiology services.

7.4.3 Communication Improvement

The trend of increasing use of teleradiology services has introduced some changes and techno-social issues to the practice of radiology, including the reduction of face-to-face contact between radiologists and their patients and referring physicians, as mentioned above, which may have some long-lasting repercussions for radiology as a specialty. Such long-term issues are beyond the scope of this chapter, so we will focus on the more immediate technical and workflow issues facing the practice of teleradiology.

In teleradiology, it is very critical to immediately communicate critical and life-threatening findings to the image-sending site verbally and in writing. It is equally important to keep a record of the acknowledgement that such communication was received. Because they seldom have person-to-person contact with the clinicians at remote sites who order the imaging studies, interpreting radiologists may communicate more effectively using added features such as notations and markings on representative images and by including their personal contact information along with the interpretation report.

Prior to teleradiology and the digitization of medical imaging, images were only available to other physicians after their interpretation by radiologists. For instance, before the broad implementation of PACS, the only copies of patient imaging studies were captured on films, which were not released to other physicians until the diagnosis was done by radiologists, so other specialists or primary physicians did not have easy access to the patient's image study data to make their own preliminary diagnosis and patient care decisions. Now, with instant access to patient imaging data through PACS, other physicians involved in a patient's care may have access to images at the same time as radiologists. This has added a sense of urgency in the radiology departments to provide the diagnosis as soon as possible. There is no doubt that radiologists, as specialty-trained experts, provide a higher quality of interpretation for imaging studies and can find more detailed pathologies in images. However, simultaneous access by other physicians to the patient images as soon as they are acquired increases the efficiency of patient care and may improve communication between radiologists and other physicians because they can view the same images while discussing them in different locations.

7.4.4 International Teleradiology Guidelines

There is a global initiative attempting to establish a framework to regulate teleradiology with defined, uniform standards. An international network of radiology organizations has drafted guidelines for the burgeoning teleradiology industry. The *Top 10 Principles of International Clinical Teleradiology* was published in 2008 by the International Radiology Quality Network. This is an evolving document that has sparked discussions and fueled debates among stakeholders, especially among those who provide global teleradiology services.

Another group, the International Clinical Teleradiology Standards Workgroup, took three years to review the literature and developed a draft based on broad standards. Their intention is to make sure that mechanisms are in place to remind people of the best clinical practices. The workgroup is composed of members of the ACR, RSNA, European Society of Radiology, WHO, and other organizations.

The workgroup realizes that teleradiology companies offer varying levels of service depending upon the individual companies' resources and the status of technical and professional development within different countries. The draft attempts to cover some image display quality, security, and medical–legal key issues. For instance, it addresses liability coverage issues at the referring and interpretation sites to be governed by international laws, but the laws for such disputes may not be well established yet. However, it is commonly agreed that the radiologist providing the final report must be a qualified specialist and meet the appropriate training, registration, certification, licensure, revalidation, credentialing, malpractice insurance, and continuing professional education and development requirements for the country of the service-providing physician. However, credentialing and licensing physicians in multiple countries presents a challenge, and these requirements would be difficult to enforce. For these reasons, some teleradiology providers have reservations about the workgroup's requirement that radiologists be licensed at both the referring and interpretation sites. Perhaps in the future appropriate certification and credentialing requirements can be established internationally.

As with many of the other guidelines, the draft is merely suggestive and has no regulatory or enforcing power. It is the workgroups hope that in the future teleradiology providers will follow these requirements. It is possible that more definitive international standards will be developed over time.

7.5 CHALLENGES AND FUTURE TRENDS IN TELERADIOLOGY

The concept of telemedicine started in the 1970s when computers were first applied to medicine. Unfortunately, the technology in those days was not sufficient for practical telemedicine applications. Some of the early attempts failed primarily because they lacked fast communication technologies such as the Internet. Today, with easy access to the World Wide Web and other Web-based technologies, the formation of the virtual world has finally arrived. Virtual communities are formed without the restrictions of geographic distance. This fundamental change occurred in the end of 1990s and early 2000s. We are just at the beginning of an information revolution whose implications are far and deep.

Telemedicine is one part of a new paradigm of communication in society today. As did the telephone, the Internet has enhanced communication between healthcare providers and patients. This separation of point-of-care from the physical presence of the patient can lead to increased access to special expertise and possibly to reduced costs, but telemedicine also enables better management of resources and manpower, thereby increasing efficiency. As with any increase of efficiency, there will be the trend of commoditization of medical services and expertise, which may lead to changes in healthcare delivery structures.

7.5.1 Potential Impact to Radiology

As pointed out by several visionary leaders in the radiology field, teleradiology may bring about a major consolidation in radiology practices to achieve economies of scale, faster turnaround times, and access to subspecialty clinical expertise to match the changing demands of imaging studies in the future. In other words, the application of teleradiology and increasing globalization may bring about the commoditization of radiology services. Regulatory restrictions may slow down this trend, but the bigger economic environment does not bode well for such a move. Universal IT standardization of medical imaging and patient EHR and PHR as promoted by the U.S. government will accelerate inevitable changes.

In order to face such challenges, the radiology community as a whole may have to stand up to join the inevitable changes. Insisting that the practice of radiology be limited to the interpretation of imaging studies has profound risks, which may lead to commoditized image interpretation services going to the lowest bidder. Therefore, radiology practice must redefine itself to become an even more active participant in patient care, making it much more difficult to outsource services to a cheaper bidder. In the end, however, the increased use of teleradiology and the associated increase in competition will improve patient care, which is and should be the ultimate priority of any medical specialty.

7.5.2 Faster Image Transmission

Nowadays, advanced and more sophisticated imaging studies generate large amounts of image data. Transmitting these data to interpretation sites may take a long time. Therefore, new paradigms of fast image transmission must be implemented. For instance, allowing the radiologist to first browse through lower-resolution images for significant findings while waiting for full-resolution images to be sent across the network may be very helpful in some emergency scenarios. In this case, an application feature of the JPEG2000 image compression format discussed in Chapter 5 can be used. The JPIP compression streamlining protocol works with JPEG2000 to download only the requested region of a picture in high spatial resolution or to download a large image in low resolution for quick viewing. This allows efficient browsing of large, 3-D volumetric image data at a reduced spatial resolution and high speed, without using too much network bandwidth. Additionally, intelligent image compression schemes may be very useful for faster image transmission in teleradiology.

7.5.3 Teleradiology Service in Multienterprise and Multivendor Environments

As in most information-sharing projects, the really difficult challenges for interoperability in teleradiology tend to lie with data exchange and information sharing, since these data are usually generated by equipment from different vendors and stored at different healthcare institutions. Imaging studies coming from various healthcare enterprises for teleradiology service often use different patient identification systems. Coordinating these different imaging studies into a single worklist for teleradiology workflow is challenging but critical for the efficiency of a teleradiology operation.

7.5.4 Cross-Enterprise Image Sharing of EHR and PHR Integration

The accuracy of a diagnosis using teleradiology across institutional boundaries may be dependent upon prior patient history and prior imaging studies for comparison, progression assessment, evaluation of treatment responses, and so on. At the present time, an automated and standard interfacing mechanism does not exist to meet such telemedicine and teleradiology needs. A seamless transmission of patients' prior medical records, including prior imaging studies, from one healthcare institution or enterprise to the next would improve the efficiency and accuracy of diagnosis. This is the fundamental driving force behind the RSNA/National Institute of Biomedical Imaging and Bioengineering (NIBIB) image-sharing initiative, which is essential for the broader scope of EHR implementation among multiple healthcare enterprises. Until such standards are broadly adopted by the healthcare industry, users will need to manage some of these tasks manually. Therefore, a systematic and error-proof procedure, such as the prefetching and transmission of prior medical records in advance, may have to be used.

7.5.5 RSNA/NIBIB Image-Sharing Initiative

There have been efforts at both local and national levels to enable the electronic exchange of healthcare information by the adoption of EHRs to improve and reduce the cost of healthcare in the United States. Sharing medical imaging data across healthcare enterprise boundaries will improve the quality of care and reduce costs by avoiding redundant imaging studies and by making imaging studies and associated reports from other institutions readily available to physicians. Timely access to patients' prior medical information can also reduce patients' exposure to the ionizing radiation and contrast agents associated with imaging studies. Furthermore, if images and other medical information are easily accessible from both a rural hospital and a tertiary medical center, physicians from the rural hospital can provide timely treatment for patients by obtaining specialty consultation from the tertiary hospital remotely.

The RSNA and the NIBIB have initiated a proof-of-concept image-sharing project designed to complement the functions of regional health information organizations (RHIO) and EHRs. Using XDS and XDS-I profiles from the IHE initiative, this project plans to create an Internet-based alternative to the physical exchange of DICOM images via portable media such as CDs. The system will have a mutually agreed implementation of interoperable technologies, a set of policies, and legal agreements dealing with security and privacy, which will allow the sharing and transmission of medical data. The RSNA/NIBIB initiative promises easy implementation of image sharing with Internet-based technology, which enables patient control and access. In the next few years, this initiative will develop and test the project at several pilot sites to study the feasibility of the proposed architecture for both clinical and research applications.

7.5.6 Future Growth of Telemedicine and Teleradiology

There are several joint initiatives in which health insurance groups and technology firms such as teleconferencing companies are collaborating to build a nationwide telehealth network. Such a network would provide patients with access to physicians and specialists remotely, as an extension of PHR patient portals (see Chapter 11). These new programs can combine audiovisual technology with other health resources to expand physicians' reach into underserved areas where access to specialists is limited, rare, and difficult. As part of telemedicine, teleradiology will be an integral part of these systems. These mostly Internet-based networks will enable the open and secure exchange of medical records, allowing them to connect with disparate telemedicine technology systems from multiple equipment vendors. These projects hope to achieve real-time connectivity for consultations among healthcare providers across the country, creating a more integrated system of healthcare. However, the keys to their success include focusing on people-to-people interactions, making the system easy to use, and connecting patients with their care providers and insurance carriers. These networks will use multimedia and medical information to create an experience similar to a face-to-face visit with a doctor.

At the present time, Medicare only covers telehealth visits and services in certain rural and underserved areas. Some of these telehealth initiatives are funded by private insurers and industry pilot grants. Other applications of telemedicine may include patients' remote monitoring at their own residences, such as the applications defined by the IHE Implantable Device Cardiac Observation (IDCO) Integration Profile. Since patients' pacemakers, implantable cardioverter defibrillators (ICD), and cardiac synchronization therapy devices need to be interrogated after they are implanted, remote follow-up may allow daily data collection for the early detection of trends and problems or provide other event information about the current status of patients. Reports may then be sent to recipient systems where the data are converted into HL7 messages and summary data are communicated to EHR/PHR systems.

7.5.7 Teleradiology and Telemedicine in Developing Countries

So far, this discussion of telemedicine and teleradiology has been focused on models in developed countries. As the world is getting "smaller," propelled by the broad use of the Internet and computer networking, access to the expertise of subspecialty physicians at affordable prices is possible from anywhere in the world. In fact, developing countries may benefit more than others from telemedicine applications. In developing countries, especially in remote regions and villages, access to specialists' expertise may be difficult to come by. Immediate access to such expertise and consultations from a foreign country may become a reality via applications of telemedicine. However, current healthcare systems worldwide are not equipped to handle such demands, not because of lack of technology but because of other socioeconomic issues and roadblocks. In order to provide sustainable teleradiology or telemedicine services to developing countries, the economic aspects of the services and the financial incentives for the specialists who possess the expertise have to be determined. Infrastructures have to be put in place for such services to be done in an economically viable way. There are several critical technical, socioeconomic, and cultural ingredients that must be in place in order for this to happen.

With the advancement of Internet use and the associated unstoppable trend of globalization, the need for teleradiology and telemedicine in general will continue to increase. For large healthcare enterprises with multiple institutions and community clinics, the use of teleradiology and telemedicine is a fact of life. For hospitals that do not have in-house specialists and off-hour coverage, there will be an increased need for telemedicine and teleradiology services. Physicians' remote/home reading of image studies and the subsequent distribution of reports to nonradiology medical specialties have come to the forefront in many healthcare institutions. Additionally, telemedicine needs for international and domestic travelers will be an area of fast growth in the next decades.

Web-based applications of teleradiology and telemedicine in general will be the future. The main issues concerning these Web-based applications are cost, speed, security, and privacy. There are numerous other technical issues that need to be resolved before a broad adoption takes place, although specific and niche applications are already creeping into the field. Telemedicine and teleradiology will experience a dramatic increase in use during the next decade.

BIBLIOGRAPHY

American College of Radiology. 2005. ACR practice guideline for communication of diagnostic imaging findings. www.acr.org/SecondaryMainMenuCategories/quality_safety/guidelines/dx/comm_diag_rad.aspx. Accessed January 17, 2010.

American College of Radiology. 2006. Revised statement on the interpretation of radiology images outside the united states. www.acr.org/MainMenuCategories/media_room/FeaturedCategories/other/PositionStatements/statement_interpret.aspx. Accessed January 17, 2010.

American College of Radiology. 2007. ACR technical standard for electronic practice of medical imaging. www.acr.org/SecondaryMainMenuCategories/quality_safety/guidelines/med_phys/electronic_practice.aspx. Accessed January 17, 2010.

American Telemedicine Association. 2008. Medicare reimbursement for telemedicine. www.americantelemed.org/files/public/policy/Medicare_Payment_Of_Services.pdf. Accessed January 17, 2010.

Andriole, K. P. 2008. Teleradiology. *The Society for Imaging Informatics in Medicine 2008 Annual Meeting*.

Andrus, W. S. and K. T. Bird. 1972. Teleradiology: Evolution through bias to reality. *Chest* 62: 655–657.

Berger, S. B. and B. B. Cepelewicz. 1996. Medical legal issues in teleradiology. *AJR* 166: 505–510.

Berlin, L. 1998. Malpractice issues in radiology: Teleradiology. *Am. J. Roentgenol.* 170: 1417–1422.

Caryl, C. J. 1998. Malpractice and other legal issues preventing the development of telemedicine. *J. Law Health* 12: 173.

Field, M. J. 1996. *Telemedicine: A Guide to Assessing Telecommunications for Health Care*. Committee on Evaluating Clinical Applications of Telemedicine, Institute of Medicine. National Academy Press, Washington, D.C.

Geis, R. 24/7 Teleradiology in private practice. *The Society for Imaging Informatics in Medicine 2008 Annual Meeting*. Seattle, WA.

Goldberg, A.S. 1998. *Telemedicine: Emerging Legal Issues*. American Health Lawyers Association, San Diego, CA.

Hoffman, T. 2005. Teleradiology: An underdeveloped legal frontier. www.acr.org/SecondaryMainMenuCategories/BusinessPracticeIssues/Teleradiology/TeleradiologyAnUnderdevelopedLegalFrontierDoc4.aspx. Accessed January 17, 2010.

Huang, H. K. 1996. Teleradiology technologies and some service models. *Comp. Med. Imag. Graph.* 20: 59–68.

Johannes, N., J. N. Stahl, and W. Tellis. 2000. Network latency and operator performance in teleradiology applications. *J. Digit. Imag.* 13 (3): 119–123.

Kumar, S. and E. Krupinski. 2008. *Teleradiology*. Springer, New York, NY.

Massat, M. B. 2008. Trouble in multi-PACS paradise. *Imaging Technol. News*. www.itnonline.net/node/29463/. Accessed January 17, 2010.

Mehta, A. 2003. *The Internet for Radiology Practice*. Springer, New York, NY.

Moore, A. Van, B. Allen, and S. C. Campbell, et al. Report of the ACR task force on international teleradiology. www.acr.org/SecondaryMainMenuCategories/BusinessPracticeIssues/Teleradiology/ReportoftheACRTaskForceonInternationalTeleradiologyDoc3.aspx. Accessed January 17, 2010.

Mullick, F. G., P. Frontelo, and C. Pemble. 1996. Telemedicine and telepathology at the armed forces. Institute of Pathology: History and current mission. *Telemed. J.* 2: 187–194.

Ontario Association of Radiologists. 2007. OAR teleradiology practice standard. www.oar.info/pdf/NewOARTeleradiologyStandard.pdf. Accessed January 17, 2010.

Royal College of Radiologists. Teleradiology—A guidance document for clinical radiologists. www.rcr.ac.uk/publications.aspx?PageID=310&PublicationID=195. Accessed January 17, 2010.

RSNA News July 2009. Teleradiology ushers in new, subspecialized era. www.rsna.org/Publications/RSNAnews/July-2009/Teleradiology_feature.cfm. Accessed January 17, 2010.

Sodickson, A. 2008. 24 × 7 Radiology: Academic radiology perspective. *The Society for Imaging Informatics in Medicine 2008 Annual Meeting*. Seattle, WA.

Thrall, J. H. 2007. Teleradiology Part I. History and clinical, applications. *Radiology* 243: 613–617.

Wrong, Y. 2008. Global group creates framework to regulate teleradiology. www.diagnosticimaging.com/imaging-trends-advances/pacsweb/teleradiology/article/113619/1181532?verify=0. Accessed January 17, 2010.

8 Practical PACS Implementation

This chapter discusses practical issues and considerations in the implementation of clinical PACS, including common issues involved in planning, installation, acceptance testing, and post-installation maintenance. The chapter will particularly focus on the steps taken in implementing a PACS and provide the potential PACS buyer practical guidance throughout the process.

It is helpful to realize that the impact of implementing a PACS is not limited to the radiology or imaging department alone. Besides, the process involved is unlike planning and installing a piece of complex imaging equipment, such as an MRI or CT scanner. Rather, it more closely resembles the installation of a telephone or IT system throughout a hospital (even though, in most cases, much of the costs will be borne by the radiology department). A PACS should be treated as a part of the EHR system of a healthcare institution. Consequently, the implementation of a PACS involves many interaction and interfacing issues with other existing information management systems, such as the HIS and RIS.

It is expected that every PACS implementation will increase the efficiency of patient care delivery and that the increase in efficiency will mostly be reflected in the workflow changes made possible by the PACS. However, the increase in efficiency and change in workflow are not limited to the radiology department. In fact, most of the efficiency improvements may be brought to other clinical departments. Regardless of the result, in today's environment of increasing managed healthcare and pressure to reduce costs, reducing patient hospitalization times by increasing the efficiency of imaging studies and shortening turnaround times is one of the most important benefits of PACS implementation.

As the U.S. and other governments have begun to push for unified EHR that can be accessed anywhere at any time, all healthcare providers may need to implement PACS and other patient EHR systems. Enhancing this pressure is the broad adoption of computer technology and Internet-based communications among the general population, even in regard to health-related issues. Combined, these pressures are prompting healthcare providers to implement PACS and EHR at an unprecedented pace.

8.1 FINANCIAL ANALYSIS

Since each institution's workflow and cost structure is different, financial analysis for PACS implementation cannot be simply copied from one to another without looking into specific clinical operations in detail. For instance, a large enterprise healthcare delivery system must consider the impact of PACS implementation on all subcomponents, whereas a small clinic may only need to be concerned with the limited impact of workflow change. Additionally, the technical support cost of a PACS may require significant involvement of the IT department in a large institution, or an additional expense related to contract support in the case of a smaller clinic. Data traffic increases by several magnitudes after the implementation of PACS. The additional infrastructure requirement or the upgrade to the existing network infrastructure may be substantial and may not have been considered as part of the cost of PACS implementation itself. That being said, some of the general principles laid out in the following text can serve as a guide to institutions of any size that are considering implementing a PACS.

8.1.1 Cost Analysis

Like RIS and HIS, PACS costs a lot to implement and does not generate obvious and direct additional income. Rather, it is a necessary tool that improves the efficiency and accuracy of patient care. Therefore, when doing the cost-benefit analysis, the buyer should consider it in broader terms than just simply the dollar amount needed to acquire, implement, and maintain a PACS. By the same token, when evaluating potential benefits or savings, the buyer should also look beyond the imaging department and look at a broader scope than just eliminating film costs and reducing staff in the film library and other areas of radiology.

The major costs for a PACS implementation can be broken down into the following components: PACS workstations; archiving and data storage systems; interfaces with RIS/HIS and imaging modalities such as MRI, CT, and digital x-ray equipment; any additional PACS support technical staff needed; and network infrastructure upgrades, which may or may not be counted as direct costs of the PACS.

Before selecting a PACS vendor, the institution must first analyze and determine the storage and accessibility requirements based on its specific workload. It is important to plan for sufficient but not excessive storage capacity since the per-unit cost of storage continues to decrease. Some third-party vendors offer data storage services; this option has a certain appeal since the vendor is responsible for the upkeep of the storage device, archiving maintenance, and, more importantly, for the currency of the storage technology. This fee-per-service option should be evaluated and compared with the alternative of owning and maintaining the storage devices outright and taking the responsibility of future data migration and upkeep of the storage technology. (See Chapter 5 for details of data storage in PACS.)

The number of workstations required is another major factor in determining the overall cost of a PACS project, since these workstations are most likely equipped with one or more high-end (and thus expensive) medical-grade display monitors. Although the cost of medical-grade monitors has declined somewhat over the past few years, it remains much higher than that of consumer-grade monitors used for office desktop applications.

Another major cost of PACS implementation may not be as obvious. That is the cost of upgrading the existing computer network infrastructure. For instance, it may be necessary to add or reconfigure the network cables for the reading room where PACS workstations are to be located and upgrade or build new communications closets throughout the healthcare institution. Since the cost of network upgrades to accommodate the PACS may fall to the hospital-wide IT department, rather than to the radiology department, it is often overlooked. However, for this reason, it is absolutely necessary to get the IT department involved in the planning phase of PACS implementation.

Even though PACS promises to increase clinical operation efficiency, it still may be necessary to add personnel to manage and maintain it after installation. A PACS administrator/manager is most likely needed to run routine maintenance, such as creating new user accounts, providing user training, implementing quality control, monitoring archive and storage backup, and adding or deleting DICOM nodes to the PACS, among other tasks. Last but not the least, the cost analysis should also include the service contract to be drawn up for the PACS system's upkeep and maintenance.

8.1.2 Benefit Analysis

The second part of the financial analysis looks at the benefits of implementing a PACS. There are two major justifications: one financial and the other the improvement in clinical service, which in turn has an indirect financial impact on the institution as a whole. In both cases, the financial benefits accrue from the increase in workflow efficiency; however, such improvements will be optimized or realized only if the PACS's impact on workflow is well planned before the system is implemented, and the benefits will be realized faster and more readily if proper training and other staff preparation has been planned and conducted before the "go-live" moment.

The most common financial benefits that can be realized from PACS implementation are as follows:

- Workflow improvement for higher efficiency and better patient care

- Simultaneous access to the same patient images at different physical locations by physicians in various departments

- Reduced delays in the availability of radiological images and associated diagnostic reports

- Teleradiology service to remote clinics or satellite sites

- Key component for EHR

- Reduced film loss

- Reduced need for repeat radiology studies due to poor image quality and film loss

- Reduced cost of film and chemicals

- Reduced image file-room cost

Most institutions require an economic justification analysis, and the scope of benefits identified may determine the scope of PACS implementation. Sometimes it is difficult, though not absolutely impossible, to attach a dollar amount to the benefits brought about by the PACS. Some savings are more obvious (e.g., elimination of film use and film-processing chemistry, reduction of staff associated with film processing and handling, cost of replacing lost patient radiographs, and the elimination of repeat imaging). Because PACS are so expensive, however, economic justification on the basis of film savings alone is rarely possible, unless the department performs a very large number of examinations and can go filmless virtually overnight.

Although the radiology department is usually the driving force in the implementation of PACS, the PACS is really an integral part of the hospital-wide patient management system. Radiological images are a part of the patient's EHR, and less easy to capture in the benefits analysis is the fact that PACS is a system that enables the radiology department to provide better and faster service to clinicians for improved patient care. Time savings for medical and administrative staff; remote image reading and consultation using teleradiology; improvement of radiology services to other departments who benefit from zero film loss; faster, simultaneous access to the radiology images; and improved patient care reduce costs for the hospital, especially within the managed-care environment. For example, if primary care physicians can have instantaneous access to radiological images while the patient is still in the hospital, than a second patient visit for the same medical problem may be avoided. This example illustrates well the fact that it is the primary care physicians and ultimately patients who benefit the most from PACS.

Another factor that makes a general analysis difficult is that the benefits and associated economic savings are different for each institution. Thus, each healthcare institution should perform the benefit analysis based on its own specific situation.

8.2 PLANNING AND DEVELOPMENT

Planning is always the most important phase in implementing any project. Very detailed planning is required to successfully implement a PACS because it is a very complex system. Even the installation of highly sophisticated radiology equipment, such as MRI or CT scanners, is relatively simple compared with installing a PACS in terms of planning. First, the buyer must justify the purchase of PACS, which as we have just seen can be difficult to do based on a simple financial analysis. Second, implementing PACS involves multiple departments, such as the information service, the facility service, and other clinical departments served by it. Obviously, a well-planned project results in a much smoother and more cost-effective implementation. The remainder of this section documents the steps to be taken in the planning phase.

8.2.1 Forming the Right Team

The buyer first needs to organize a PACS strategic planning team. This team is tasked with identifying problems or areas requiring improvement and studying the potential benefits PACS would provide. At the beginning, the team typically includes only a small number of people so that it can stay focused and thus clearly define the goals and expectations of the PACS project. It is recommended that the following staff members be included in this team:

- Radiology administrator
- Radiologist
- Physicist
- Information service (IS) administrator

This small group of people will also serve as the steering force of the expanded team, whose members will increase as the implementation progresses. At the purchasing and installation stage, the following personnel are included to form a site and vendor joint implementation team:

- Site IT project manager
- Site PACS administrator
- Site scheduling department
- Site network engineer
- Facility manager
- Radiology supervisor and lead technologist from each area
- Patient scheduling representative
- Vendor team, including project manager, installation engineer, or service support engineer

After the team is organized, members can meet and discuss the goals for and the expectations from PACS. When the PACS's initial implementation is complete, the site team will continue to meet regularly not only to search and coordinate solutions for any problems that may have arisen related to clinical applications, but also to identify and plan for future upgrades. PACS is still evolving at a fast pace, both in hardware (following the IT industry in general), and in software, which has increased functionality (following the development of new standards in DICOM, HL7, and IHE), because PACS's role goes well beyond just storing and transmitting digital medical images. New software updates are regularly released and the typical life span for a PACS system without a major hardware/software upgrade is approximately 5 years. All these matters require routine discussion at team meetings.

8.2.2 Defining Goals and Vision

In the early days of PACS, several projects around the world made the mistake of not defining specific goals before system implementation. The lack of clearly defined goals and expectations by the project team often resulted in unrealistic expectations among PACS users, who ended up feeling strong disappointment. Like all other technologies, PACS cannot solve all problems in a radiology department. It only provides the technical means to improve some aspects of patient care. Still, the human factor is the most important factor in PACS, and the acceptance of a PACS system by all related medical staff is critical to its success. Thus, a strategic PACS plan must be defined; such planning is likely to include the following steps:

- Form a steering group for strategic planning.
- Determine the reason to implement PACS.
- Determine the institution's vision and goals to be realized by implementing PACS.
- Determine areas to be served by PACS.
- Set high-level milestones and timelines for PACS implementation.
- Define benchmarks for measuring the success of PACS.
- Estimate equipment, infrastructure, staff, maintenance, and service costs.
- Estimate direct cost savings from using PACS.
- Define the number of implementation phases.
- Set goals to achieve at each phase.
- Determine the budget for each phase.
- Determine the budget for infrastructure investment.
- Determine purchasing leverage.

The goals defined by the team directly determine the scope of and steps toward implementing the PACS; therefore, they can serve as guidelines in contract negotiations. The goals and expectations also need to be communicated with all users as soon as the implementation of the PACS begins. In this way, there is no surprise for users following PACS deployment.

8.2.3 Coordinating Efforts and Consolidating Resources among Various Departments

Traditionally, PACS implementation is based on departments. Thus, at large healthcare institutions, it is not unusual to find the radiology, cardiology, gastrointestinal disease, and pathology departments at various stages of pursuing their own departmental PACS. However, in strategic planning at a large healthcare institution, it is never too early to plan for enterprise-wide multi-department PACS implementation and to consolidate both financial and human resources, if possible. Though each department has its own unique PACS implementation requirements, they will also have many common needs: for instance, network infrastructure within the institution and a redundant tiered data storage system for recovery after a disaster. These individual PACS requirements when planned strategically as an enterprise solution can almost always result in improved IT support and efficiency and a lower total combined cost for the institution. Individually managed and supported department- or modality-wide PACS, with their own upgrade and data migration needs, are a nightmare for the IT department responsible for systems support.

8.2.4 Project Planning

After strategic planning is completed, the process advances into the actual implementation phase, at which time project planning comes into play. Project planning for the healthcare institution typically includes the following items:

- Form a site implementation team expanded from the strategic planning team.
- Define team members' roles, responsibilities, and task assignments.
- Arrange regular meetings to communicate progress.
- Analyze and generate a flow chart for the current workflow.
- Based on the workflow, determine the estimated amount of data to be transferred during peak hours.
- Determine existing network infrastructure for the institution and the upgrades required to meet the PACS requirements.
- Determine short- and long-term data archive capacity requirements.
- Analyze the capabilities of the existing RIS and HIS; if no RIS is in place, implement an RIS first.
- Determine how long read images will be archived in the patient folder and whether images modified by clinicians will be archived or not.
- Track issues encountered.
- Select proper sites for a visit to see an implemented PACS.
- Analyze potential issues found during site visits.
- Define strategy for user training.
- Determine user and vendor responsibilities and roles during installation and support after acceptance.
- Perform acceptance testing for PACS subcomponents and system integration.

Next comes benchmarking of the installation at its various phases:

- Determine types of equipment to be installed and times when the equipment must be fully functional.
- Determine measurable deliverables and timelines.
- Determine if clinical operation is to be interrupted by installation and its duration.
- Determine the time to start user training.

All these processes must be undertaken with the understanding that PACS implementation is becoming increasingly about departmental or institutional workflow improvement, rather than about electronic storage and transmittal of images for softcopy reading only.

8.2.5 Workflow Analysis

Workflow analysis is one of the most important steps in the planning phase of a PACS. It determines the layout and archive configuration of the PACS. It also determines the network's bandwidth requirement. In the workflow analysis, the site implementation team studies the current workflow pattern. Based on clinical needs, the team also defines what locations images will be sent to and when. The team establishes rules about when, where, and what images can be retrieved and about when, what, and how images should be archived. Based on the number of images and their size and the frequency with which images are transferred among workstations, the team can define network speed and capacity requirements.

Each institution's workflow will differ depending on the services it provides and the amount of image data needing to be transported. There may also be differences in policies as to whether images should be released to clinicians without a radiologist's report and, if so, when they can be released. The team cannot overlook the impact of PACS on these policies and the consequences or issues involved. Customs and traditions tend to be very difficult to change and directly impact PACS acceptance by users.

8.2.6 Facility Planning

Like any big project, successful PACS implementation requires much coordination among the various departments at the healthcare institution. Facility planning involving the site implementation team is one essential component. Facility planning to meet required changes for PACS implementation typically includes the following:

- Floor space, reading room remodeling

- Electrical power supply, emergency power supply

- Special lighting for the reading room, ergonomic desktop furniture for PACS display monitors/workstation

- Simplified lightbox for occasional film reading

- Additional special ventilation and air-conditioning requirements

- Appropriate network and communication ports for all work areas

- Communications closet upgrade or construction, optional separate PACS-only LAN segment

Some of these items require frequent review because the needs may change over time.

8.2.7 Storage Requirements

Database and storage management is probably the most important component of a PACS. This management process controls the data archive and flow among workstations around the hospital in addition to data management in the archive. The archive usually includes both short-term, fast-access, and long-term slower-access components. It is crucial that users decide correctly how much archive capacity to purchase and when. Because the price of archive media decreases constantly, it makes little sense to spend a sizable amount up front for storage media that will be available for a lot less in the near future. On the other hand, the initial installation of a PACS needs enough capacity to handle operational requirements. In other words, the user should initially have enough archive capacity to keep newly acquired data stored for a relatively short duration of typically 2 years and add required media such as magnetic tapes or Blu-ray optical disks as needed.

The site implementation team needs to estimate the amount of imaging data that will be generated per year by reviewing the data from the previous year for all imaging modalities involved and adding an annual volume increase estimate; the team must also decide what kind of archive media to purchase and the initial archive capacity. To make sure the estimate is accurate, the team should consider the compression ratio used by the PACS archive system, as stated by the vendor. The long-term scalability of the archive portion of the PACS selected also needs to be addressed. Currently, the short-term archive that keeps the data online and provides fast access is usually a RAID, as discussed in Chapter 5. As the price of spin disks decreases, this format is becoming more popular for long-term archiving, together with tape or optical disk libraries/jukeboxes. However, redundant long-term data backup for disaster recovery usually uses only tape or optical storage media. Some vendors provide PACS data storage services, with one copy of the data stored on spin disks in one location and a replicate of the data backed up and stored on slower tapes or optical disks in another location. For institutions with limited in-house IT resources and a fairly small volume of data to be stored, this is an option worth considering: the user is freed from the responsibility of keeping the data safe, and the vendor is responsible for migrating the data when storage technology becomes obsolete and new technology is available. Details about PACS data storage were covered in Chapter 5.

8.2.8 Network Infrastructure

Depending on the amount of data flow, the existing network infrastructure may or may not be adequate for the successful operation of PACS once installed and in clinical operation. In fact, in the early days of implementing PACS, a common mistake made was improper network analysis, resulting in insufficient network capacity. Inadequate analysis was one of the main reasons why several PACS projects failed, which created a general resistance to PACS. It is not difficult to imagine that if a network based on 10 Base-T Ethernet created a decade ago is used to handle traffic from a PACS that includes computed radiography (CR), ultrasound, MRI, and CT images all together at the same time, the result would be a predictable failure. Even though the network issue can be resolved by upgrading after the PACS is installed, the damage may already have been done in terms of the overall medical staff's forming a bad impression of PACS. Network analysis during implementation planning must involve an experienced network or IS engineer. Most of the time, the PACS network can be segmented from the hospital-wide network so that the heavy traffic within the PACS network will not slow down the functionality of other departments.

8.2.9 Imaging Modality Assessment

The status of DICOM conformance and IHE integration of image-acquisition modalities in radiology changes constantly. Radiology equipment manufacturers are at different stages of implementing DICOM and IHE, and the degree of DICOM conformance and IHE integration depends on the equipment type and vendor. For example, the latest MRI scanner from a manufacturer may support typical DICOM service classes and some IHE integration profiles, whereas the latest CT scanner from the same manufacturer may not support any IHE integration profiles at all. At present, no single piece of commercial equipment on the market fully supports all DICOM functions and all IHE integration profiles. This lack is partially due to the fact that the DICOM standard and IHE are constantly evolving. However, it is expected that all new radiology equipment will be DICOM conformant in some basic functions, including storage, querying/retrieval, modality worklist, and DICOM printing. It is foreseeable that, increasingly, more equipment will support IHE integration profiles.

Currently, major image-acquisition modalities in radiology include MRI, CT, ultrasonography (US), nuclear medicine (NM), radiography, mammography, fluoroscopy, and angiography. All the latest devices in these modalities acquire and store images in digital format. At present, all imaging modalities support the DICOM image format. However, the image matrix size varies from modality to modality. In addition, DICOM conformance status varies from vendor to vendor. The following is a brief summary of the DICOM conformance status across modalities, without regard to differences between vendors.

8.2.9.1 Magnetic Resonance Imaging

MRI images are inherently digital. Most images are stored in a 256×256 data matrix, some in a 512×512 data matrix. The average patient has about 200–500 images, but the range is wide, with significantly more images stored for patients receiving some MR angiography or FMRI studies. All the latest MRI scanners sold today support the aforementioned DICOM basic functions.

8.2.9.2 Computed Tomography

CT images are inherently digital as well. The image matrix size is usually 512×512. The average patient has about 700–1500 images when the number of CT detector rows are increased from 4 to 16, images, significantly more if the images were acquired with even more detectors or if the patient underwent CT angiography. All of today's CT scanners support all the basic DICOM service classes.

8.2.9.3 Ultrasonography

US equipment has advanced rapidly in the last decade. Nearly all the latest high-end US scanners are inherently digital and support basic DICOM service classes. However, some legacy US scanners may still need an external video frame grabber to capture and convert the video signal into digital images. Then, the digital images are converted into DICOM format and communicated to a PACS on a separate interface box. The US image matrix size is usually 256×256 or smaller, with 50 images per patient on average, significantly more in the case of dynamic or vascular studies.

8.2.9.4 Nuclear Medicine/Positron Emission Tomography

The latest NM cameras acquire images in digital format. The image matrix size is usually 128×128, with about 2–4 images per patient, except in the case of positron emission tomography (PET), for which the number of images can be of the order of 100 per patient. Current NM cameras support the basic DICOM service classes.

8.2.9.5 Computed Radiography and Digital Radiography

Radiographic images traditionally have been captured with an analog screen/film system. CR, however, captures the radiographic image on a photostimulable phosphor plate and then uses a laser to read out the image digitally. Digital Radiography (DR), conversely, uses micro-semiconductor devices to convert a radiographic image directly into a digital signal. A typical CR/DR radiographic image is approximately 2000 × 2000 pixels large, yielding a per-image size of around 8–10 MB. In a busy radiology department, several hundred examinations may be performed each day. Therefore, to store all the CR/DR images requires considerable archive capacity. All the latest CR/DR products support basic DICOM service classes.

8.2.9.6 Digital Fluoroscopy and Vascular Imaging

There are two types of digital fluoroscopic images. One is the series of images taken at 15–30 frames/s (often called the scene image). Another is the digital spot image. All the latest digital fluoroscopy systems support basic DICOM service classes; however, it is important to note that legacy fluoroscopy units may not support all the DICOM service classes.

8.2.9.7 Mammography

In the past decade, digital mammography equipment has become broadly available. Digital mammography images are captured in matrix sizes of 4000 × 4000 or larger. All digital mammography systems support the basic DICOM service classes.

In summary, advances in DICOM conformance of radiology equipment have made directly connecting imaging modality scanners to a PACS much easier now than it was a decade ago, when the first generation of commercial PACS became available; thus, today's systems provide physicians access to radiological images as soon as they are acquired and improve the radiology workflow significantly, resulting in better patient care. However, there is still a long way to go before all patient data, including lab results, patient histories, and radiological images, are integrated seamlessly into a master database with the EHR. Nevertheless, with the rapid developments in PACS, associated standards, computer technology, and network communications technology, a more efficient, computerized healthcare system is foreseeable in the near future, and the PACS will become only a small component of the whole.

8.3 PACS EQUIPMENT SELECTION AND CONTRACT PREPARATION

In the equipment selection and contract preparation phase of the planning, the scalability and future expansion of the PACS must be considered. Since PACS involves primarily computer hardware and software, it is critical to keep the system open to future upgrades and to the integration of equipment. Ideally, the buyer would not be locked into having to purchase any specific vendor's product. It is only because of the DICOM standard and conformance to it that such open PACS are possible. Therefore, the buyer should insist on DICOM conformance of the system purchased. In fact, the buyer should not waste time looking at a proprietary system if its functions can be performed by a DICOM-conformant one.

DICOM is quite a complex and evolving standard. It has been discussed in Chapter 6, and the detailed documents may be obtained at NEMA (www.nema.org). It includes many functional specifications called service classes. Each service class specifies a particular function. For example, the storage service class specifies the function of sending the image data in the standard way (by the storage class user) so that the receiving party (the storage class provider) can receive it properly. At present no vendor claims that it supports all service classes specified by the current DICOM standard. Further, ACR/NEMA has specific guidelines vendors must follow to claim DICOM conformance. The buyer should bear in mind that a sales representative from a vendor may not necessarily understand the full meaning of DICOM conformance. The system's specifications should, therefore, be studied very carefully, perhaps in consultation with an expert.

8.3.1 PACS Selection

Selecting the PACS to be implemented in the healthcare institution is no doubt the single most important decision in the entire planning phase. There are well over 30 PACS vendors in the marketplace today. Each has a different set of technical and support strengths; they target different user groups, but with some overlap; and they have various reputations. These factors, together with the constant corporate mergers occurring in the IT industry that raising the possibility of discontinued product support, and the different price tags associated with the product, selecting the right PACS system, one that will meet most needs of the healthcare institution, is not a trivial task.

8.3.1.1 Hardware Considerations

Hardware options for PACS follow advances in the IT industry in general, including servers, network, data storage, display monitors, and so on. The following factors must be considered during hardware selection for a PACS:

Facility requirements for PACS:

- Physical floor space

- System heat dissipation and ventilation

- Electrical, emergency power, and uninterrupted power supply

- Lighting for the primary diagnostic reading room and clinical reviewing area

Network capability and protocol:

- Proposed infrastructure, configuration, and bandwidth of the network for server, storage, and workstations; for example, Ethernet with 100 Mbps, 1 Gbps, or FC as discussed in Chapters 4 and 5

- Accessibility of the data

- Speed of retrieving PACS data from the archive to the workstation and of transferring it from workstation to workstation

Database and storage system:

- Centralized or distributed archive

- Unit cost for short-term online storage per TB of data, RAID level, and hot swap disk standby

- The number of maximum simultaneous disk failures allowed before unrecoverable data losses occur, the time it takes to rebuild a failed RAID disk for either the server RAID or short-term storage RAID

- Short-term and long-term data storage configurations; for example, DAS, NAS, SAN, and the level of RAID they employ

- Long-term storage types; for example, spin disk arrays, tape, or Blu-ray disk libraries/jukeboxes, the number of tape cartridges or disks that can be held, the total capacity, and unit storage cost per TB of data

- Data retrieval and write speed at 0%, 25%, 50%, 75%, and 90% capacity

- Short-term and long-term data storage scalability for future expansion and migration

- Database management used

- Data storage failure rate and redundancy in the archive and tiered storage options

- Type of data compression and compression ratio used, where it is compressed and decompressed

- Security measures for accessing the central storage system

Image display monitor configuration on PACS workstation:

- Display monitor type, size, and frame rate

- Display graphics card and video memory

- Display resolutions and pixel bit depth

- Bank of display monitor numbers supported; for example, two- and four-bank display monitors

- Luminance range, for example, minimum black versus maximum bright

- Luminance consistency among multiple monitors

- Stability of display monitors, characteristic variation over time, and recalibration

- Uniformity of luminance

- Life expectancy of monitors

- DICOM LUT or GSDF calibration capability
- Display QA/QC monitoring services
- Warranty on display monitors
- Antireflective coating on monitors
- Color of the display monitor case

Digitizer devices:

- Film digitizer for films, and frame grabber for analog imaging modalities
- Dynamic range of the digitizer, for example, pixel bit depth
- Optical density range of the film digitizer
- Image format, resolution, and compression
- Built-in SMPTE or AAPM TG 18 test patterns as discussed in Chapter 3
- Interface of the digitizer with the PACS
- Preview station for the digitizer

Software and hardware upgrades:

- Proprietary or off-the-shelf hardware
- Scalability and upgrading of hardware
- Future hardware upgrade guarantee and associated cost

8.3.1.2 Software Considerations

In contrast to hardware options, software options are very PACS-oriented and unique to individual PACS vendors. In addition to bug fixes, new software releases often provide new features of PACS by adopting more DICOM and HL7 standards and IHE integration profiles. The following is a general list of items to be considered when selecting software components for a PACS:

Standards and their versions supported:

- HL7 v2.x or HL7 v3.x, CCOW, and so on
- DICOM service classes supported by the workstation and archive
- IHE integration profiles supported, for example, SWF, Patient Information Reconciliation (PIR), and so on

File format and standards supported:

- The image matrix size
- Pixel bit depth; for example, 8, 10, 12, or 16 bits
- Raw and processed image management
- CR/DR stored image file format, for example, processed or raw
- Type of file compression and compression ratio, lossy or lossless, used by the PACS

PACS workstation software and user interfaces:

- Workstation requirements, for example, PC or Macintosh-based, CPU types and frequency, operating system (Windows, Linux) and version
- Window and level auto/manual
- Hanging protocol support
- Image manipulation tools provided, for example, magnifying, roaming, zooming, rotation, flips, distance measurement, text notation, ROI measurement
- User interface, for example, ease of learning and use, short-cut keystrokes

- Intuitive graphic patient list, worklist, and ability to sort studies by name, date, modality, clinical number, type of examinations, reason why examinations ordered

- Adequate magnification range with continuous zoom

- Postprocessing capabilities or options supported, for example, edge enhancement, surface rendering, 3-D reformat, maximum intensity projection (MIP), image fusion, histogram analysis, computer-aided detection (CAD) and their cost

- Acceptable speed of the software, fast switching between function windows

- Easy configuration on PACS workstation to add additional servers

- Capability for creating teaching files

- Screen capture capability

- Software support and customization, upgrade, and cost

- Web distribution capability

- Capability to interact with other vendors' products

  Custom configuration capability:

- Hanging protocol tailored for individual user with customary settings

- Workstation display of prior examinations and reports

- Querying the archive and other servers conveniently

- Adding image send or query destinations conveniently

- Adequate speed of bringing up images

- Adequate speed of roaming and moving images with acceptable latency

8.3.1.3 System Integration Issues

Unlike in the early days of PACS adoption, during which much effort focused on hardware and software developments with very basic DICOM capabilities, recent and future advances concentrate more on integration of the PACS with RIS/HIS/EHR systems, adoption of newer DICOM standards, integration of IHE integration profiles, and realization of the potential of fully digital, filmless operation to significantly improve workflow and efficiency and ultimately benefit patient care. Attention should be paid to the following issues related to system integration:

  Hardcopy output capability:

- Hardcopy print capability

- Dynamic range of the hardcopy print, for example, maximum and minimum optical density for film, pixel bit depth, and DICOM print service class supported

- CD/DVD recording, embedded DICOM image viewer on CD/DVD, and similar capabilities

  Integration with RIS/ HIS:

- Presence/lack of a built-in RIS function

- Compatibility of the RIS reports with HL7

- Interface between the PACS and the existing RIS/HIS of the hospital

- Known capability to interface with various RIS

- Worklist arrangement, auto data entry from RIS/HIS versus bar code entry, and similar features

- System throughput and the maximum number of simultaneous users allowed

  Security issues:

- Password and associated privileges

- Regular password update

- User password and associated access privilege setup by PACS administrator
- Appropriate warning for attempts of unauthorized break-ins
- Regular user log-on monitor and record
- Record for access history to the patient image files

Downtime and uptime:

- Average downtime and guaranteed uptime of the system
- Downtime procedure
- Emergency backup mechanism
- Credit for excessive downtime
- Remote reboot capability for the server

QC program with the system:

- System self-check capability and frequency
- Self-QC items
- Self-QC duration
- Automatic alert for file loss
- QC workstation and function
- QC on monitors and workstations
- Storage space check

Training:

- User training length
- User training cost arrangement
- Staff training plan at different implementation phases

8.3.1.4 Image Display System Requirements

The PACS display system is not discussed in the hardware section because PACS workstation display monitors most likely will be selected separately from the rest of the hardware. The use of the individual PACS workstation dictates the requirements of the display monitor. Although the differences between monitor types and their capabilities are covered more thoroughly in Chapter 3, the following requirements for various display monitors are recommended.

For primary diagnostic workstations (nondigital mammography):

- Minimum monitor luminance level greater than 400 cd/m^2
- Luminance stable (decrease of less than 5%) over a 3-year period
- Ambient light level controlled at lowest possible
- Minimum monitor resolution of 3 megapixel (MP), preferably greater than 5 MP
- Antiglaring or antireflection on monitor
- Ergonomic workstation design

For clinical review workstations (the floor workstations):

- Minimum monitor luminance level at least 100 cd/m^2
- Luminance stable (decrease of less than 5%) over a 3-year period
- Ambient light level controlled at lowest possible
- Minimum monitor resolution of 2 MP
- Antiglare or antireflection coating on monitor preferred

For office PC-based workstations:

■ Statement such as "Not used for diagnostic purpose" posted on the workstation

■ 21-in. monitor preferred

■ No display QA required

■ Department housing the monitor to comply with network requirement for these computer systems

■ Department housing the monitor responsible for computer maintenance

8.3.2 Writing Specifications for the Request for Proposal

The Request for Proposal (RFP) can serve as the basis for the final purchase contract. It is therefore important to specify the system requirements in detail. Unfortunately, accomplishing such a goal is not easy because PACS are very complicated systems and are still evolving. Therefore, hiring a professional to help with the RFP may be a reasonable option, especially for smaller institutions or clinics without in-house expertise in this area.

Two approaches are common in preparing a PACS RFP. The first approach, which is also the most common in radiology equipment purchases, is to specify every technical detail about the system. Government agencies use this method often. This approach works well for a known, mature product such as an x-ray or CT unit. However, it may not work very well for institutions considering PACS purchases. The second approach is to ask very specific technical questions and let vendors answer them. For example, instead of specifying the type and the speed of network that should be used with the PACS, the buyer can ask how long the PACS takes to prefetch certain images from one point to another. Once the information is obtained from all vendors, the buyer can compare these competitive PACS products.

Since conformance to DICOM standards and IHE integration are being adopted incrementally, a vendor should not be excluded solely based on lack of support for a single DICOM service class or IHE integration profile. However, buyers may ask that specific DICOM and IHE functions be provided at no or minimal cost if these functions become available within a certain period of time from the initial purchase.

In summary, major areas to consider in selecting a PACS from a vendor are:

■ Conformance to DICOM standard

■ IHE integration profiles implemented

■ System openness and network topology

■ Database distribution

■ Workflow management capability

■ User interface design of PACS workstations

■ Interfacing capability to RIS and HIS

■ System function and performance

■ Reliability

■ Service and technical support

■ Existing installation sites or customer base

■ Reputation and years in the PACS business

The minimum DICOM functionality requirements for very basic DICOM service classes for image transfer and storage should be:

■ DICOM storage service class

■ DICOM query/retrieve service class

■ DICOM modality worklist service class

■ DICOM print service class

As for IHE, preferred IHE integration profiles to be supported are:

- SWF

- PIR

- CPI

- PGP

- KIN

- SINR

- PWF

Details about the functions of these IHE integration profiles have been covered in Chapter 6.

8.3.3 RFP Evaluation

Once bids and quotes from vendors are received, the site implementation team meets to analyze each bid carefully. The main functionality of the various PACS products should be compared. At the time of this writing, there are still major differences in PACS products. Some use centralized archives, some distributed. Some conform with many DICOM service classes and IHE integration profiles, others with only a few. It is critical to have an expert who has PACS experience to analyze the bids. Sometimes a consultant may provide the necessary help.

8.3.4 Site Visit

After careful evaluation of bids, the buyer can narrow down a few vendor products for site visit. During a site visit, the buyer can see the PACS in actual clinical operation. The buyer may find that the clinical operation at the site differ from their own. Unique operational issues will need to be addressed to assure PACS success. In addition, during a site visit the buyer may identify conditions or issues not mentioned by the vendor. It is always a good idea to discuss a vendor's system performance with friends and colleagues who are currently using it. During the site visit, the buyer should pay special attention to workflow, archiving, and the user interfaces at the PACS workstations.

8.3.5 Contract Negotiation

As in any negotiation process, the buyer of PACS can fully utilize their purchasing power. Often the buyer has more leverage than he or she realizes. For example, the buyer may be in a geographical location that one PACS vendor wants to use as a demonstration site to compete with other vendors who are already doing business in the same area, or the buyer may belong to a multihospital purchasing group, or the vendor may be approaching the end of its fiscal year and so be inclined to offer some concessions in order to make the sale and improve the bottom line.

In the contract, the purchaser should address clinical workflow issues so that there are no last-minute surprises or excuses for the vendor's team to delay installation. It is recommended that the criteria by which the PACS will be considered acceptable be specified. In addition, the buyer must specify some of the critical functions that the PACS will have to perform, such as whether a user is able to print from a given workstation to an arbitrary printer. The vendor typically provides such information in its response to the buyer's RFP. Because there are so many details to the PACS and so many options available, putting technical details into the contract can avoid many misunderstandings and frustrations in the future. For example, the buyer should have the vendor lay out all hardware and software requirements to perform a certain function to avoid incurring costs to add so-called "optional software" so that the PACS can perform a certain critical clinical function.

Further, the buyer can insist that the price of PACS components be itemized, though PACS vendors may try to sell the system as a whole package. An itemized list gives the buyer pricing information to consider in the context of future expansion, such as adding workstations. Many PACS vendors will agree to give the buyer an itemized cost list when requested.

In the contract negotiation, the buyer also must address the servicing of PACS. In the first few weeks or months after installation, users frequently want to reconfigure it to meet some specific needs. For example, the user may decide to send images to somewhere other than originally planned. It is important to put some service clauses in the contract so that such requests are met in a timely fashion. It is mutually beneficial to both the buyer and the PACS vendor to build a good, long-term relationship so that the future expansion of the PACS is smooth and painless.

It also may be necessary to have the legal department involved early in the contract negotiation process, depending on the institution's internal policy. It is always a good idea to put some penalty and payment clauses in the contract so that the vendor has some incentive to finish installation on schedule and maintain the guaranteed up-time for the system. One common practice is to pay 80% at shipment of equipment and the final 20% after completion and acceptance by the user.

The purchaser should also be aware of contractual issues related to PACS functionality and future upgrades. As noted above, because the nature of PACS is quite different from that of any other single piece of medical imaging equipment, software releases are issued frequently, not only for fixing bugs, but also for adding functionalities to further improve workflow. Therefore, during the contract negotiation phase, it is recommended that the buyer include the following:

- Vendor guarantee to keep PACS software up-to-date for 3 years at no or minimal cost to the purchaser; the purchaser and the PACS vendor must agree on the definitions of "software upgrade" and "new software release" in advance

- As part of the service contract agreement, that the vendor will provide free system software reconfiguration, such as adding more DICOM acquisition devices (e.g., US, CT, MRI, CR/DR, NM)

- System up-time requirement and service response time

- The cost of expanding short-term and/or long-term archive capacity

8.4 IMPLEMENTATION

When the purchase order is close to being issued, the healthcare institution forms the site implementation team as soon as possible. This team includes network engineers from the hospital, project engineers from the vendor, the facility manager from the hospital, and anyone else who is involved in the PACS installation. Others who may be involved in the installation may be invited to the implementation meeting at least once so that they will understand the scope and timeline of the installation. The team should meet regularly or as needed during the installation, because good communication between all the departments involved is one of the key elements for a successful and smooth PACS implementation, especially when workflow is changed significantly.

8.4.1 Implementation Team and Responsibilities

The installation team usually includes personnel from both the healthcare institution and the vendor. Project engineers from the vendor coordinate most of the configuration work and software uploads onto the site's servers and workstations. The healthcare institution appoints a coordinator to coordinate the work within the institution, such as network cabling and facility modification, and assigns clinical or equipment field engineers to configure the imaging modality scanner and related tasks. The installation team also serves as a channel for communication among all parties involved. When the members of this team hold regular meetings, it is much easier to resolve issues that come up during installation and beyond.

8.4.2 Timelines and Clinical Area Downtime Planning

Careful planning of the PACS installation is important for a busy clinical setting because there must be minimal interruption of clinical operations. Usually, installation of PACS workstations does not interfere too much with routine clinical operations. PACS workstations can be configured at prestaging, when the software is loaded into the system. Installation of the archive and storage device can be time consuming, but it causes minimal interruption to clinical operations.

However, interfacing image-acquisition devices such as MRI, CT, and other imaging modality scanners with the PACS may require scanner downtime. To minimize the interruption to clinical operations, the team may opt to do the interfacing over a weekend or late in the evening. Usually, DICOM configuration of CT and MRI scanners takes less than 2 hours if they are already in DICOM conformant.

The radiologist's routine suffers the most during interfacing work. If in the past many scans had been taken on film-based scanners, the radiologists may have a difficult time comparing images, as the radiographs from previous scan will have to be viewed on a light box, but the images from the new scan must be viewed on a PACS workstation display monitor. Even if the light box and monitor are close to one another, because of the different luminance levels of these two displays, the radiologist's eyes will be constantly trying to adapt. In addition, it is very tiresome to switch back and forth from a light box to a computer display monitor. To resolve this issue, the institution may consider scanning films with a digitizer and then preloading prior examinations into the PACS before the patient comes in for a new one, so that radiologists can read from the PACS workstation from day one. Other challenges may occur when new features of RIS, such as speech recognition for report dictation, discussed in Chapter 2, are incorporated into the RIS and interfaced with the PACS.

8.4.3 Acceptance Criteria and Testing

The criteria for acceptance testing must be defined prior to PACS installation and written into the contract. All those responsible must be satisfied with the PACS before the final acceptance is signed and the last payment made. Issues related to acceptance testing are:

- Acceptance criteria setup
- Final sign-off procedure
- Function of PACS workstation software
- Function of optional software
- Function of QA/QC workstation
- Proper function of interface with HIS/RIS
- Workflow configuration
- Speed of image transfer between modality scanners to PACS, query, and archive
- DICOM function of interfaces between image-acquisition devices and PACS
- IHE implemented integration profile capabilities
- Image quality on the monitor
- Conformance of each display monitor to quality specifications
- Image file integrity
- Image printing and quality
- Functionality of the archive and database management (e.g., prefetching, autorouting)
- Server remote reboot capability
- Proper staff training
- Completion of staff training

8.4.4 Transition and Continuous User Training

User training is a very important factor that may well determine the acceptance of PACS by various user groups in the healthcare institution. Different user groups require different levels of training, and user training is an ongoing process. Since PACS includes all imaging modality scanners and is interfaced with the RIS/HIS, any PACS software changes and release of new functions related to integration with RIS/HIS and any modality scanner additions will cause a change in the workflow. Benefits from changes to workflow and efficiency improvements can only be realized when all staff are properly trained and communicated with about any updates. Especially for first-time PACS implementation, the following decisions are to be made before the PACS is installed:

- Determine the responsibility for user training at various levels (e.g., who will bear the cost and ensure that prerequisites are fulfilled?)
- Determine who should be trained and at what level
- Determine the PACS vendor's responsibilities (e.g., to provide an on-site application specialist for a specific period of time)
- Determine a super-user group which consists of expert users

8.4.5 Teleradiology Considerations

PACS implementation within a healthcare institution is the first step in enabling the institution to perform teleradiology services. Similar to the planning of a PACS, implementing a teleradiology service requires analysis based on economic and technical common sense. There may not be enough need for such a service in some areas; likewise, sites with very few patients cannot justify the cost of setting up a dedicated workstation. One additional difficulty teleradiology initiatives have to face in the United States, as discussed in Chapter 7, is the legal issues involved in practicing medicine across state boundaries. Because of turf issues, technical difficulties, and some legitimate concerns about the quality delivered by this technology, teleradiology as part of the telemedicine effort has not progressed as far as some analysts predicted when PACS began to gain in popularity in the 1990s and 2000s.

For teleradiology, it is crucial to implement a QA program on all workstations to make sure that images are presented in a consistent manner and with optimal quality.

8.4.6 Routine QA

After installation, some minor bugs and software glitches in the PACS may require service. Once these problems are resolved, the PACS usually operates very smoothly. Maintaining the system may or may not require an occasional reboot; for some PACS, this can be done remotely at any place with Internet access. The user should follow the instructions provided by the vendor. There may also be some minor changes in the PACS configuration over time, such as autorouting images to a different location. Well-trained PACS administrators should be given the tools and training to make these changes themselves. They should be trained by the vendor prior to the system's installation; they can then train new users within the organization. Administrators are responsible for the regular maintenance of PACS, including QA and data integrity checks. In addition, they are responsible for maintaining the QC program on the PACS, which includes, at a minimum, QC of workstation display monitors, maintenance of the archive device, and monitoring and testing of the data integrity in the archive.

8.5 CASE STUDIES

The following sections present two actual PACS implementation cases, one at a children's hospital with 400 beds and the other at a tertiary care hospital with 950 beds; the latter is also the flagship hospital of an enterprise healthcare delivery system with many hospitals and clinics.

8.5.1 PACS Implementation in a 400-Bed Children's Hospital

This children's hospital was among the pioneers in the early adoption of PACS. Since most of the imaging modalities did not fully support DICOM service classes such as storage and query/retrieve, several third-party DICOM conversion boxes were used in order to create interfaces between the image-acquisition equipment and PACS. Although some of the technical issues that arose at the time are no longer relevant, the principle of planning and implementing the PACS is still valid and applicable today. The case is presented here partially to add some historical perspective to the development of PACS and to provide a point of comparison to today's state-of-the-art.

8.5.1.1 Defining Goals and Expectations

The project started with a small group of people as the strategic planning team: a radiologist, a medical physicist, the vice president in charge of the information service, and the radiology administrator. This group defined goals and identified the areas in which a PACS might most improve the efficiency of the radiology service. An internal analysis found that the ER, ICU, and neurology and neurosurgery units constituted ~70–80% of the CT/MRI services. In addition, CT and MRI were inherently digital and DICOM conformant. These factors made CT and MRI ideal candidates for the first phase of the PACS implementation. It was decided that the ER, ICU, and neurology and neurosurgery departments would be included in the first phase of PACS implementation. Based on each of the department's needs, the team decided on the numbers and types of workstations to be installed at various locations. This team was the steering force behind the later, expanded site implementation team, which included a technologist from each clinical area in radiology, the patient scheduling manager, an IT engineer, facility management personnel, and others.

8.5.1.2 *Justifying PACS*

The PACS project was not justified based on the possible financial savings within the radiology department. Rather it was justified on the grounds of improved radiology services to other departments, from which the hospital as a whole could be improved.

8.5.1.3 *Workflow Analysis*

The planning team met numerous times to map out the workflow of CT/MRI services. The team carefully studied the needs of the various departments that used the CT/MRI service. Then the team decided on the configuration for the physicians' workstation, analyzed the physicians' needs, and decided how and when physicians would be able to access images. The team decided to make CT/MRI images immediately available to physicians as soon as they were acquired.

The simplified diagram in Figure 8.1 indicates the image data flow in the first phase of PACS implementation. CT and MRI send images to the cluster server. New CT and MRI studies are stored on the diagnostic reporting workstation server of the neurosurgery node for creating reports, and after sign-off are sent to the archive. Images are also sent immediately to the neurosurgery workstation for viewing. Should neurosurgery wish to print images on film, the images are sent to the laser camera. The same basic procedure was followed for the neurology department.

New CT and MRI studies for ER and ICU were also stored on the diagnostic reporting cluster server during reporting, and after sign-off were sent to the archive. Selected images from the CT and MRI studies were manually sent to the ER department workstation or the ICU workstation.

8.5.1.4 *Network Requirement Analysis*

Based on the workflow, the strategic planning team calculated the amount of image data to be transferred from one place to another. The workflow analysis found that the network infrastructure at the institution was adequate to handle the data flow. The strategic planning team decided to use the existing network as the backbone for the PACS network but to segregate the PACS LAN from the main hospital-wide LAN.

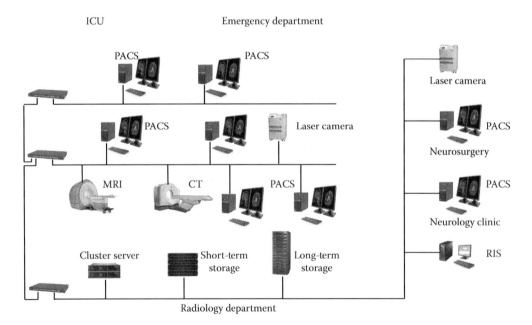

Figure 8.1 CT and MRI integration with PACS, so that all new studies are sent to PACS workstations in the neurology clinic and to neurosurgery with the option to print, and selected studies were sent to PACS workstations in ER and ICU department. All studies are archived on both short- and long-term storage systems.

8.5.1.5 Estimation of Archive Needs

Based on a study of the amount of data generated from the previous few months, the team determined the size required for the short-term RAID and the long-term tape library and purchases were made accordingly.

8.5.1.6 Writing Specifications for the RFP

The strategic planning team did careful research on all major PACS vendor products. Instead of issuing an RFP, however, the team found a PACS product that met all essential needs and then negotiated directly with the vendor. However, not every institution or clinic can take this approach. Therefore, an RFP may still be critical. The children's hospital did not go through the bid process, as the PACS was selected based on research of the existing products and the identification of a PACS product that met the institution's needs.

8.5.1.7 Site Visit

Several site visits were conducted to look at various PACS products. During these visits, the team paid close attention not only to the PACS product itself, but also to the workflow and operational issues. Site visits are very beneficial to the decision-making process in selecting the PACS vendor. From these site visits, it was found that most problems or issues related to interfacing between HIS/RIS and PACS, as well as the interfaces between image-acquisition modalities and PACS.

8.5.1.8 Contract

Once the vendor was selected for the PACS, a contract was negotiated for a multiphase, multiyear project. Itemized prices for all PACS components were requested. The hospital agreed to be a demonstration site for the vendor in exchange for a price discount.

8.5.1.9 Installation and Timelines and Clinical Interruption Planning

The initial strategic planning team was expanded to include facility management and network engineers from the information service. HIS/RIS representatives were also included to coordinate interfacing issues. Occasionally, imaging modality vendor representatives were invited to coordinate solutions for interfaces involving their equipment.

Installation and configuration of the CT and MRI scanners was planned to occur over weekends and in the late evenings. In-patients were rescheduled and scanners kept available with 30 min of notice for emergencies.

Past MRI/CT images of each patient were preloaded onto scanners and then pushed into the PACS workstation and archive the evening prior to the patient's new examination date, according to the schedule. This was very labor intensive, but it made the reading process go much faster.

8.5.1.10 Maintenance and QA

A PACS system administrator was hired from among the technologists of the radiology department. The administrator became responsible for training new users, maintaining the PACS workstations, and archiving images efficiently. To ensure image quality consistency across all workstations of the same class, the administrator implemented a QC program on the workstations. This step was critical in controlling image quality because the hospital's computer display monitors varied greatly in size, quality, brightness, and color tones.

8.5.2 PACS Implementation in a 950-Bed Tertiary Care Hospital

The second example of PACS implementation is somewhat different from the first one. This is a community tertiary care center with over 950 beds. It is part of a healthcare enterprise consisting of several large flagship hospitals and many clinics, with several radiology groups. The history of PACS used in the radiology department can be traced back to 1996 with the implementation of an US mini-PACS, with mobile US scanners sending images during examinations to the central server for access by workstations located in four central reading areas. The initial benefit of filmless operation for the US area was much appreciated by the medical and technical staff. A few years later, a department-wide PACS was implemented to include CT, MRI, US, and NM, which were then digital modalities with DICOM capabilities in the department. Past digital US images stored on the US mini-PACS were migrated to the department PACS system. During this time, filmless operation was expanded to the above mentioned imaging modality areas, and the need for a PACS that was integrated with the RIS to improve workflow was widely appreciated by the department staff.

8.5.2.1 Planning

The enterprise-wide adoption of PACS and its integration with the HIS and RIS started in 2003. The hospital was selected to be included in the first phase of the enterprise's PACS implementation. A site implementation team was formed that was made up of a strategic planning team to define the enterprise's PACS vision and develop the implementation strategy, a project leader to coordinate the various departments and areas within the hospital, clinical engineering staff to provide imaging modality support, and a corporate PACS IT team fully engaged in PACS implementation for the enterprise. The hospital further appointed a full-time PACS coordinator who had a radiology technology background; this person was dedicated to implementation and management of PACS at the hospital. The enterprise healthcare system, with its own fully staffed IT department, was responsible for all IT-related issues.

During the implementation, the PACS vendor assigned a dedicated project manager as a liaison to coordinate activities requiring collaboration between the site, the vendor, and the various imaging modality manufacturers. In addition to the project manager, the PACS vendor also formed a vendor team including engineers and clinical and application specialists to work on-site or remotely, as necessary, to accomplish the tasks required for a successful installation.

The switch to the new enterprise-wide PACS started from the portable digital radiography area, with the workflow illustrated in Figure 8.2. Clinical engineers and the PACS vendor configured CR and C-arms to send images to the PACS server. The vendor also configured the server such that short-term storage located in the hospital's main building complex sent data to the central archive in the enterprise data center for long-term storage. The short-term storage was based on RAID-5 and was initially planned to hold all modality scanner data for 12 months; the capacity was later expanded to hold 24 months' worth of data.

8.5.2.2 Facility Planning

Facility planning was the institution's responsibility. The institution finished all facility preparations complying with the vendor's functional and environmental requirements. The site renovated the reading room to provide the proper lighting conditions and furniture for the PACS workstations. Some simple light boxes were also kept in the main reading area so that PACS images could be compared with those on film from previous exams. All infrastructure modifications, including creating floor space for PACS equipment, performing network communications closet upgrades or construction, as shown in Figure 8.3, and providing electrical power with emergency backup were prepared by the institution. Also the institution ensured the physical security of the delivered PACS equipment against tampering, damage, or loss. During the installation and configuration period, the institution provided the vendor with office space for its on-site work crew and the privilege to access areas where the installation was occurring. The vendor and its subcontractors were required to follow general safety guidelines when on-site, observe security policies, and follow any special state, local municipal, or site regulations and codes that were applicable.

8.5.2.3 Network

The institution was responsible for the network infrastructure, except that the vendor was allowed to arrange for network evaluation and assessment from a third party when necessary. Thus, the institution's responsibilities included:

- Administration of hospital LAN and enterprise WAN configurations
- Reliable network connection where PACS devices were located
- Network performance
- Management of secure remote user access and configuration
- Provision of VPN or modem dial-in access so that the vendor can remotely support and monitor PACS performance

8.5.2.4 Imaging Modality DICOM Connectivity

The imaging modality scanners' proper DICOM functionality was a prerequisite for the scanners to be integrated into the PACS. However, ensuring the appropriate DICOM connections was the responsibility of the site. The institution's clinical engineers or field service engineers who would be servicing the imaging modality scanners, together with the PACS vendor's engineers, assessed

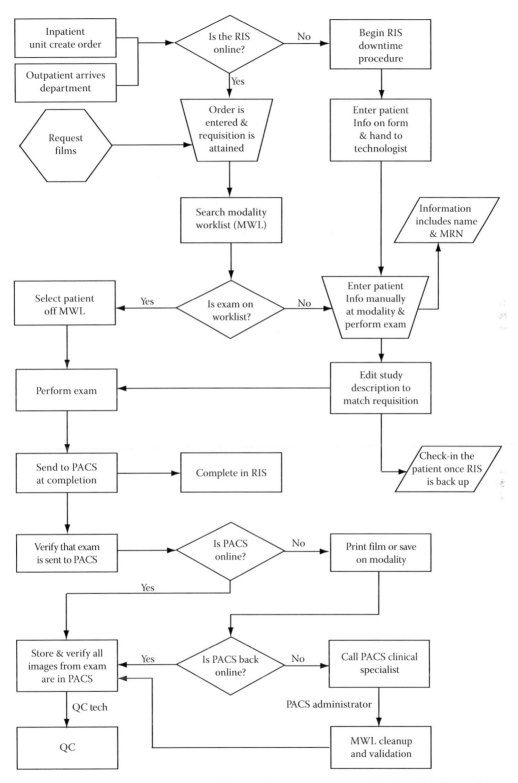

Figure 8.2 This is a workflow analysis diagram for integrating portable digital radiography into PACS. Similar workflow analysis can be performed for other imaging modalities when they are integrated into PACS.

Figure 8.3 A new communication closet was under construction to meet the increased network needs when PACS was expanded to different areas of the institution.

the DICOM service classes supported by the scanner manufacturers. The institution purchased and installed additional DICOM compliant hardware upgrade or software for modality scanners as needed based on the PACS vendor's recommendations. Then the site team and vendor team collaborated to configure the scanners in preparation for the PACS integration.

8.5.2.5 Installation

The vendor provided the site with a written list of all environmental requirements needed for the vendor to install the PACS equipment and to support their proper functioning. During this phase of installation, the following actions were taken:

- Server installation
- Short-term storage installation
- Short-term storage configuration and verification
- Review workstation installation
- Server and workstation function verification
- HL7 interfacing for modality worklist and structure report implementation

8.5.2.6 Training

The PACS vendor trained the institution's PACS administrator at the vendor's headquarters, with the institution being responsible for the trainee's transportation and lodging. The institution had designated the administrator as one of its super-users, to be trained by the vendor extensively with in-depth knowledge so that the administrator could then provide in-house training to task-specific users at the institution (with support from the PACS vendor as needed). It was agreed that the vendor would provide training materials to super-users for end-user training, and that the institution would coordinate its own training schedule. The training itself was also a collaborative effort; although the institution's super-users were always on-site, initially their technical skills did not match that of the vendor application specialists. Therefore, though most end-user training was conducted by site's super-users, the vendor application specialist was frequently required to be on site during the early stages of PACS implementation, right after the system went "live."

Enterprise-wide PACS implementation is an ongoing project. While more hospitals and clinics have completed the initial PACS implementation, those that already have PACS in place are in the process of upgrading to add functionalities and thereby further improve workflow. After five years in service, a major upgrade to the system is imminent; the upgrade process is discussed in Chapter 9.

In conclusion, successful implementation of a PACS requires careful planning, teamwork, and cooperation from many parties, including the buyer's and the vendor's team. The approaches discussed here do not represent the only or even necessarily the best way to implement a PACS, but only serves as examples for PACS implementation.

BIBLIOGRAPHY

Ball, M. J., C. A. Weave, and J. M. Kiel. 2004. *Healthcare Information Management Systems: Cases, Strategies, and Solutions.* Springer, New York, NY.

Carrino, J. A., P. J. Unkel, and I. D. Miller, et al. 1998. Large-scale PACS implementation. *J. Digit. Imaging* 11 (3 Suppl 1): 3–7.

Cohen, M. D., L. L. Rumrelch, and K. M. Garriot. 2005. Planning for PACS: A comprehensive guide to nontechnical considerations. *J. Am. Coll. Radiol.* 2: 327–337.

Dreyer, K. J., D. S. Hirschorn, and J. H. Thrall. 2006. *PACS: A Guide to the Digital Revolution.* Springer, New York, NY.

Gross-Fengels, W., C. Miedeck, and P. Siemens, et al. 2002. PACS from project to reality: Report of experiences on full digitization of the radiology department of a major hospital. *Radiology* 42: 119–124.

Hickman, G. T. and D. H. Smaltz. 2008. *The Healthcare Information Technology Planning Fieldbook: Tactics, Tools and Templates for Building Your IT Plan.* Smaltz. Health Information and Management System Society (HIMSS).

HIMSS Enterprise Information System Steering Committee. 2007. Clinical perspectives on enterprise information technology. www.himss.org/content/files/clinicalperspectives_whitepaper_052907.pdf. Accessed June 5, 2010.

Huang, H. K. 2004. *PACS and Imaging Informatics: Basic Principles and Applications.* Wiley-Liss, Hoboken, NJ.

Huang, H. K., K. P. Andriole, and T. Bazzil, et al. 1996. Design and implementation of a picture archiving and communication system: The second time. *J. Digit. Imaging* 9 (2): 47–59.

Integrating the Healthcare Enterprise. 2005. *IHE Radiology User's Handbook.* www.ihe.net/Resources/upload/ihe_radiology_users_handbook_2005edition.pdf

Liu, Y. 2003. Experience in implementation of PACS in a multi-site tertiary health care center. *Med. Phys.* 20 (6): 1439.

Locko, R. C., H. Blume, and J. C. Goble. 2002. Enterprise-wide worklist management. *J. Digit. Imaging* 15 (Suppl 1): 175–179.

Nagy, P. 2008. PACS technology assessment. *The Society for Imaging Informatics in Medicine 2008 Annual Meeting.* Seattle, WA.

Nagy, P., E. Siegel, and T. Hanson, et al. 2003. PACS reading room design. *Semin. Roentgenol.* 38: 244–255.

Ong, K. R. 2007. Medical informatics. *Health Information and Management System Society (HIMSS).* 2007.

Samei, E., A. Seibert, and K. P. Andriole, et al. 2004. AAPM/RSNA tutorial on equipment selection, PACS equipment overview: General guidelines for purchasing and acceptance testing of PACS equipment. *Radiographics* 24 (1): 313–334.

Siegel, E. L., J. N. Diaconis, and S. Pomerantz, et al. 1995. Making filmless radiology work. *J. Digit Imaging* 8: 151–155.

Thrall, J. H. 2005. Reinventing radiology in the digital age. I. The all-digital department. *Radiology* 236: 382–385.

Thrall, J. H. 2005. Reinventing radiology in the digital age. II. New directions and new stakeholder value. *Radiology* 237: 15–18.

9 Upgrading and Replacing Legacy PACS

PACS systems are constantly changing following the pace of changes in the IT industry. The life expectancy of PACS is much shorter than that of other medical imaging equipment. Despite healthcare institutions' or enterprises' desire for IT systems stability, typically in 3–5 years, institutions face a major upgrade requirement. PACS/RIS or other department information system vendors also prefer their customers to upgrade their latest software release to streamline their responsibility of service and support. Additionally, PACS has been around long enough that healthcare institutions may want to replace legacy or current PACS systems. The issues institutions or enterprises face when upgrading on switching to different PACS/RIS are quite different from those faced when PACS is implemented for the first time, especially when the replacement system is from a different vendor.

9.1 LEGACY SYSTEM AND IT INFRASTRUCTURE ASSESSMENT

Before any decisions are made to replace the existing PACS/RIS system, the implications to the current department workflow, and the general clinical operations of the healthcare institution must be assessed. It is also necessary to assess the current status of the IT network infrastructure, the database used interfaces among various information systems, the overall needs of the institution, and the performance criteria for the replacement system.

The IT department must assess how the PACS/RIS replacement impacts the existing interfaces, because, in many cases, they may not be from the same vendor. The time and resources required to develop and test the interfaces, which is essential for an optimal workflow in the department, should be considered as well. Additionally, the impact of the replacement PACS/RIS on other information systems such as EHR and PHR should also be evaluated. The assessment should be made in terms of:

- The urgency and time frame within which the system must be replaced

- Budget for the replacement

- Benefits of the replacement system

- Limitations of the IT network infrastructure and storage

- Path to integration with EHR and cross-enterprise image exchange

These are primary elements based on which proper planning for the replacement PACS/RIS can be done.

9.1.1 Reasons for Replacing PACS

There are many reasons why a healthcare institution or enterprise chooses to replace the current or legacy PACS systems. Typically, these reasons include:

- End of the service life of the system (3–5 years)

- Higher expectations from PACS/RIS

- The legacy system needs to be scaled up for increased clinical volume

- The legacy system is not DICOM, HL7, IHE, or HIPAA compliant. (Health Insurance Portability and Accountability Act, discussed in Chapter 12)

- Additional PACS functionality is needed

- Uneasy graphic user interface

- Unsatisfactory product service support or relation with the vendor

Additionally, it is not uncommon that within a healthcare enterprise, different institutions use different radiology or cardiology PACS systems, because multiple physician groups selected their own PACS during the initial implementations. This requires complicated support from the IT department, and it is not cost effective. When the PACS reaches the end of its service life, it is logical for the enterprise to implement a single PACS product for all its institutions.

9.1.2 Differences between First-Time and Replacement PACS Implementations

Compared to initial PACS implementation, replacement system implementation is different, because healthcare institutions have first-hand experience of PACS/RIS and know their priorities and the associated costs. Even with the experience gained through initial PACS deployment, there are differences in implementation processes and decisions involved in the replacement implementation that require significant internal resources. The radiology department and IT department have the experience to comprehend both department clinical needs as well as enterprise requirements. However, the problem of buying replacement PACS from a different vendor is that images and data residing on the legacy system need to be accessed after the switch over. Another matter for consideration is that some of the hardware could possibly still be reused if they are relatively new.

If for any reason the selection of PACS for the initial implementation had proved not to be the right choice, replacing the system gives the healthcare institution a second chance to get it done right. Educating, engaging, and involving radiologists and other users in the replacement PACS decision-making process increases the likelihood of a successful implementation. Regardless of the reasons for unsatisfied clinical users, IT issues, or other integration issues, the implementation process of a replacement PACS differs from that of the initial deployment, especially when the vendor has been changed. The change may not be easy; however, with good planning the end result usually proves it well worth the effort.

9.2 PACS/RIS UPGRADE CONSIDERATIONS

A new software release from a vendor typically has solutions for the existing problems of the previous version and with additional new features. The healthcare institution consequently must decide if the existing PACS/RIS should be upgraded to the new release based on financial and nonfinancial considerations. Even though some software upgrades are indicated as "free," and at no additional cost beyond current valid service contract, the software itself may require additional hardware that is usually not provided free, or it may be associated with additional installation service charges. Nonfinancial considerations include impact of the upgrade to the current operation of the PACS/RIS, the IT environment, and additional training of clinical staff.

9.2.1 Major Upgrade Planning

A major PACS/RIS upgrade should be evaluated based on both financial and nonfinancial aspects. Associated costs, benefits, training needs and resources, and timing of the implementation should be carefully assessed before the decision is made. A major upgrade to a large PACS/RIS system will inevitably have an impact on all users and on the routine clinical operation of the department. A few large healthcare institutions have both PACS and RIS systems from the same vendor. Thus, the software version and upgrade status of both systems has to be assessed separately. If both systems have to be upgraded within the same time frame, it is desirable to upgrade the RIS and complete staff training before moving on to PACS upgrade.

To minimize any possible negative impact on the clinical operation of the department immediately following a major upgrade, the upgrade should be carefully planned to occur at a time when both the department and clinical staff are least busy. Additional staff support should be arranged to accommodate unexpected training requests and software technical issues, which are common for any new software release in the IT industry. These additional requirements on the resources should be counted as indirect financial cost. Possible reduction in the patient volume caused by the upgrade, with revenue loss should also be considered.

A PACS/RIS upgrade may affect physicians in other medical specialties as well. Their access to the PACS/RIS may not be the same after upgrade. The planned upgrade should therefore provide at least the same levels of service to those medical staff, and provide all necessary training.

9.2.2 Independent Enterprise Long-Term Archive

Many current PACS systems have a two-tiered storage system. Shown in Figure 9.1 is a long-term archive that is separated from the workflow of PACS. It not only improves disaster recovery for the PACS, but also provides a vendor-independent archive for future PACS. The healthcare institution or enterprise should request from the vendor a migration plan for export of data in future if any new enterprise PACS storage archive is purchased, preferably with the image data stored in DICOM media file format and including only standard objects. When these DICOM images are stored in a long-term archive separated from the PACS workflow, data migration will be simpler in the future.

Figure 9.1 Aurora Healthcare long-term enterprise PACS archive for all PACS systems used at various institutions. For the flagship hospital (St. Luke's Medical Center) of the healthcare enterprise, it is also used for downtime access when the hospital's short-term PACS archive is down.

9.3 PACS-REPLACEMENT PLANNING

There are many PACS systems available on the market today. Choosing a replacement PACS/RIS suitable for a healthcare institution is significantly different from an imaging equipment replacement. In the past, the PACS was purchased by the radiology department, and the system was supported in part by the IT department. However, PACS is now becoming an enterprise-wide system. It is interfaced with a variety of information systems, departments, and institutions supported by the IT department of the healthcare enterprise. Therefore, the decision to replace PACS should be made jointly by the radiology and IT departments, and physicians and senior administration staff should be engaged to buy into the replacement plan.

The process of replacing a legacy system can be as time consuming or complicated as the initial PACS /RIS deployment. In addition to contract negotiation, product and vendor selection, the entire implementation project can take about a year to complete. Regardless of whether it is an initial deployment or a subsequent replacement, planning is a vital part of a smooth and successful implementation. During the planning phase, all key personnel from the administration and radiology departments, radiologist groups at each institution of the healthcare enterprise, technologists in various imaging modalities, the IT department of the enterprise, and the large-volume referring physician groups should be involved in the enterprise expansion considerations. At the planning stage, communication with all current and potential users is always the key to success. Feedback from all involved user departments is very important as well, and staff here can provide opinions and be informed of decisions. After the system is launched, all should be informed of any unexpected issues to shorten the learning phase.

9.3.1 Replacement System Selection and Evaluation

Locating a replacement PACS product and vendor is different from locating a vendor of an initial legacy system. The healthcare institution or enterprise has already had first-hand positive and negative experiences of digital imaging management using PACS. Therefore, the selection of the replacement PACS is much more specific and realistic than the initial search. Typically, initial PACS deployment is centered on radiology imaging of a single healthcare institution only, whereas the replacement system search is based more on:

- Enterprise-wide approach including multiple departments and their imaging modalities

- Multiple institutions within the same enterprise

- Higher degree of data security with disaster recovery capability

- System integration with information systems such as RIS, and speech recognition reporting

- Integration with EHR

- Image exchange among all institutions in the enterprise

These issues should be addressed by the replacement system if they have not been covered in the initial PACS/RIS deployment.

9.3.1.1 Enterprise PACS

A replacement PACS implementation in a clinical setting is no longer limited to the radiology department of a single healthcare institution. Replacing legacy PACS is a good opportunity for the healthcare enterprise to plan its overall strategy for PACS implementation for the entire healthcare enterprise. Very often, there are multiple radiology PACS used in different institutions by various physician groups because, historically, the purchase decision was made at the radiology department level. The current opportunity can be used to upgrade or replace individual PACS/RIS with an enterprise PACS, so that enterprise IT support can be done in a coordinated and cost-effective manner.

As discussed in Chapter 10, departments such as cardiology, pathology, and radiation oncology that use digital imaging modalities are in the process of integrating with PACS. A common enterprise-wide image archiving system for all those imaging specialties has many advantages. An enterprise PACS is one of the essential steps towards the establishment of fully integrated EHR/PHR. It includes medical images drawn from various specialties during patients' encounters with all healthcare institutions within the enterprise. For physicians working at multiple institutions of the healthcare enterprise, it improves workflow, and for patients seeking healthcare services at different locations of the healthcare enterprise, it improves the quality of care they receive, as physicians need to access fewer computer workstations or applications. Since radiology PACS is evolving into enterprise solutions ahead of any other departmental PACS, IT experience and expertise involved in radiology workflow and enterprise distribution can be valuable resources. If the replacement PACS is capable of fulfilling the role of an enterprise PACS, the associated storage and network requirements should be evaluated accordingly to reflect the expanded usage.

9.3.1.2 Replacement System Vendor Selection

Currently, there are many PACS vendors on the market. The healthcare institution or enterprise should first research products available on the market and understand what the replacement PACS must deliver. In addition to well-established traditional imaging equipment vendors that now also offer PACS products, there are many PACS vendors that are specialized only in PACS/RIS/HIS products. The PACS industry is part of the IT industry, and it is not unusual for these companies to merge or go out of business. Large healthcare institutions should take special care that the PACS vendors selected should be able to support their products for the next few years and are less likely to go out of business or be bought in the near future. Healthcare institutions should look for well-established companies with expertise in small or large institutional settings, whose products meet their specific needs. For large healthcare institutions with multiple institutions and different radiologist groups, the product selected must be accepted by all groups. Sometimes a single product/vendor that can be utilized for the enterprise may be hard to find. In general, vender selection should be based on:

- Details of how services are provided

- The number of similar-sized sites where the vendor has installations

- How long the vendor has been in the business

- Product integration with EHR

- Image accessibility through EHR

- Image accessibility through the Web

- Imaging modalities interfaced with the system

- Consistent user interface for diagnostic and clinical workstations

- Integration of advanced image processing and display functions

- Image format and compression techniques used for archiving

- Reference installation sites to check the track record of the vendor, that it is capable and actually delivers on its commitment and promises

- Training methodology and plan

Last but not the least, the healthcare institution or enterprise should spend time studying the purchase contract well, to understand all the details covered in the contract, especially the responsibility of the new vendor during data migration from the legacy system, and the support or plan for data export from the current replacement to future PACS systems a few years later.

9.3.1.3 Replacement System Evaluation

When the legacy system of an institution was originally selected, database migration, paperless departments, speech recognition systems, and other such newer technologies may not have been considered. However, these are essential features of a PACS/RIS implementation today. The criteria for new PACS selection and evaluation must therefore include all that is required for initial PACS/RIS implementation, as discussed in Chapter 8. Special care must be taken to ensure that:

- The product from a single vendor can fulfill all the radiology imaging needs of the healthcare enterprise.

- The product does not use proprietary databases.

- The product uses the latest technologies.

- If possible, some additional functionalities, such as 3D/4D viewing, multiplanar reconstruction, or other advanced postprocessing capabilities, should be incorporated, because it will be inconvenient to add additional workstations just to add those postprocessing capabilities.

- The product uses standard equipment with open architecture.

- The interface or broker used between the PACS and third-party RIS is also recognized as a key criterion.

- Speech recognition, and its interface with RIS, is not an optional feature, but essential to the replacement system.

- An uptime of more than 99.99%, which translates into less than 0.01% downtime, or 53 minutes downtime per year, is highly desirable, since in a paperless and filmless radiology department, the clinical operation relies entirely on the PACS and RIS.

- The product is in compliance with the relevant standards. It is not enough for the product to claim to be HL7, DICOM, and IHE compliant. Although HL7 and DICOM are adopted by almost all vendors, even on some legacy PACS systems, compliance to these standards remain important today. Specific DICOM functions, the HL7 version, and the types of IHE integration profiles that the system supports or conforms to should be verified with vendors' conformance statements in writing.

In addition to the above criteria, a factor that is not significant when selecting a PACS for the first time, but is important for replacement systems is the PACS user interface. A user-friendly and intuitive PACS workstation interface may be more important in replacement implementation. A similar user interface between the legacy PACS and the replacement PACS will make user training and learning easier and result in a smooth transition for end users. While the initial PACS deployment was primarily a radiology department-centered system with users exclusively remaining within the department, the replacement system is likely to be used by radiology staff and referring physicians from all across the healthcare enterprise. A user interface easy to navigate through will thus make the training process for all staff much simpler.

9.3.1.4 RIS Replacement System Evaluation

Since replacing RIS may involves workflow change, representatives of all user areas from adminis-tration, scheduling, radiologists, technologists, file room, and billing need to be included to evaluate the advantages and disadvantages of the potential replacement RIS under all possible workflow scenarios and to coordinate any future changes. This will improve the communication of workflow change to those most involved, and obtain feedback from users, so that the workflow can be optimized to fit the needs of the department. The team should consider all possible, complex but realistic workflow scenarios and request RIS vendors to present solutions during site visits. Attention should also be paid to the scalability, and the maximum number of concurrent users of the RIS system.

During system evaluation, details of interface with other clinical information systems such as HIS, order, billing systems (technical and professional charges), dictation, EHR, PACS, and imaging modality scanners should be checked and validated, because interfacing and systems integration is vital in any new systems launch. A speech recognition reporting tool interfaced with the RIS is necessary as well, because the workflow will have to be adjusted to adapt to the new RIS system if the legacy system does not have this feature. Since the workflow of an independent imaging center is obviously different from that of a multi-institution healthcare enterprise, the workflow and RIS will need to be customized to meet the specific needs of a particular healthcare institution.

9.3.1.5 Single Vendor PACS and RIS Considerations

Since an outdated RIS may limit the full functionality or workflow of the new replacement PACS, RIS may typically need significant upgrade or replacement when the PACS is replaced. Another decision to be made about PACS/RIS replacement is whether to choose products from a single or multiple vendors and if both PACS and RIS should be replaced simultaneously. There are both benefits and limitations to replacing PACS and RIS at the same time with products aquired from a single vendor. The benefits are:

- Potentially more efficient workflow with better integration between PACS and RIS
- Easier and simplified implementation
- Lower total cost because of economies of scale
- Simplified and synchronized product upgrades for new functionalities
- Convenience of dealing with one vendor for the resolution of any issue
- Simplified product service support and maintenance
- Convenience of one-stop shopping

The limitations of choosing a single vendor PACS and RIS solution are:

- Potentially limited functionalities provided by the vendor
- Historically, few vendors developed both PACS and RIS products. Many single-vendor PACS/RIS products are actually PACS from a large vendor combined with RIS from a smaller vendor through corporate acquisition, and hence the PACS/RIS from a single vendor may not be developed in unison
- Single-vendor PACS/RIS solution may be in name only and based on multiple architectures

Therefore, a single-vendor PACS/RIS solution should be evaluated carefully in terms of its architectures. Otherwise, there may not be significant advantages in selecting such a solution.

9.3.1.6 Multivendor PACS and RIS

Compared with a single-vendor approach for replacement PACS/RIS products, the benefits of a multivendor solution include:

- Best-of-breed solution to maximize the desired functionalities of PACS and RIS individually
- Flexible timeline to rollout major upgrade or replacement of either PACS or RIS

- Less financial burden because of the option of replacing either PACS or RIS one at a time

- Easier adoption of advanced image analysis applications

  The limitations of a multivendor solution are mainly as follows:

- Potential integration issues between PACS and RIS, which may result in an inefficient workflow, and inconvenience and nonoptimized productivity for users

- Requiring close collaboration between PACS and RIS vendors for their seamless integration

- The need to negotiate, manage, and maintain system support separately

Before the decision is made to select either a single- or multivendor PACS/RIS, solution, the healthcare institution should clearly understand the goals, evaluate the products based on their track record of system integration and compliance with standards, and approach the process based on specific needs and limitations.

9.3.1.7 Replacement PACS Storage Design

The storage design of the replacement PACS requires clinical operational data, such as imaging volume and the data retention period, both in the short term as well as the long term. Knowing these factors allows the healthcare institution or enterprise IT department and the vendor to optimize the design of the replacement PACS storage infrastructure.

Since, typically, it takes more time and cost to add additional image archive at a later time, sufficient storage should be installed when the PACS is replaced. It is not unusual that the initial projected storage usage is exceeded and more storage is needed. The healthcare institution or enterprise may reevaluate the image archive need with a strong vision for enterprise imaging, not just a single PACS system. If the new storage can also be used by enterprise PACS of other medical specialties, the possibility of consolidating all PACS (such as radiology and cardiology PACS) storage together should be seriously considered. Compared with department-wide specialty mini-PACS systems, a common PACS image repository has many benefits of scale, including IT support, capacity expansion, universal data security policies, disaster recovery strategies, and facilitating EHR/PHR implementation. It also allows the enterprise to extend the workflow and productivity benefits already seen in radiology into other medical specialties across the enterprise. The data storage in this case is essentially part of the IT infrastructure; the cost should be considered on an enterprise basis instead of a single PACS, and the total cost of all current PACS storages together with their maintenance must be compared with future consolidated storage and maintenance expenses. The healthcare enterprise can leverage all internal resources and assets to reduce the total cost of ownership, while at the same time providing better and safer healthcare.

9.3.1.8 System Uptime and Downtime Issues

When PACS, department information systems, and EHR are widely adopted and integrated, the daily clinical operation becomes increasingly dependent on these systems. If the PACS system is claimed to guarantee an uptime of 99%, it means that the system may have up to 1% downtime, or the system can be down for up to about 3.5 days per year, which is not acceptable to many healthcare institutions or enterprises. For a 99.99% uptime-guaranteed system, the downtime is no more than 53 minutes a year, and for a 99.999% system, it is about 5 min per year. The healthcare institution or enterprise should consider this reliability issue to see if the system meets the business operation requirements, and set out any compensation or penalty for additional downtime other than that indicated in the contract.

9.3.1.9 Replacement System Integration Considerations

Many legacy PACS were designed to be used within the radiology department, and some of them are either not linked to a department's information system (RIS), or they lack reporting tools. Since the integration of PACS with RIS can significantly improve department workflow, when the legacy PACS is replaced, it is the time to integrate the new PACS with the RIS. When referring physicians are able to easily access patient images and reports online anytime, anywhere, shortly after the imaging studies are completed, their collaboration with other clinical specialists can result in better and more efficient patient care. Easy access to both PACS and RIS information also enables physicians to take informed and timely patient care decisions much more efficiently and without relying on phone calls or fax as communication tools. Therefore, it is necessary for operational success that all replacement PACS/RIS hardware and software are integrated with other existing clinical information systems.

Setting up replacement PACS is complicated, because it involves more than just configuring imaging equipment, the PACS server, and workstations. The integration between PACS and RIS, the way images are linked with orders, reports, and data mapping at patient or study level also need to be considered. Replacement PACS with capabilities to be interfaced with RIS, HIS, and EHR may require interface engines or brokers. Since each information system generates, passes, and stores data, proper systems integration allows data flow among disparate systems without requiring users to enter the same data repeatedly. Therefore, systems integration is the most important and challenging part in any clinical information system implementation.

9.3.2 Feasibility of Reusing Workstation Display Monitors

If the PACS system to be replaced is still fairly new, it is possible that workstation hardware and display monitors can be configured to be re-used by the replacement PACS. The feasibility of such reuse should be discussed with the replacement PACS vendor, so that the initial investment may still be leveraged. Depending upon its condition and the extent of use, display monitors of the legacy PACS workstations may be reused. As discussed in Chapter 3, there are two types of display monitors for diagnostic and clinical purposes. Since diagnostic display monitors are of a higher quality and very expensive compared with clinical display monitors, if they are maintained properly and meet quality specifications with a valid service contract or warranty, they can be reused in the replacement PACS, or they can be reused as clinical display monitors. Depending on the number of workstations deployed in the replacement PACS, the savings on display monitors may be significant.

9.4 NETWORK INFRASTRUCTURE UPGRADE

Just as during the initial PACS installation, the network infrastructure may need to be reevaluated in terms of image data generation, network traffic loads, and system response time because any major PACS upgrade or replacement will likely impact the network infrastructure. Since new replacement PACS are most likely equipped with faster network interface cards, to fully realize the improved speed of the new PACS, the improved data access time from the legacy system data archive media to the new spin disk media, and to meet the demand for expanded data storage needs from imaging modalities, the network speed should match that of the interface. Even without replacement PACS implementation, the network traffic itself may demand an upgrade to the existing network infrastructure because of the increased use of digital mammography, multidetector CT, functional MRI, EHR by various medical specialties, the access to radiology PACS by more referring physicians, and the increased radiology imaging volumes.

9.4.1 Network Upgrade

The network infrastructure design requires clinical operational data such as daily imaging volume, the speed that clinical staff need to retrieve patient images, and so on. These operational requirements determine the bandwidth requirement of the LAN linking various areas within a healthcare institution, and the WAN linking different sites of the healthcare enterprise. Accurate information about these two factors allows the healthcare institution or enterprise IT department and the replacement PACS vendor to design the network infrastructure efficiently. For smaller healthcare institutions without their own IT department, IT contractors may be used and should be informed of all PACS upgrades or replacement plans.

For a healthcare enterprise with institutions at multiple geographic locations, the architecture of the replacement PACS also determines the requirement of the network infrastructure. If the healthcare enterprise switches from distributed multiple storage systems located at various institutions of the enterprise to a centralized PACS storage architecture, it may require different specifications of WAN that links these healthcare institutions involved in the replacement. In this case, images generated by each institution are all archived to a central location, and they are transferred over WAN for both primary archive and second- or third-tier storage as backup for disaster recovery. The central PACS storage is later queried by institutions for any subsequent retrieval requests. The network traffic on WAN is higher than that of a distributed PACS storage architecture, as images generated by each institution are archived locally, which limits traffic load on the WAN. If different compression techniques and compression ratios are used by the replacement PACS, or digital pathology images with a much higher compression ratio compared to radiology images are included in the new PACS, the network upgrade assessment should include these considerations as well.

9.4.2 LAN Network Cabling

The network interface cards equipped on servers and workstations are usually capable of a much higher data rate compared with those on legacy systems. The PACS server's and workstations' potential will not be realized if the LAN network they connect to is slow. Therefore, it may be necessary to upgrade the LAN network infrastructure. One important aspect of the LAN upgrade is its cabling system. Because of the overall lower cost of technical expertise and equipment required for installation and maintenance, copper-based cabling is still widely used for LAN applications. Although cabling only represents a small percentage of the overall network cost, it is the most difficult and potentially most expensive component of the network to replace. It is expected to outlive most network components, as its useful life span is in excess of 10 years whereas Ethernet applications such as PACS typically have a useful life of 5 years. It is recommended, therefore, that the upgraded cabling system be able to support two generations of network applications. Based on lifecycle considerations, the strongest cabling system capable of supporting both present and future network application should be selected to upgrade an existing LAN network.

9.4.3 Next Generation Ethernet WAN

The success of Ethernet as a data network technology of choice in the LAN is primarily due to the tremendous Ethernet network connection speed, simplicity, interoperability among different vendors, and steady decrease of associated installation and operational cost of Ethernet equipment.

Ethernet LAN is currently widely used by healthcare institutions and enterprises. Traditional WAN data connection speeds are essentially limited by the private leased line service speeds of T1/DS1 (up to 1.5 Mbps) or T3/DS3 (up to 45 Mbps) circuits, as discussed in Chapter 4, which are primarily used for point-to-point connection and Internet access. At the time of PACS/RIS replacement, the network infrastructure used to link the various geographically separated institutions should be reevaluated to check if it still meets the requirement of the new replacement PACS. Although Ethernet-based networks are already deployed by healthcare institutions for LAN, traditional as well as nontraditional network service providers have also recognized the benefits of Ethernet and have begun to utilize their high-capacity fibre network infrastructures to offer high-bandwidth Ethernet-based WAN or metropolitan area network (MAN) services.

9.4.4 Benefits of Ethernet WAN

Ethernet WAN allows much higher data connection speeds compared with leased-line technologies. The benefits of deploying Ethernet WAN technology for healthcare institutions and enterprises are:

- The bandwidth limit of Ethernet is up to 10 Gbps.

- The cost of connection is low compared to leased lines in terms of unit cost per megabit per second.

- Matured standards-based Ethernet technology ensures interoperability among the equipment of different vendors.

- There is no requirement for specially customized equipment; typical Ethernet devices can be used.

- The setup is relatively easy and simple with widely available Ethernet expertise.

- The infrastructure cost for vendors is low because of the availability of the widely used switch equipment.

- Capability of multipoint connectivity simplifies the management of multiple point-to-point connections on the equipment with a leased-line service.

Ethernet-based WAN is becoming an alternative to traditional leased-line WAN service to business customers, and it will be increasingly used in the healthcare environment.

9.4.5 Ethernet WAN Services

Several key services or a combination of key Ethernet WAN services may be packaged and marketed under various names by service providers. The major advantage of these services is that without any hardware change or software configuration modification, existing customer equipment with a standard Ethernet interface can connect to the network of a service provider and use the subscribed services. These services are:

- *Ethernet Wire Service*: This is the Ethernet equivalent of leased-line service for a point-to-point connection between two geographically separated sites using standard routers, switches, or hosts. This type of service ensures that all data exchanged from the customer's equipment at the source site through the network to that at the destination site are unaltered.

- *Ethernet Multipoint Service*: This is the WAN equivalent of the multipoint Ethernet functionality in the LAN environment. Using this service, multiple physical sites of the enterprise can be linked just as in LAN, and data are delivered without alteration.

- *Ethernet Relay Service*: In the Ethernet relay service, although customers' routers are used to establish a single point-to-point physical connection between two sites of the enterprise, multiple logical connections can be multiplexed on a single physical connection to the network of the service provider, which enables connection to multiple remote sites.

- *Ethernet Relay Multipoint Service*: This service is the multipoint capability of Ethernet switches extended to WAN. It allows both multiplexed logical multipoint-to-multipoint and point-to-point WAN connections among multiple sites, where routers are used by these sites.

At each site where the Ethernet WAN service is provided, the interface between the network infrastructure of the service provider and the institution site as the user is based on standard Ethernet. Figure 9.2 is an actual interface of the Ethernet WAN inside a healthcare institution. It is typically specified by its physical medium, bandwidth, and operational mode. The service provider is responsible for the maintenance and management of any equipment beyond the interface, while the institution is responsible for all infrastructures within the boundaries of the institution.

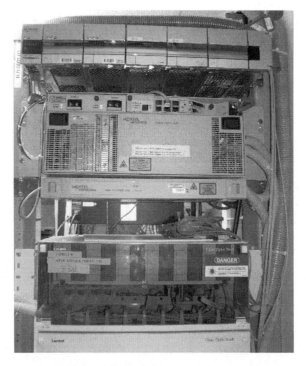

Figure 9.2 Interface inside St. Luke's Medical Center, Aurora Healthcare, which links the institution's LAN to the WAN with an Ethernet connection.

9.5 LEGACY PACS DATA MIGRATION

When institutions replace legacy PACS systems, there are several new challenges. One of them is to decide if the legacy system should be maintained so that prior studies can be accessed from the new replacement system, or if the image database should be migrated to the replacement system, because access to historical image data is vital for the continuity of patient care. After the switch over to the replacement PACS, many institutions do not keep the legacy system for access to prior studies because administrative and technical support is needed to service the legacy system, the service contract for maintenance and repair has to be maintained, and additional space has to be allocated for the system. In terms of service and maintenance of the legacy system, the original vendor may not be able to support the legacy system any more without significant expense to upgrade, which may be necessary to keep the legacy system operational. Additionally, hardware, software, and engineering expertise for support of the legacy system may not be available a few years later. In later years, service support may even be impossible, because the original vendor may no longer exist. In short, keeping the legacy PACS operational after switching over to a new replacement system may not be a good option for the healthcare institution or enterprise.

If it is decided that the data on legacy PACS are to be migrated, the data migration should be planned early in the system replacement process. Decisions should be made concerning the migration method, the extent of migration, and the timing of migration (before, during, or after the launch of the replacement system). Depending on the image format used for archive by the legacy PACS, complex data migration issues can emerge during the transition to a replacement PACS, because it is necessary to access data stored on the legacy system after the operations switch over to the new system. The most prominent issue is that images stored in the legacy PACS archive may or may not be in DICOM format, and most PACS databases are not generic with only DICOM objects. They frequently have proprietary components such as key images, image annotations, or image overlay. If data were archived in DICOM format, although the data migration may still take a long time depending on the archive size, it will be easier if data can be accessed using the DICOM query and retrieve.

The additional challenges of data migration from legacy PACS to the new replacement system is the operations during the transition period. Typically, for a short transitional period, both the legacy and the replacement PACS systems may simultaneously be operational. This may add complexity for the service and support team, and the respond time in the new replacement and/or legacy is not optimized because of the added load of data migration.

9.5.1 Data Retention Period

Images and reports of a PACS/RIS system are part of patients' medical record. Depending on where the healthcare institution or enterprise is located, there may be different medical record retention requirements by the local or national government. The healthcare institution must check with its legal or risk management office about the data retention policy, and use its film retention policy as a reference to determine the image data retention period. Pediatric, mammography, and oncology imaging study data are required to be kept significantly longer than other diagnostic studies, and many institutions keep these data much longer than the minimum time required. The size of the replacement PACS data archive should be large enough to store data accumulated on legacy PACS.

9.5.2 Scope of Data Migration

Since legacy PACS storage may use various storage media such as optical disks, tapes, or magneto-optical-disks (MOD), and they may be on-line in the "jukebox," or some of them off-line on the shelf, the institution must decide the extent of data migration: should it include the entire database or should it be a partial migration, with only limited most recent data being sent to the new replacement system? Although images on legacy systems may be claimed by vendors as "DICOM Compliant," many systems are not plug-and-play, because of their "unique" implementation of the DICOM standard. Depending on the type of legacy system, data migration from some mini-PACS may need more correction and take longer, and it is not unusual to have lost some data after migration.

The timing of the "go-live" of the replacement system should also be decided in relation to how much data has been migrated from the legacy system, to ensure continuity of patient care, and older images may need to be retrieved regardless of whether they are on the legacy system or on the new replacement PACS. In addition, the integrity of migrated data has to be validated to make sure that the migration was done correctly, and the inevitable data discrepancies have to be reconciled.

9.5.3 Data Migration Planning

After the decision is made to migrate the data from the legacy system, it is time to plan the data migration. Data migration also requires careful planning. Without a well-planned strategy, data migration from the legacy to the new replacement PACS can result in an unexpectedly high hidden cost. In addition to the selection of the data migration service provider and budget preparation, the following are to be considered in the planning, because there are unique issues associated with it. The data migration planning includes:

- Defining objectives, goals, and measurable outcomes of the data migration.

- Evaluating all options available.

- Custom hardware and software requirements.

- Determining the amount of data to be migrated and the amount that can be left behind legally and practically.

- Assessing the size of the archive. The replacement system's long-term archive size is much larger than that of a first time PACS implementation, since it includes both the legacy system's archive plus that required for future use.

- Data migration and preload of new studies to the replacement system. Often, 1–2 years of most recent data are migrated to the new replacement system by the go-live date of the new system.

- Planning data migration, since it may take up to 30% of the time during which, depending on the specifics of the legacy and replacement systems, data accumulation occurs.

- Correction of legacy data. Some data from the legacy system may need to be corrected for inconsistent data format, incomplete exams, misplaced exams, and test studies.

- Minimizing department workflow interruption.

- Analyzing possible technical challenges and creating a solutions roadmap.

- Evaluating the likelihood of the replacement system being used for enterprise PACS, and the need for a separate long-term archive for disaster recovery.

Data migration usually starts from the most recent data first. Together with data preloading, data within 1–2 years should be on the new replacement system when it is launched. Among all of the data migrated, it is not unusual that up to 10% of the data may be flagged as exceptions requiring some corrections, especially data from mini-PACS systems, and some data loss is inevitable. Before data are stored on the new system, they have to be manipulated to correct errors and to be updated to current DICOM standards, and reconciled at patient and examination level to map patient demographics. When data are migrated, key image functionalities, such as image selection, grayscale image window and level setting (image display state), proprietary image annotation and markup, may not be exported to the new system properly. Depending on the nature of the migration, it can take longer to complete the entire data migration. Therefore, the legacy system should be maintained clinically operational with a valid service support contract until after the data migration has been completed and all images have been migrated to the replacement PACS. To minimize issues, the replacement PACS purchase contract should specify that image data stored on the new system be retrievable by any third-party application using standard DICOM services. The selection of the replacement system also directly determines the current and future data migration costs.

9.5.4 Data Migration Approaches

From the operation and maintenance point of view, the best option is to migrate the entire database residing on the legacy system to the new system. Regardless of who is going to perform the data migration, it may take a long time to finish the project. Data storage media on legacy systems are most likely on MODs, DVDs, or tapes. They may be in the jukebox system or off-line, and thus loading, searching, and accessing data are much slower than spin disks. Data migration from these legacy systems are usually arranged to occur during the time of the night that there is not much clinical activity, because the process takes computation and network resources from the new PACS, and the performance of the new system may not be optimized if data migration occurs during the day. In addition, often the data on the legacy system are not in the same format as data on the new replacement system, and customized software and tools have to be developed to migrate the database. Since most PACS are closely linked with order and result data from RIS, order and result data should be sent to the replacement PACS before images from the legacy PACS are migrated.

9.5.4.1 Migration Using Long-Term Archives

One data migration approach is to use disaster recovery data backup either from the healthcare institution's second- or third-tier long-term archive or from the third-party disaster recovery application service provider. Since data were already loaded to these storages for disaster recovery purposes, they can be exported to the replacement PACS to avoid any direct data migration issues from the legacy to the new PACS, because the long-term archive is separated from the workflow of the PACS. The data migration can occur without any interruption to the operation of the legacy PACS.

9.5.4.2 Self-Supported Data Migration

In-house data migration from legacy to replacement PACS can be time consuming and technically challenging. For a relatively small database, or a large database with strong in-house IT support, data can be migrated by the institution with limited assistance from the legacy system vendor. The data on the legacy PACS can be migrated using one of the following methods: a "Send" utility application software on the legacy system, an "Import" utility application software on the new replacement PACS, an in-house developed script file, or a data migration utility. However, for a large legacy PACS database, without institutional expertise in the area of data migration, it might be more practical to hire a third-party data migration service provider to transfer the data.

9.5.4.3 Data Migration Service

Another option for data migration is using the services of a vendor with expertise in managing the actual data migration, data analysis, PACS and RIS data correlation, and providing the required software and hardware for migrating data from the legacy system to the replacement system. Using the services of a reputable vendor can minimize the support from the legacy system vendor. If the data migration service vendor is recommended by the new replacement PACS vendor, the challenge of interfacing to the new system is also minimized. However, there is an additional cost that can offset the convenience of using the data migration service from a service provider. For example, if an institution originally had an annual volume of 100,000 radiology exams 5 years ago, at an average study volume increase rate of 5% per year, the PACS database would hold approximately 553,000 studies. Depending on the nature of the data migration and contract terms, the cost to migrate this can be $0.25–0.5 per study, and thus the total cost to migrate data from the legacy to the new replacement system is not insignificant. The healthcare institution or enterprise should include the data migration cost at the financial planning stage.

9.5.4.4 Data Migration Service Considerations

Data migration services can be provided by vendors of the legacy or replacement systems, or an independent third-party service provider, either in the form of data migration service itself or as data migration software with training and support services. Before contacting a service provider who specializes in data migration, the healthcare institution should prepare detailed technical information about the legacy system:

- Hardware platform of the legacy PACS

- Software version and upgrade history of the legacy system

- Total database size in terabytes

- The total number of examinations and images

- All imaging modalities involved

- The data range

- Data cleansing options

- The data to be migrated

In addition to information about the legacy system, that of the new replacement system should be given to the data migration service provider as well, especially the hardware and configuration of the data storage system associated with the replacement system, and its sustained data through-put rate, because there may be a period of time that the new PACS/RIS is fully operational, while the data migration is still going on in the background from the legacy to the replaced system. After the collection of information about both the legacy and new replacement systems, the healthcare institution or enterprise can contact a data migration service provider.

9.5.5 Vendor Collaboration for Data Migration

Usually, during the data migration, technical assistance from both the legacy and replacement system vendors may be needed, especially if the images are in a proprietary format or use a nonstandardized version of the DICOM format, to facilitate the migration. However, this could prove to be difficult and they may not work together well, because it is not to the best interest of the legacy system vendor, while the new replacement PACS vendor is very motivated in the data migration project.

It is necessary to put PACS database migration support terms in the new replacement purchase contract, because it is equally important now to get support from the new PACS vendor to migrate data from the legacy PACS, and in the future when the current replacement PACS becomes another "legacy" system.

9.5.6 Data Cleansing and Error Correction

An important aspect of data migration is data cleaning, or data matching. Regardless of whether in-house IT resources or an external service provider is used to migrate the data, errors such as incomplete studies, data in wrong patient folder, or patient demographic data in wrong format must be corrected. Also in the PACS, there are data that is generated for tests, services, and demonstrations that are associated with particular identifiers and naming conventions. These processes require manual correction during the migration to make the replacement PACS database "cleaner," and this process can be very time consuming.

9.5.7 Proprietary Data on Legacy System

Other area of concern is to capture proprietary grayscale window and level setting, and key image selection or annotation. Grayscale settings may have been inserted in individual images, or included as DICOM grayscale presentation state objects (GSPS) (discussed in Chapter 6) stored together with image data. Key image selection and annotation for reports in proprietary format cannot be exported to the replacement PACS. Although image objects stored in many PACS are in DICOM format, the database of the PACS systems that manage images are usually proprietary.

Since the life span of a PACS system is limited to a few years, it is very much possible that a healthcare institution or enterprise is going through the data migration process for the second or third time. Although data migration is neither economical nor efficient, currently there is no long-term PACS data archive that can be seamlessly integrated with all other systems without any customization. The healthcare institution should understand the technical approach used by the service provider, negotiate the contract, and hope this is the last data migration.

9.6 REPLACEMENT SYSTEM GO-LIVE AND TRAINING

A well-planned replacement system go-live and training is a must for any smooth and successful implementation. Usually the go-live date is selected on a midweek day, when the clinical operation of the department is not the busiest. Before the go-live date, the key users of all involved areas should be well trained, so that they can be the front-line help during the initial phase of system launch. Training of these trainers must start well before the go-live date. A few weeks before the launch of the replacement system, training sessions for all users can be offered on several new workstations in each area, which are not just demonstration workstations. During the initial "go-live" period, key users can effectively feedback any unexpected significant or nonsignificant issues encountered. They also need to train many users, because few PACS users are seriously involved in the training until their daily work relies solely on the new system, and it is proven that this type of "on-the-job" training is the most effective. A few weeks following the launch of the replacement system, a more in-depth training session can be offered to those who want more advanced applications than just the basic ones.

9.6.1 Transition Planning and Going Live

The launch of the new replacement PACS/RIS can either be in gradual phases, or in a "big-bang" fashion. Although the gradual rollout has an advantage in that users adapt to the new system a lot easier, the disadvantage is that this period of learning is stretched out as new components are constantly being added, and it sometimes requires the new systems to be configured to be back compatible with legacy systems. Another disadvantage associated with this approach is that the healthcare institution's infrastructure may be a limiting factor. Space is usually never enough in any department. In this case, additional space is necessary to accommodate two systems at the same time. Network access and bandwidth can be a limiting factor as well, because more network access will be needed for two systems to be operational at the same time. Some imaging modalities may have the capability of sending images automatically only to one destination. Probably the most difficult is to coordinate workflow on two systems, which is complicated compared to one system.

9.6.2 Replacement PACS Data Preload

To have approximately 1–2 years of most recent historical data on the new replacement system by the go-live date, both data migration and preloading are needed. Preloading is the process of sending new data from imaging modalities to the replacement PACS while preparing for the launch of the replacement system. During this time period, it is necessary for both the legacy and the new replacement system to be operational. There are several approaches for PACS data preloading.

9.6.2.1 Dual-Send Approach

Dual-send is the configuration that all imaging modalities send newly acquired images to both the legacy and the new replacement systems, as described in Figure 9.3. This requires assistance from clinical engineering or field engineering from imaging modality vendors, because some imaging scanners do not support the configuration of automatically sending data to more than one destination without reconfiguring the imaging modality scanner.

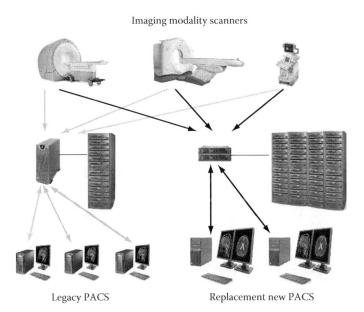

Imaging modality scanners

Legacy PACS   Replacement new PACS

Figure 9.3 Using the dual-send approach, newly acquired images from modality scanners are sent to both the legacy and the new replacement PACS during data preload for the replacement PACS.

9.6.2.2 Relay from Legacy PACS

The relay using the legacy PACS is essentially an operation of migrating the most recent data to the new replacement PACS, as illustrated in Figure 9.4. After the legacy PACS receives images from imaging modality scanners, they are forwarded to the new replacement PACS, so that the new system always has the latest images. Since one of the reasons for the legacy system to be replaced may be its unsatisfactory performance and poor response time, this operation takes additional computational power and resources from the legacy system, and the performance of the legacy system may be further reduced.

9.6.2.3 Relay from Intermediate Server to Replacement System

Intermediary relay uses an additional server that first receives all newly acquired images from all imaging modality scanners, and then forwards them to each of the legacy and replacement systems, as shown in Figure 9.5. Although each imaging modality scanner is configured to send images only to this server during the preload period, this setup requires additional equipment. When the new replacement PACS is launched, all imaging modality scanners need to be reconfigured, so that images are automatically sent to the new system. Depending on the particular needs of the healthcare institution or enterprise, any combination of the above three approaches can be applied for preloading the new replacement PACS before it goes live.

9.6.3 Testing and Simulation before Replacement System Launch

Some healthcare institutions or enterprises prefer the "big-bang" approach, so that all components of the new replacement PACS/RIS system are all configured and go live the same time; this makes it faster for users to adapt to the new system and simplifies the interface setup among various information systems. The approach requires that the test system is setup and operational simultaneously with the legacy system for about 1–2 months before the go-live of the new replacement system. Before the replacement PACS/RIS goes live, the functionality of the system has to be fully tested to find the possible pitfalls of system under normal clinical conditions, and as many as possible workstations to be installed in each area with full access to the new replacement PACS/RIS. Users are encouraged to test the entire workflow process using the replacement system as much as possible and communicate any issues encountered, while the system administrators can duplicate the process on the legacy system if necessary. With the assistance and support from the vendor, all technical and workflow problems should be solved before the go-live date, because there will be enough issues to be taken care of after the replacement system is launched.

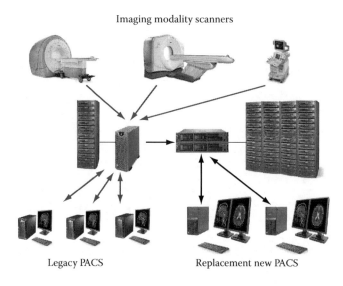

Imaging modality scanners

Legacy PACS   Replacement new PACS

Figure 9.4 Using the relay approach, most recent images acquired from modality scanners are forwarded from the legacy PACS to the new replacement PACS during data preload.

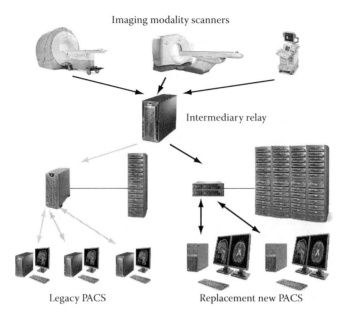

Imaging modality scanners

Intermediary relay

Legacy PACS   Replacement new PACS

Figure 9.5 Using the intermediary relay approach, an additional server is used to receive all
newly acquired images from modality scanners. These images are then forwarded to each of the
legacy and replacement PACS.

9.6.4 Replacement System Training

The importance of training cannot be overestimated for a successful replacement implementation.
From the training point of view, supporting staff of healthcare institutions prefer to have stable
systems, especially in large healthcare institutions or enterprises, where there are many staff
users. If PACS and RIS undergo significant upgrades, or if systems are completely replaced, new
training modules need to be developed tailored to the new systems, so that staff can learn with
minimum time, effort, and cost. Since on-the-job training is the most effective, and it is impossible
to train all users at the same time, it is important to develop a strategy of priority for the training
to minimize any possible disruptions to users after the replacement system has gone live. During
this training period, PACS/RIS support staff need to reprioritize other projects to meet increased
training and service demands in connection with the launch of the PACS/RIS replacement system.

A PACS/RIS replacement for a large healthcare institution or enterprise involves many clinical
staff and users. During the planning stage a training team can be formed to include PACS administrators, PACS team from the IT department, image library or file room personnel, and radiology
technical staff. The team not only serves the needs of planning and providing training, but also
provides an effective mechanism for receiving feedback from end users. This team may be trained
by the vendor's application specialists to become experts in using the system. In addition, vendor's
on-going application support is also important, because many technical difficulties encountered
by users after the replacement system is launched are actually application training issues. The
training plan should also include workflow change and not be limited to just the operation of
PACS/RIS. Understanding the specific needs and workflow of the institution, the team can develop
its own training tools to facilitate the training process, and engage clinical staff and nurses beyond
the radiology department.

9.6.5 Replacement System Acceptance Testing

Just as in initial PACS/RIS implementation, acceptance test planning is also a crucial component to
ensure a fully functional replacement system that meets the goals, and protects both the finance
and time invested by the healthcare institution or enterprise. Very often, acceptance testing is
overlooked in the entire implementation process. In addition to checking the vendor-provided
acceptance test list, the healthcare institution should look into the following areas:

- Define key elements of a replacement acceptance test plan

- Define test condition and test criteria

- System completeness test

- All functional element test

- Performance test

- Clinical acceptance

- Vendor communication for the resolution of any discrepancies

After the acceptance test has been completed, it is important to have all issues of installation and discrepancies resolved before the last payment is made to the vendor.

9.7 FUTURE DEVELOPMENT

The future development of PACS is enterprise-wide integration, where medical images from various departments and imaging modalities can use a common image data repository, and different departments can select best-of-breed systems to work with the enterprise image archive. Currently, medical images are mostly stored in their own archives and detached from EHR in clinical patient care, so that they are not easily accessed by clinicians. However, the issue is that diagnostic imaging studies with several hundred images are not uncommon, and clinicians need to easily access the right images. It is therefore crucial to have processes in place for the enterprise image archive, clinical or department information systems, hospital information systems, EHR systems, and many other clinical databases to be seamlessly linked together, so that clinicians can access the right images from a single point of interaction for patient care.

To entirely replace an existing PACS from a different vendor is a significant undertaking that is no easier than the initial PACS implementation. The entire process is lengthy even if the transition is smooth. Although there are similar concerns in the implementation of PACS storage, network requirement, training all clinical staff and users, budget and cost planning, and available IT resources planning, there are unique challenges for the second, third, or fourth replacement implementation, which involves data migration and enterprise PACS integration. In areas not addressed in the initial PACS implementation, such as multitiered data storage for disaster recovery or voice recognition technology-based reporting tools for RIS, this is the time to review and implement these measures. However, there is no perfect PACS system. It is not realistic to expect that all undesirable characteristics of the legacy PACS will be gone, while all desirable features and functionalities will be automatically carried over to the new replacement system, because they are different systems from different vendors. Therefore, the new replacement system may still not please every user or supporting staff. The ultimate success of implementing a major upgrade or replacement PACS/RIS depends on defining clear goals and undertaking a thorough cost analysis, timing selection, and cost and benefit analysis. It also requires close cooperation and communication among the various imaging departments, the administration, the IT department, and all clinical staff and users, so that the value of the upgrade or replacement can be maximized while clinical activities during the process are not interrupted.

BIBLIOGRAPHY

Avrin, D. 2007. PACS/RIS migration. *The Society for Imaging Informatics in Medicine 2007 Annual Meeting.* Providence, RI.

Behlen, F. M. 2008a. Data migration: Plan early and sleep better. *The Society for Imaging Informatics in Medicine 2008 Annual Meeting.* Seattle, WA.

Behlen, F. M. 2008b. PACS operational policies & procedures: Policies surrounding migration & long term data management. *The Society for Imaging Informatics in Medicine 2008 Annual Meeting.* Seattle, WA.

Behlen, F. M., K. R. Hoffman, and H. MacMahon. 1996. Long term strategies for PACS archive design. *Proc. SPIE,* 2711: 2–8.

Cavanaugh, B. J., H. T. Garland, and B. L. Hayers. 2000. Updating legacy systems for integrating the healthcare enterprise (IHE) initiative. *J. Digit. Imaging* 13 (2 Suppl): 180–182.

Cisco Systems, 2003. Metro ethernet WAN services and architectures white paper. www.cisco.com/warp/public/cc/so/neso/meso/ns227/mtrwn_wp.pdf. Accessed January 16, 2010.

Crowe, B., L. Sim, and V. Whitter. 2008. PACS Upgrades require careful advance planning. www.diagnosticimaging.com/imaging-trends-advances/pacsweb/teleradiology/article/113619/1179352?verify=0. Accessed January 16, 2010.

D'Arcy, S. and J. D'Arcy. 2003. Securing data from disaster. *Imaging Economics*. www.imagingeconomics.com/issues/articles/2003-06_02.asp. Accessed January 16, 2010.

Examine PACS: A conversation with … Alex Jurovitsky: The cost-effectiveness of PACS. 2007. www.rt-image.com/Examine_PACS_A_Conversation_with_Alex_Jurovitsky_The_cost_effectiveness_of_P/content=8904J05C4856B690409698764480A0441. Accessed January 16, 2010.

Fratt, L. 2006. Changing PACS vendors: getting it right the second time around for radiology and the enterprise. www.healthimaging.com/index.php?option=com_articles&view=article&id=4235. Accessed January 16, 2010.

Hagland, M. 2009. Parsing the PACS market: Evaluating a PACS upgrade or replacement decision is all about understanding need. http://www.healthcare-informatics.com/ME2/dirmod.asp?sid=&nm=&type=Publishing&mod = Publications%3A%3AArticle&mid=8F3A7027421841978F18BE895F87F791&tier=4&id=B9177ADDF4B244BD9FBD8A264B234EB8. Accessed January 16, 2010.

Healthcare Informatics. 2007. PACS grows up: As imaging technology matures into its second generation, the industry begins to feel the shift forward. 2007. www.healthcare-informatics.com/ME2/dirmod.asp?sid=&nm=&type=Publishing&mod=Publications%3A%3AArticle&mid=8F3A7027421841978F18BE895F87F791&tier=4&id=6D7809D1B8D846F6974230A2F146BD32. Accessed January 16, 2010.

Jones, A. 2009. PACS/RIS replacement: A PACS administrator's perspective. *The Society for Imaging Informatics in Medicine 2009 Annual Meeting*. Charlotte, NC.

Liu, Y. 2004. Experience in integrating old PACS to new PACS in a multi-site clinical center. *Med. Phys.* 31 (6): 1735.

Massat, M. B. 2009. Key considerations for replacement PACS. www.itnonline.net/node/32511. Accessed January 16, 2010.

Oliva, V. 2002. Ethernet—The next generation WAN transport technology white paper. www.10gea.org/ethernet-wan.htm. Accessed January 16, 2010.

Orenstein, B. W. 2008. Moving to the next PACS. *Radiology Today*. www.radiologytoday.net/archive/rt_072508p14.shtml. Accessed January 16, 2010.

Page, D. 2008. Breaking up with PACS vendor can be hard to do. www.diagnosticimaging.com/imaging-trends-advances/pacsweb/article/113619/1180047. Accessed January 16, 2010.

Page, D. 2008. Choosing next PACS gets complicated, especially where RIS is concerned. www.diagnosticimaging.com/imaging-trends-advances/pacsweb/article/113619/1343356?verify=0. Accessed January 16, 2010.

Philips Healthcare 2008. Considering replacing your PACS? www.healthcare.philips.com/pwc_hc/main/shared/Assets/Documents/Healthcare_informatics/enterprise_imaging_informatics/Pacs_replacement_white_paper.pdf. Accessed January 16, 2010.

Philips Healthcare. PACS replacement strategy. www.healthcare.philips.com/de_de/products/healthcare_informatics/products/enterprise_imaging_informatics/isite_pacs/PACS_replacement.wpd. Accessed January 16, 2010.

Walsh, B. 2008. Replacement PACS: How to plan, purchase & put it in place. www.healthimaging. com/index.php?option=com_articles&view=article&id=11454. Accessed January 16, 2010.

Wiggins, R. H. PACS/RIS replacement: A radiologist's perspective. *The Society for Imaging Informatics in Medicine 2008 Annual Meeting.* Seattle, WA.

10 Enterprise PACS

10.1 INTRODUCTION

A PACS originating from the radiology domain is capable of providing digital medical image storage and retrieval anywhere at any time across the healthcare enterprise. Enterprise imaging is a concept referring to image acquisition and archiving in cardiology, pathology, gastroenterology, radiation oncology, and other specialties because PACS can potentially be expanded to serve all medical specialties in which imaging equipment is used. Diagnostic images or images acquired during therapeutic procedures across these medical specialties can be archived on PACS and integrated with department information systems to be reviewed by clinicians and used for workflow improvement, just as radiology images are. With the development and broad adoption of the DICOM standard, many imaging devices now incorporate native image formats that are DICOM compliant and suitable for PACS. Various nonradiology specialties are at different stages of adopting PACS and integrating it with hospital or department information systems. When PACS is integrated with these systems, its benefits can be fully realized to improve the department's workflow.

10.2 CARDIOLOGY PACS

Cardiology is a complex medical specialty covering many areas in patient care. The various modalities in a cardiology department typically include catheterization, echocardiography, hemodynamics and physiological monitoring (for blood pressure and flow), electrocardiography (ECG or EKG), nuclear imaging, and reporting. In addition to the modalities within a cardiology department, the chronic nature of cardiac diseases means that patients will typically have many encounters with healthcare institutions and enterprises in their lives. To ensure the continuity and timeliness of patient care, it is critical for cardiologists and referring physicians to collaborate efficiently and communicate through shared access to patients' prior clinical histories, images, and clinical data.

As patients move through the cardiac care process, from initial clinical screening and diagnostic procedures through therapeutic interventional procedures to post-treatment health condition monitoring, a variety of medical records documenting these procedures is generated. These records may include images from echocardiography, intravascular ultrasonography, vascular angiography, and nuclear medicine along with reports of laboratory data and measurements of electrophysiology, cardiac function, and hemodynamics. Traditionally, these images and reports were produced and stored on films or paper, limiting their accessibility and effective use. The redundant procedures that are necessary to ensure against inaccessible records consume professional and financial resources and delay patient care.

With the successful implementation of PACS and RIS in the radiology domain, it is natural to expect digital evolution to occur in cardiology. Although some cardiac imaging and measurement equipment have become digital, the images and data can be difficult to access because imaging archives are often proprietary, report systems for measurements are standalone systems, and cardiology information systems (CIS) and integrated solutions for multivendor, disparate systems are not common. To achieve the goals of improved workflow and interoperable EHR systems, cardiac care imaging and measurement equipment must be able to exchange data electronically.

10.2.1 Cardiology Workflow

In a routine cardiology workflow, orders are generated for procedures, procedures are scheduled, and patients are preadmitted and registered in the HIS. When patients are admitted to the cardiology department, it is critical that the demographics be entered correctly and efficiently when patients go through the various steps in the course of care in the department. Without systems integration, patient information must be manually entered at each step; this method is prone to error and inherently inefficient, wasting valuable human resources and potentially impairing the quality of patient care. In a fully integrated department, the patients' demographic data and orders for their procedures are automatically sent to the appropriate imaging modality scanners when patients are admitted to the healthcare institution, and the information is populated to all equipment involved in patient care. After procedures are completed, the acquired images and associated measurements are securely archived and are ready to be displayed on reviewing workstations for report dictation.

The scheduling of cardiology procedures differs from that of radiology examinations in that it is not unusual to have cardiology procedures ordered without requests being made through a department information system. When patients with symptoms are admitted to a cardiology department, typically ECG waveform and blood laboratory data are acquired first. If the results are positive, additional diagnostic echocardiography or nuclear imaging are performed. When these tests reveal arterial blockage, patients may be sent to catheterization laboratories for therapeutic interventional procedures, which almost always require diagnostic cardiology data and reports from the previous steps. These therapeutic interventional procedures generate additional angiographic and hemodynamic monitoring data and reports. All these data and reports are essential components for the planning of subsequent patient care by cardiologists.

10.2.2 Comparison between Cardiology and Radiology Workflow and PACS Requirements

The clinical workflow of cardiology is similar to that of radiology, but there are significant differences. Understanding each department's workflow can help plan the integration of cardiology information systems. They have similarities in the areas of patient admission, patient demographics input, imaging diagnosis, and report generation. Additionally, echocardiography, various NM studies, cardiac CT, x-ray angiography, and cardiac MRI may be performed in either the radiology or cardiology departments. Images and data from these imaging modalities can all be included in a cardiology PACS. For radiology procedures, the workflow is usually complete when diagnostic reports are generated for referring physicians. For cardiology, a diagnosis may be done with nonimaging equipment that generates waveforms, but patient care continues beyond diagnostic reports and may be followed by therapeutic procedures.

The PACS workstations for cardiology and radiology have many common characteristics and requirements. They both need large image data archives, multimodality and multivendor interface capabilities; they have to interface with department information systems and HIS, and integrate with EHR systems. However, the display, manipulation, and measurement of cardiology images are different from those of radiology images. Instead of viewing mostly static images as in radiology, images from catheterization and echo images are usually acquired in cine, or movie mode (with a large file size) because dynamic image display is very important to cardiologists. Image analysis tools for wall motion analysis, ventricular stroke volume and ejection fraction measurements, and stenotic indexing of the coronary arteries are also required for cardiology PACS workstations. Other electrophysiology, hemodynamic, and ECG measurement systems or devices also need to be interfaced with cardiology PACS.

Because of the different needs of the cardiology and radiology departments, most existing cardiology PACS are dedicated for echo and interventional studies only. From the perspective of a healthcare enterprise, the radiology PACS is evolving into an enterprise PACS. With a true multidepartment PACS approach, financial and human resources, IT resources, and technical expertise can be used much more efficiently.

10.2.3 Challenges and Requirements for Cardiology PACS

Although it is common to have PACS, RIS, and HIS integrated in the radiology department, it is rare for a cardiology department to have completely integrated imaging modalities, lab equipment, and department information systems. For most cardiology departments, there is a mix of legacy and new equipment from various vendors on heterogeneous platforms. The integration challenge is compounded by large volumes of cardiology-specific images, numeric measurements, and unique clinical data.

10.2.3.1 Data and Dynamic Display

In cardiology, grayscale, still-frame images are frequently used, just as in radiology. Additionally, color data for cardiac function or viability depiction and cine image loops of echo and angiography are used extensively, and the processing and display of dynamic motion images are required and emphasized more than in other medical specialties. Shown in Figure 10.1 are dynamic cine displays of x-ray angiography on a cardiology PACS workstation. These image data sets are much larger than those of radiology images, and they require a cardiology workstation for effective dynamic image display and review.

Another challenge for multimodality cardiology departments is the coordination and synchronization of data from various cardiology imaging systems and waveform measurement devices. These imaging and waveform data in cardiac examinations are more comprehensive and include more extensive quantitative measurements than basic radiology systems' images and numeric

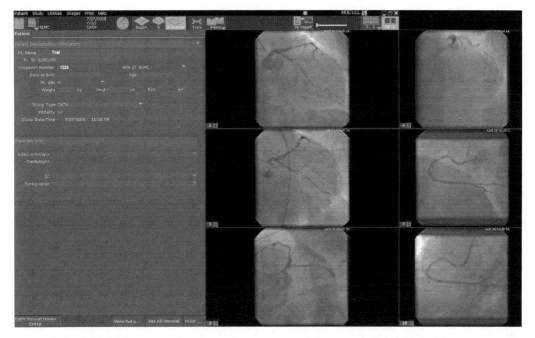

Figure 10.1 Dynamic cine display of x-ray angiography on a cardiology PACS workstation.

reports. All these challenges result in different approaches for integrating cardiology PACS and information systems.

10.2.3.2 Data Mining

Because cardiology relies heavily on outcomes analysis, cardiology departments mine their clinical databases extensively for clinical, administrative, accreditation, business, marketing, research, and other purposes. To mine data effectively, cardiology departments have to integrate various information systems to allow specific outcome queries. Currently, integration solutions are limited to the single-vendor approach from large vendors, which integrate their measurement, monitoring, and report systems so all the systems are automatically updated when data are entered into one system.

10.2.4 DICOM Cardiology Extensions and Reports

DICOM for cardiology started in 1995 as a joint effort by the American College of Cardiology, the European Society of Cardiology, and NEMA. The DICOM Cardiology Workgroup initially defined a standard digital medium for the exchange of x-ray angiography data between healthcare institutions. Since then, multiple DICOM supplements developed by the DICOM standards committee and Workgroup 1 have been released to handle cardiology-specific requirements. The majority of these supplements are related to specific cardiology reports.

Traditionally, cardiology reports have been generated with dictation systems, which are expensive and time-consuming to transcribe. The implementation of structured reports for cardiology procedures allows reports to be generated without transcription services. DICOM has developed several structured report standards for specific cardiology procedures. These reports are built for text with the addition of images, and the trend is to integrate all relevant patient data—test results, patient histories, procedure data, and hemodynamic monitoring results—to follow patients throughout their continuum of care. Using structured report software tools such as pull-down menu selections and predefined data fields for quantitative measurements, transcriptions of procedures can be built easily to include key personnel involved, procedure start times, equipment and supplies used, and other relevant data. All, including free-text descriptions, can be completed while patients are still in the procedure laboratories or immediately after procedure. Because dictation and transcription can essentially be eliminated, physicians and other healthcare professionals can view patients' structured reports, static images, and dynamic cine sequences almost instantaneously after the procedures are performed. With structured reporting from CIS integrated with PACS, more clinical care data are included in the reports than would be included in reports made from transcriptions, and cardiologists can efficiently access patients' images and other clinical data to make informed clinical decisions.

Users should check all relevant statements and measurements shown on their reports prior to implementing any structured reporting tools because customization is usually required to use reporting software packages. In addition to improving cardiology workflow, the integration of structured cardiology reports with cardiology PACS also provides referring physicians with easy access to cardiology studies.

10.2.4.1 Echocardiography Procedure Report

The DICOM echocardiography procedure reports define the structure and codes for the exchange of echocardiography information, which includes the most commonly used quantitative measurements and calculations from images acquired from ultrasound scanners.

10.2.4.2 Cardiac Stress-Testing Structured Report

Cardiac stress tests evaluate chest pain using ECG and echocardiography or NM imaging. The cardiac stress-testing report includes the patient's physical characteristics, medical history, technical descriptions of the examination, and measurement sets defined for stress monitoring and imaging. At the end of the report is a summary of significant clinical findings, measurements, clinical conclusions, or subsequent recommended diagnostic or therapeutic procedures.

10.2.4.3 Cardiac Catheterization Laboratory Structured Report

Cardiac catheterization is used to conduct interventional procedures with image guidance. During these procedures, medications are delivered and therapeutic procedures performed. Cardiac catheterization procedures require collaboration among many medical specialties and involve the acquisition and analysis of images and waveforms from various imaging modalities and equipment. Since a variety of imaging modalities and equipment exists in the catheterization laboratory environment, integrating these systems requires a uniquely structured data exchange system. The structured DICOM cardiac catheterization laboratory report is a comprehensive but basic report template used to document a catheterization procedure's overall clinical results in a format that is accessible to a variety of information systems used by clinicians involved in the patient's care. This report template has a large and useful set of coded concepts suitable for data mining and automatic data extraction for submission to various outcome registries if needed.

10.2.4.4 Intravascular Ultrasound Structured Report

The intravascular ultrasound (IVUS) is a variation of the ultrasonography modality frequently used during vascular imaging or angiography procedures in catheterization laboratories. Shown in Figure 10.2 is an IVUS image displayed on a cardiology PACS workstation. It helps in the selection and deployment of stents during these procedures. The DICOM IVUS structured report standard defines specific structured report templates and context groups, which enables the exchange of both qualitative and quantitative evaluations and measurements during catheterization procedures so that IVUS images and measurements can be incorporated into the catheterization laboratory report. In the report, vessels with lesions are associated with ultrasound multiframe movie loops, qualitative evaluations, and quantitative measurements.

10.2.4.5 Procedure Log

In catheterization laboratories, there are many systems and devices generating data that need to be recorded in the procedure log. These data need to be time stamped because they concern events that may span several hours of the procedure. Since past and current diagnostic and therapeutic procedures are all related, it is critical to log all procedural events—image acquisition, waveforms, and data measurements—during the procedure to enable retrospective clinical analysis. The templates for procedure logs consist of fields for time-stamped free-text comments and comprehensive structured data fields for coded and numeric values. Data mining of procedure logs containing these structured data can yield a variety of administrative, clinical, or scientific information.

10.2.4.6 Quantitative Arteriography and Ventriculography Structured Report

The DICOM quantitative arteriography and ventriculography structured report is used to convey information about image contour analyses of blood vessels and heart chambers acquired by

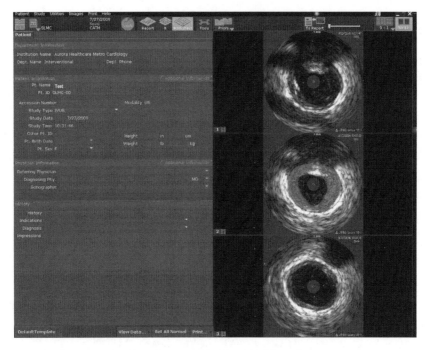

Figure 10.2 IVUS images displayed on a cardiology PACS workstation.

projection x-ray imaging. This structured report includes various qualitative and quantitative measurements, logs of procedures, and clinical findings.

The quantitative arteriography analysis is used for clinical diagnostic or research purposes to assess the condition of the blood vessels, such as the coronary arteries. When quantitative arteriography analysis software is used, the contours of these vessels can be evaluated by measuring the vessels' diameter from projection x-ray images, which enables stenosis measurement.

The quantitative ventricular analysis assesses cardiac output, ventricular contraction, and ventricular wall motion. Cross-sectional ventricular areas, at both end-diastole and end-systole, can be measured using projection x-ray images to determine cardiac volumes (e.g., stroke volumes, ejection fractions).

10.2.4.7 Hemodynamic Measurements Report

Hemodynamics refers to blood flow and pressure measurements and monitoring at various locations in a patient's cardiovascular system. The DICOM hemodynamic measurements report is defined to document observations and findings from waveforms and measurements acquired during catheterization laboratory procedures. Data indirectly derived from these measurements are also included. However, clinical interpretations and conclusions derived from these data and measurements are not part of this report; they are included in other DICOM-structured report documents.

10.2.4.8 Electrocardiography Report

An electrocardiography report is generated by ECG equipment or an ECG management system. In the report, clinical findings from 12-lead, resting, stress, and other ECG acquisitions are documented. This report and the ECG waveform service object pair (SOP) class are essential for the ECG waveform and the clinical observations derived from the waveform to be effectively exchanged under DICOM.

10.2.4.9 CT/MR Cardiovascular Report

The DICOM CT/MR cardiovascular analysis report is designed for noninvasive CT and MR cardiovascular analysis report storage and exchange. It defines report templates including those for reference images, numeric data and associated graphs, and the examiner's narrative text.

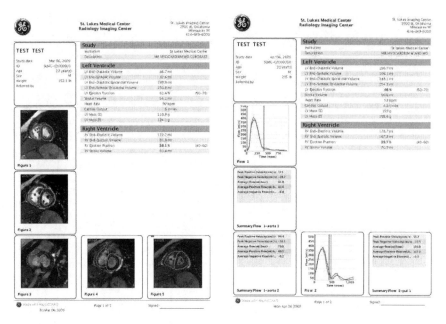

Figure 10.3   An MRI cardiovascular analysis report generated using the ReportCard software on an Advantage Workstation (GE Medical Systems, Milwaukee, WI).

Shown in Figure 10.3 is an MRI cardiovascular analysis report. Different CT/MRI cardiovascular report-generating systems may add optional items or produce different contents as needed.

10.2.5 IHE Cardiology

IHE is an initiative to implement existing standards for efficient patient care information exchange. Since 2004, industry members and professional societies such as the American College of Cardiology, the American Society of Echocardiography, the American Society of Nuclear Cardiology, the Society of Nuclear Cardiology, the Society for Cardiovascular Angiography and Interventions, and the European Society of Cardiology have worked extensively to create the IHE Cardiology Technical Framework. Clinical use cases and technical specifications for the use of existing standards such as DICOM and HL7 are documented in the technical framework as integration profiles. The IHE cardiology initiative is still evolving, with new integration profiles being proposed.

10.2.5.1 Benefits of IHE Cardiology Implementation

The benefits of implementing IHE in cardiology departments include:

- Reducing errors and time lost from identical data entry into multiple systems in the cardiology department

- Ensuring automatic and correct entry of procedure codes and patients' demographics and identifiers

- Allowing automatic synchronization, which is important when all systems are integrated and all events during a procedure are logged

- Presenting procedure status feedback in real time, which can improve department management by streamlining workflow

- Providing structured reports for measurement and procedure logs

- Reducing the number of duplicate procedures

- Providing cardiologists and other clinicians with timely and accurate images and reports

- Ensuring secure image and data archiving before its deletion from imaging modalities

- Allowing "best-of-breed" system selection from various vendors

- Simplifying the equipment procurement RFP process

- Providing building blocks for integrated EHR/PHR systems

In the cardiology environment, the implementation of various IHE integration profiles can provide efficient and automated clinical information interchange, ensure the security and privacy of patient information during these interchanges, and develop the foundation for comprehensive EHR/PHR systems.

10.2.5.2 Cardiac Catheterization Workflow

Multiple modalities and devices are simultaneously used in a cardiac catheterization procedure. While x-ray angiography dynamic cine images are acquired, an ECG system records ECG waveforms, and a hemodynamic recording system captures intracardiac pressure through a pressure sensor on the catheter. The DICOM standard for waveforms not only allows the description of waveform temporal characteristics, but also of specific waveform parameters' measurements and locations. When these systems are coordinated and data acquisitions are synchronized, image acquisitions correlated to end-diastole and end-systole cardiac phases can be determined for volumetric measurements and calculations, such as stroke volume and ejection fractions. Nevertheless, most of these systems manage their own data as isolated systems. These data typically are acquired independently without time synchronization, and they are stored in proprietary formats. Their measurements and analysis cannot be shared electronically, and they require multiple reentry into clinical reports or submission to mandatory data registries.

The IHE Cardiac Catheterization Workflow (CATH) integration profile deals specifically with the consistent handling of patient identifiers and demographics. It allows basic patient data to move through the complex workflow of catheterization procedures, as shown in Figure 10.4, which includes ordering, scheduling, and image acquisition, storage, and viewing. Because many cardiac catheterization procedures are performed during emergencies, orders often are not placed before procedures begin. Patient identities may not be known until after the procedure has started or even some time after the procedure is completed. During a procedure, there may be many steps including a variety of diagnostic and therapeutic equipment. The CATH integration profile facilitates workflow scheduling and the central coordination of data flow through various imaging, measurement, and analysis systems. It also provides information about the completion status of each stage of a multiphase diagnostic or therapeutic procedure and makes data available to support subsequent workflow. Following the completion of a procedure, the CATH integration

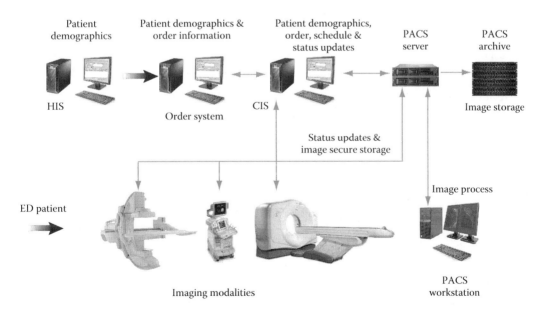

Figure 10.4 A typical cardiac catheterization workflow.

profile coordinates the working of the measurement, analysis, reporting, and data archiving systems. The integration profile eliminates the need to manually find all the catheterization data results and update them according to corrected patient identifiers. The CATH integration profile has elements similar to the IHE radiology SWF, PIR, and Evidence Document (ED) integration profiles, with data requirements specific to catheterization laboratories in a multimodal environment.

Using the CATH together with other IHE integration profiles can ensure continuity, integrity, and integration of basic patient, clinical order, and procedural data across all involved imaging modalities and information systems.

10.2.5.3 Echocardiography Workflow

Digital echocardiography includes transthoracic echocardiography (TTE), which uses probes on the surface of the chest; transesophageal echocardiography (TEE), which uses small ultrasound probes inserted through the esophagus behind the heart; and resting and stress echocardiography Figure 10.5a and b show the worklist and the dynamic echocardiography images displayed on a cardiology PACS, respectively.

Stress echocardiography is a multistage examination. It acquires images and measurements before and after exercise or pharmacologically induced stress. Just like the CATH integration profile, the IHE Echocardiography Workflow (ECHO) integration profile handles order entry, patient identifiers, scheduling, status reporting, and data storage associated with digital echocardiography, as shown in the Figure 10.6. However, IVUS, intracardiac echocardiography, and 4-D echocardiography (Figure 10.7) are not supported in this ECHO profile. Instead, IVUS and intracardiac echocardiography used in cardiac catheterization are supported in the CATH profile, and a DICOM standard for 4-D ultrasound images is still under development.

The ECHO integration profile also emphasizes modality- or independent workstation–based image acquisition processes and measurements. For example, stress echocardiogram image display is unique in that the order in which the images are displayed (baseline, prestress, mid-stress, poststress, and recovery stages) is important. For workstations supporting this integration profile with the Stress Echo option, the image sets are consistently displayed without manual rearrangement. The ECHO profile also addresses the use of digital echocardiography devices in portable configurations and during possible intermittent network disconnection.

The ECHO profile has components similar to those of other integration profiles from the IHE Radiology Technical Framework, such as the radiology SWF, ED, and PIR (discussed in Chapter 6). Although the Modality Worklist requirements are similar to those of radiology ultrasound scanners, the DICOM storage requirement are different because there are measurements made during echocardiography.

10.2.5.4 Retrieve ECG for Display

The IHE Retrieve ECG for Display integration profile is used for the review of ECG documents both enterprise-wide and outside the enterprise by cardiologists, clinicians, and referring physicians. The ECG waveforms were previously acquired during diagnostic 12-lead ECG tests and have already been archived in an information system. ECG documents typically include measurements, diagnosis, reports, signatures, and high-quality ECG waveforms and are usually managed on ECG management systems. The primary purpose of the Retrieve ECG for Display profile is to simplify and standardize the ECG access and viewing process to allow ECG documents to be selected, retrieved, and reviewed on general-purpose computers in diagnostic display resolution without specialized cardiology workstations and software. Since the display computer does not manage or archive ECG documents, the documents are not saved on the display computer and a new request must be sent each time there is a need for ECG document display. When they are displayed, these ECG waveforms are faithfully represented with diagnostic quality, together with their measurements and interpretations. This profile is intended only for retrieving stored ECGs and not for ordering, acquiring, interpreting, or storing ECG processes or documents.

10.2.5.5 Stress-Testing Integration Profiles

Currently, many steps are involved in ordering and performing cardiology stress tests. Typical cardiology stress-testing workflow involves treadmill- or drug-induced stress under which patients' cardiac functions are assessed using ECG, echocardiography, or nuclear cardiology. When a cardiac stress test is ordered for a patient by a physician and scheduled, the appropriate room, equipment, healthcare professionals, and radiopharmaceuticals (if nuclear imaging is

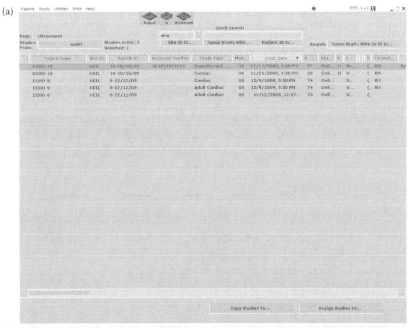

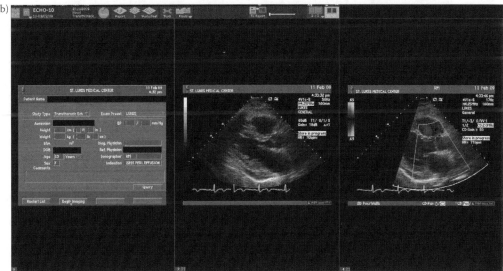

Figure 10.5 (a) The worklist for echocardiograph displayed on a cardiology PACS workstation. (b) Dynamic echocardiography images displayed on a cardiology PACS workstation.

ordered) are prepared before the test. A physician or a healthcare professional may perform the test. If nuclear imaging is indicated in the order, the technical parameters of the radiopharmaceuticals' administration need to be recorded. There are several progressive levels of exercise on treadmills, during which ECG waveforms, the patient's performance, and exercise equipment settings are documented. When a predefined heart rate is reached, or if the patients' symptoms of arrhythmia, hypertension, or angina prevent the continuation of the stress test, the procedure is completed. The stress tests and images acquired are interpreted and reported by cardiologists. Without system integration, the process from ordering the test to the final report involves many manual steps. ECG and imaging systems are usually separate, requiring different systems and workstations for data storage and display. To fully integrate all stress-testing equipment and imaging scanners into the stress lab workflow for synchronized patient orders, schedules,

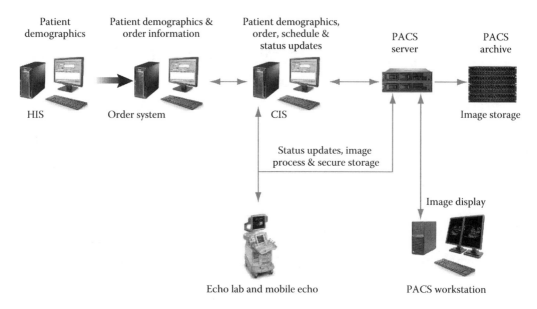

Figure 10.6 A typical echocardiography workflow.

demographics, and identifications, several IHE cardiology and radiology integration profiles can be used.

The Stress-Testing Workflow (STRESS) defines details of managing the workflow for cardiac stress tests procedure management and for reliable data storage in an archive for subsequent reporting support. Messages between hospital and cardiology department information systems about orders, scheduling, admissions, discharges, and transfers are exchanged in the HL7 format across a variety of systems. These include ECG systems (for acquisition, measurement, and analysis) and imaging modality scanners. Various DICOM service classes are used to synchronize and automatically link patients' information systems with imaging modalities and stress-testing devices, eliminating the redundant entry of patient demographics into multiple systems and reducing the risk of error. DICOM-defined objects such as waveforms, structured reports, and encapsulated PDF files are also used to exchange waveform and stress-test data. These objects facilitate the convenient display of stress-test results and their incorporation into the EHR system by allowing results from imaging and ECG to be managed as a linked data set.

Other useful cardiology-specific integration profiles include ECHO with the Stress Echo Option (for consistent image labeling and viewing at various stress levels), ED with the Echo Option for

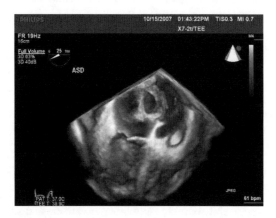

Figure 10.7 A 4D echocardiography.

standardized stress echocardiography measurement templates, Retrieve ECG for Display (ECG), NM–Cardiology Option, and Displayable Reports (DRPT). These integration profiles allow patients' stress test laboratory data flow across various systems without repeated data entries at each step. When tests are completed, data and images are stored and available to cardiologists and clinicians with easy access for viewing and reporting, which lead to significant improvement in patient care.

10.2.5.6 Displayable Reports

The DRPT profile defines transactions for the creation, revision, inter- and intradepartmental exchange, and viewing of display-ready cardiology clinical reports, which can be distributed enterprise wide when other profiles are used. It links images, reports, and other evidence of the imaging procedure to report creators, managers, repositories, and displays. PDF is the format of choice for encoding a full range of documents with both text and graphic images generated by various systems. When used with other integration profiles defined in the IHE IT Infrastructure Technical Framework, such as Retrieve Information for Display (RID) and XDS, the report can be further distributed within the enterprise and beyond the healthcare enterprise border. The Displayable Reports profile does not have a standardized reporting worklist management function because most physicians in catheterization laboratories create reports immediately after performing interventional procedures.

10.2.5.7 Electrophysiology Lab Workflow and Useful Integration Profiles

Electrophysiology (EP) laboratories tend to have a variety of systems, often from multiple vendors, for implanting and adjusting cardiac rhythm control devices such as pacemakers and ICDs. Without systems integration, each system manages its own patient data. The entering of order, schedule, and patient's demographics is very similar to that followed in the cath and echo labs. Patients' demographic information and identifiers are entered separately into intracardiac echocardiography, fluoroscopy, recording, and special mapping systems immediately before procedures are performed. Relevant data from patients' medical records are also frequently re-entered for procedure records. The data generated by these independent, specialized systems need to be reviewed on each individual system, and the timing information of these data acquisitions must be manually sorted and entered into the records. Furthermore, these data are archived in a variety of formats and are usually not archived with patients' medical records. Several IHE cardiology-specific integration profiles can be used to integrate systems in EP laboratories: CATH for laboratory workflow management, Retrieve ECG for Display for easy access to archived ECG waveforms, and Urgent Implantable Rhythm Control Device Identification (PDQ-IRCD) and IDCO for extracting data from implanted devices.

Other useful IHE integration profiles from the IHE Radiology and IT Infrastructure domains include the ED for standard electrophysiology lab data templates, the RID, and the XDS for EHR system access both within the healthcare institution and across enterprise borders.

10.2.5.8 Implantable Device Cardiac Observation

During follow-up checks, implanted cardiac rhythm control devices are interrogated either in healthcare institutions or remotely from patients' homes. Information obtained from the devices includes device identification, device diagnostics, therapy settings, and therapy delivery. Typically, proprietary device programmers or pacing system analyzers from vendors are needed to collect information from implanted cardiac devices. The IDCO integration profile defines a standardized method to create processed data and report attachments, and to translate or transfer information from the interrogation or observation of key implantable rhythm control devices. Information from multiple vendors' interrogation systems, programmers, pacing system analyzers, or remote follow-up systems can thus be stored consistently in a central data repository and management system such as EHR. The integration profile allows implanting physicians to dictate operation notes efficiently, and clinicians involved in patients' follow-up care are able to view these data from anywhere by accessing the EHR system.

10.2.5.9 Urgent Implantable Rhythm Control Device Identification

In emergency rooms, when patients come in with implantable rhythm control devices such as pacemakers and ICDs, physicians and clinicians need to know the details of these devices, including vendor, model, and programming. However, often patients do not have or cannot communicate this information readily. The PDQ-IRCD is a simple specialization of the general

IHE Patient Demographic Query (PDQ) integration profile. The generic PDQ is used to access a list of patients matching some specific demographic query parameters, while the PDQ-IRCD allows emergency department physicians to electronically access patient device registry data from various sources, including vendors, national, or government databases. Quick identification of a patient's IRCD can be achieved by using all available patient identification information. When a single query searches all available registries, the correct information can be retrieved quickly, which is necessary for the selection of the right programmer for the particular device implanted. This process improves patient care and workflow efficiency.

10.2.5.10 NM Imaging Cardiology Option

Nuclear cardiology images and reports are critical diagnostic tools for patient care, but without systems integration it is difficult to use one vendor's workstation to view patients' cardiology images acquired by other vendors' camera systems. This causes the performance of unnecessary redundant studies or the use of only text reports as the basis for essential clinical decision making. The IHE Nuclear Medicine Imaging-Cardiology Option integration profile was developed together with IHE Radiology to define NM image acquisition and display requirements. The profile allows nuclear cardiology's raw, postprocessed/reconstructed, and quantitative images to be easily exchanged across various systems, even when the workstations are from different vendors, and it allows PACS workstations to display functions unique to NM and critical for accurate interpretation, such as simultaneous stress/rest cine movie displays, stress-over-rest perfusion displays, color maps, and image realigning within displays. If needed, images can be reprocessed to meet the specific clinical requirements of a different vendor's nuclear cardiology system. This profile not only significantly improves cardiologists' ability to access critical nuclear cardiology images and data at the point of care, but also allows images to be incorporated into comprehensive EHR/PHR systems.

10.2.6 Cardiology PACS and Information System Integration

The radiology community has been working on ways to integrate PACS with department information systems for many years, and it is a proven fact that the integration of various systems brings significant improvements to a department's workflow. Cardiology PACS used to be based in individual laboratories as mini-PACS, but with the development of digital and information technology in the cardiology domain, it is anticipated that cardiology departments are going to benefit from integration, just as radiology departments have. The trend is now toward the development of systems in which the department information is integrated with laboratory-based mini-PACS throughout the cardiology department and to add capabilities and functions for reporting support, which constitute the foundation for EHR systems. The keys to integration are IHE integration profiles and the CIS, which complements cardiology PACS as a patient and department information management tool. CIS allows imaging studies and cardiology procedure reports to be accessed throughout the healthcare institution or enterprise. The driving force for systems integration is similar to those at radiology; an integrated system meets the needs for department workflow tasks such as patient scheduling, ordering, and cost capturing while also serving as an administrative and business operational data source for department staff and resource management, administrative reporting, data mining, laboratory efficiency improvement, and efficient communication with other clinicians and third-party healthcare payers.

10.2.6.1 CIS Benefits

CIS provides cardiology departments with a tool that improves patient care and workflow. Fully integrated cardiology PACS and CIS give cardiologists immediate access to patients' images and data allowing clinical decisions to be taken faster. When the integrated PACS system is equipped with the structured report profiles mentioned earlier, enabling cardiologists to generate data-rich, organized reports almost immediately after procedures, these reports and detailed statistical data extracted from CIS can be used to improve operational efficiency and patient care. Cardiology departments with multiple subspecialties can be automated by applying IHE integration profiles, avoiding redundant data entries and improving efficiency. When PACS from all cardiology laboratories and CIS are seamlessly integrated using DICOM standards and IHE integration profiles, images and patient information can be efficiently accessed throughout the healthcare institution or enterprise.

10.2.6.2 Cardiology PACS and CIS Implementation Considerations

Before the implementation of the cardiology PACS and CIS systems, key areas to be assessed include department structure, procedures performed, cardiac imaging volumes, image data retention period, how quickly physicians need to retrieve data, scheduling and ordering workflow, clinical operations workflow, modality scanners and clinical systems, the results formats and distribution within the department workflow, and potential workflow changes for improved efficiency. General infrastructure considerations were discussed in Chapter 8. The implementation requires careful analysis and very possibly updating of the IT infrastructure, as in radiology PACS and RIS implementations.

In addition to the IHE cardiology integration profiles discussed in this chapter, the general integration profiles from the IHE Radiology and IT Infrastructure domains, discussed in Chapters 6 and 12, can also be used in the integration of cardiology department systems for security, privacy, and cross-enterprise exchange. These profiles include Consistent Time (CT), PIR, RID, Key Image Notes (KIN), Simple Image and Numeric Reports (SINR), SWF, XDS, XDS Medical Summary (XDS-MS), XDS Imaging (XDS-I), and Audit Trail and Node Authentication (ATNA) integration profiles. Among these integration profiles, SWF and PIR have been the most successfully implemented integration profiles. The use of IHE for cardiology systems procurement is very similar to that in radiology, as discussed in Chapter 6.

10.2.6.3 Cardiology Systems Integration Challenges

Compared to radiology departments, cardiology departments lag behind in integrating data into the information systems of their healthcare institutions and enterprises. Complete systems integration in cardiology is not widespread, and a truly integrated CIS does not yet exist because of multiple challenges. There is much more information to integrate in cardiology than in radiology or other medical specialties. Instead of dealing mostly with images as in radiology, cardiology deals with an array of patient data including images, ECG waveforms, and hemodynamic information. Rarely do all imaging modalities and image management and storage systems in a cardiology department use the latest technology supporting the most recent DICOM, HL7, and IHE standards. Often equipment and systems of a department are acquired and upgraded over years. Because of the subspecialty nature of cardiology, and because of legacy systems acquired at different times, the scope and cost associated with integrating the entire department are beyond the reach of many cardiology departments; as a result, specialty mini-PACS covering individual areas are common. Another challenge is that the extensive workflow changes caused by new system adoptions or integrations will require training of all involved clinicians and staff.

10.2.6.4 Development Trends

The growing number of cardiology procedures has created a need for improved workflow efficiency. The development and successful integration of radiology PACS and RIS information systems show that it is feasible for a similar approach to be used to integrate cardiology PACS and CIS. The trend of cardiology PACS and CIS development has been to move from an imaging-centric system to an information-centric system with imaging components. The continued digitization of cardiology department workflow has improved efficiency by centralizing all cardiology equipment and services. This process will lead the once-isolated islands of cardiology PACS and other systems toward integration with department information systems and ultimately into enterprise-wide PACS solutions and comprehensive EHR/PHR systems. Healthcare institutions or enterprises that accomplish complete integration of patient data and image will have improved patient care and efficiency, and can fully realize fundamental operational and financial benefits.

10.3 PATHOLOGY PACS

PACS has become commonplace in radiology departments; however, this is not yet the case in most pathology departments, which also use imaging extensively. Anatomic pathology involves providing diagnosis based on examinations of tissue and fluid specimens. These examinations may be gross, microscopic, or performed using other instruments. For over a century, high-power microscopes have been used by pathologists to study small tissue samples embedded in glass slides. Through the study, patients' diseases have been identified and diagnostic reports provided to clinicians who plan and deliver treatments. In most pathology laboratories today, pathologists still review glass slides of stained tissues under a microscope to provide diagnosis. This process requires shipping many glass slides back and forth between pathology laboratories, and delivering the right slides to the right pathologist, which may cause delays both at the initial stage and during subsequent second opinion diagnosis.

With the development of radiology PACS and department information systems, pathology departments are beginning to use new digital tools to scan glass slides rapidly and to display these scanned digital pathology images on computer workstations for pathologists to review. Although PACS in pathology requires different image acquisition modalities—which include digital gross specimen photographic stations and microscopy glass slide scanners—the essential components and functionalities of pathology PACS are very similar to those of radiology PACS and include image archives, delivery, and management as well as integration with the department's laboratory information system (LIS). This digital approach provides opportunities to improve pathology workflow in the areas of reduced turnaround time, second opinion consultation, pathologist specialization, computerized image processing, reduced need for glass slides, and incorporation of images and results into patients' EHR/PHR.

Pathology departments are in the process of transitioning to a digital imaging and electronic workflow. Commercial instruments for digital microscope slides are increasingly available, aided by standards developed for handling diagnostic high-resolution microscope whole-slide image (WSI) digitization. When this transition is completed, it will allow faster pathology services for patients and potentially reduced costs for healthcare institutions.

10.3.1 Pathology Workflow

The pathology workflow is different from that of radiology or cardiology. Anatomic pathology belongs to the pathology domain, which is divided into surgical pathology, biopsy pathology, cytopathology, autopsies, and molecular pathology. The diagnostic process of an anatomic pathology laboratory is different from that of a clinical laboratory because diagnostic reports are based on image interpretation. During operations, surgeons collect, label, and send specimens to pathology laboratories for analysis. Without a pathology PACS, the workflow following specimen collection includes the following steps:

- Label and identify tissue collections in containers
- Examine the collected specimen
- Determine and select small portions of tissue for analysis
- Prepare tissue into thinly sliced sections
- Prepare staining if needed
- Place sections on slides
- Examine slides under microscopes (done by pathologists)
- Generate reports (done by pathologists)

In the workflow, each slide is identified uniquely from all other slides and managed separately by the LIS of the pathology department.

In a digital pathology department, although all the above steps are the same, a case, or "accession," with a unique accession number is assigned by the LIS. An accession consists of analysis of tissues collected in a single procedure. Each specimen is considered a "part," and multiple parts can exist for an accession. In the pathology workflow, the part is an important logical component. According to DICOM, the accession concept in anatomic pathology is the equivalent of accession in the radiology workflow, and accession numbers at the DICOM-defined study level can be reused as in radiology. Thus, a requested procedure or imaging study in anatomic pathology is identified by its unique accession number, and the "study" is the image folder defined at the level of the accession number. When a requested procedure involves multiple modalities such as gross imaging, microscopic imaging, or more than one specimen, new imaging series are created because each series can only have images acquired from a single specimen by a single imaging modality. If new images are acquired for an existing study, a new series has to be created for these images.

The workflow of anatomic pathology also differs from that of radiology in that it is driven by specimens. A variety of imaging devices for gross imaging, microscopic imaging, whole-slide imaging, and multispectral imaging may be used when digital anatomic pathology images are acquired. In contrast, diagnostic processes in radiology are driven by patients; a single radiology imaging modality is typically used in a study, and all images of the study are acquired from a single patient.

10.3.2 Digital Pathology Benefits

Pathology image digitization enables efficient, easy, and fast workflow in the department. It eliminates the shipment and management of glass slides and allows pathologists to improve their diagnostic processes.

10.3.2.1 Improved Workflow

Improved pathology workflow is one of the major benefits of digitization, which was proven when radiology transitioned to digital. When digital virtual slides, gross images, and associated clinical data are stored and organized electronically within department information systems, patients' prior histories are readily available from digital archives. The electronic exchange of all information through computer networks streamlines the workflow for the pathologists and the laboratory staff, improves accuracy, and reduces errors.

10.3.2.2 Reduced Turnaround Time

Pathology PACS allow cases to be organized and prepared efficiently, which results in reduced turnaround time from tissue collection to final diagnostic reports as compared to systems using glass slides and microscopes. A fast, accurate diagnosis, followed by appropriate and definitive therapy, could improve the treatment outcome and lower the healthcare cost.

10.3.2.3 Eliminating Glass Slide Transportation

Because virtual slides, also called digital slides, are transmitted electronically over a computer network, the images are available to pathologists almost instantly after the slides are scanned. This process minimizes the time, cost, damage, and loss associated with the physical transportation of specimens or glass slides.

10.3.2.4 Telepathology and Distributed Workload

Digital pathology allows pathologists to be more specialized, and workloads can easily be distributed to multiple physical locations without the time and cost incurred for pathologists to travel to where the glass slides are located.

10.3.2.5 Collaboration and Second Opinion Consultation

Digital pathology makes it much easier for pathologists to collaborate and obtain second opinions on difficult cases in which timing is critical for patient care. Also, communication with oncologists and patients can be conducted efficiently.

10.3.2.6 Advanced Review and Digital Imaging Analysis on Workstations

Digital pathology platforms are capable of providing image analysis algorithms and software applications to help pathologists efficiently screen large numbers of slides in search of small nests of cells, diagnose diseases accurately, and perform tedious tasks such as counting cells of interest. The image contrast, brightness, and gamma curve of display characteristics can be adjusted interactively during image review on workstations. Multiple digital slides can be displayed simultaneously for comparison study, which is not possible when reviewing glass slides using microscopes. Advanced computer analysis software incorporating pattern recognition algorithms can automatically detect, classify, and count thousands of cells with improved efficiency. These algorithms can be used to find tumor cell regions, and for cell membrane morphometric analysis. Otherwise, it is very difficult and time consuming for pathologists to search for rare events by examining an entire glass slide under a microscope at high magnifications.

10.3.2.7 Digital Records for EHR/PHR

Anatomic pathology is an imaging-rich specialty. Digital pathology images and reports are easily integrated into a patient's comprehensive EHR/PHR. Displaying gross and microscopic images improves the assessment of the patient's disease conditions by clinicians, thereby improving patient care.

10.3.3 Digital Pathology Challenges and Considerations

Although there are many benefits of digital pathology, several factors have prevented it from being adopted on a widespread and standardized basis. The most challenging aspect of digital pathology WSIs is their file size. A digital slide is a digitized scan of an entire glass microscope slide at high spatial resolution. When pathologists review slides, usually a low resolution of 2× to 4× magnification (5–2.5 μm/pixel[mpp]) is used to pan through the slide, followed by a zoom in to a higher resolution for a region of interest and focus. Because of this, the software for digital WSI storage and review must be designed for fast panning, zooming, and focusing when multiple focus planes are acquired during WSI digitization.

A typical digital slide has a spatial resolution of 0.25 mpp, conventionally referred to as 40× magnification, and is captured in 24-bit color (8 bits for each of the three basic colors: red, green, and blue). With a typical sample size of 20 mm × 15 mm, the data matrix is 80,000 × 60,000 pixels, making a total of 4800 million pixels. Without any image compression, the digital image file size is about 14.5 GB, which is much larger than any other medical image. The size of the digital slide is going to be even larger if the sample size is larger, which can be up to 50 mm × 25 mm, or if the spatial resolution is higher, up to 0.1 mpp, or if the sample is thicker than the depth-of-field of the lens, which requires scanning at multiple focal planes. Therefore, in a typical pathology laboratory processing as many as 200,000 glass slides a year, it is a tremendous technical challenge to scan and digitize many glass slides rapidly, archive these digital slides efficiently, transmit them over computer networks, and navigate each image easily.

A digital pathology image data storage and retrieval structure is different from other medical images currently stored in the PACS systems. Pathology imaging presents another incompatibility with the DICOM standard because pathologists need frequent and fast panning and zooming during image review sessions, just like moving slides under a microscope for viewing different areas. A typical diagnosis involves much viewing at low magnifications, with occasional zooming to higher magnifications of 10×, 20×, or 40× to review small regions in detail. Figure 10.8 shows (a) a virtual slide displayed on a workstation, (b) digital slide review tools, and (c) a pathology PACS

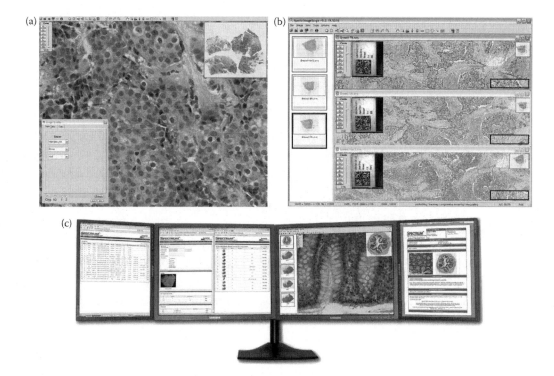

Figure 10.8 A pathology PACS system (Aperio Technologies, Vista, CA) with (a) a displayed digital virtual slide on a workstation; (b) digital slide review tools; and (c) a pathology PACS display monitor. (Courtesy of Ole Eichhorn from Aperio Technologies.)

Figure 10.9 A ScanScope CS (Aperio Technologies, Vista, CA) slide scanner. (Courtesy of Ole Eichhorn from Aperio Technologies.)

display monitor. The review patterns are quite different from those used by radiologists or cardiologists, who usually view an entire image at once and at full spatial resolution. Therefore, DICOM image viewers are not suitable for digital slides review by pathologists.

10.3.3.1 Image Data Acquisition with Glass Slide Scanning

Slide scanning is the process of transforming glass slides into digital slide files. Shown in Figure 10.9 is a digital slide scanner.

There are currently two approaches to scanning as shown in Figure 10.10a and b one is image tiling, which involves scanning and capturing many small regions of a microscopic glass slide by a digital camera, and the other is strip line scanning by employing a linear array of detectors for contiguous overlapping image stripes. The details of the slide scanners are outside the scope of this book, but in both methods, the entire WSI is composed by stitching or aligning together individually acquired tiles or strips.

10.3.3.2 Image Data Compression

Since digital pathology slides have very large file sizes, the development of digital image applications in clinical pathology is limited. The computational power and resources required for virtual microscopy WSI differ to a large extent from those of radiology and other medical imaging because of the much larger file sizes, which are 100–1000 times larger than typical radiology imaging studies. For this reason, data compression is mandatory for WSI application software, storage, and management. Various lossless and lossy data-compression techniques such as LZW, JPEG, and JPEG2000 discussed in Chapter 5 can be used. Although lossless LZW with an average

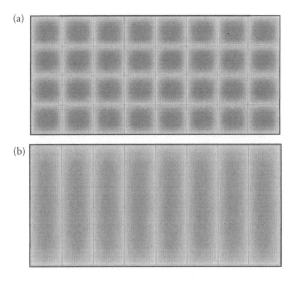

Figure 10.10 (a) Image tiling and (b) strip line approaches for slide scanning.

239

compression ratio of 4:1 can be used, lossy JPEG and JPEG2000 with compression ratios of up to 20:1 and 50:1, respectively, are frequently used. With a compression ratio of 25:1 using JPEG2000, a typical 15 GB image data file can be compressed into 600 MB, which is significantly smaller than the original and is much easier to review and analyze. Like the tiled pyramid image organization being adopted by DICOM, the JPEG2000 Interactive Protocol (JPIP) is also reported to be useful for compressing and viewing large WSIs. JPIP is capable of downloading only the requested part of the picture for a relatively quick viewing of a large image either in low or a high spatial resolution.

It has been found that data compression does not degrade the diagnostic value of WSIs; therefore, lossy data compression is widely used for virtual slide images in pathology. This is quite different from radiology PACS, in which image files are magnitudes smaller and lossless compression techniques, typically with compression ratio of 2 ~ 3:1, are commonly used. Even though JPEG2000 has a higher compression ratio and less image artifacts than JPEG, it requires more computational power to process, and it has not yet been as widely adopted as JPEG. Therefore, JPEG is still used by many vendors.

10.3.3.3 DICOM Limits for WSI

To make use of PACS's proven capabilities of image storage, retrieval, managing, and searching for other medical images, WSIs need to be acquired and formatted in a way that they can be stored in commercial PACS systems using DICOM standards. This will accelerate the adoption of digital pathology by healthcare institutions and laboratories, one of the prerequisites for establishing a unified EHR/PHR system in which medical images from radiology, pathology, and other imaging modalities are managed together. However, one of the technical challenges is that the current DICOM standard does not have a provision to accommodate such a large pathology WSI 2D image; both the spatial resolution and the image size exceed the maximum value limit set by DICOM. The DICOM image spatial resolution is limited to an unsigned 16-bit integer, which constrains the row and column sizes of the 2D image data matrix to no more than 65536×65536 ($64\text{ K} \times 64\text{ K}$) pixels, and the data file size of each image cannot be larger than 2 GB. Since these two constraints are not going to be changed, and since a WSI image data file can easily exceed these two limits, the DICOM-compatible storage of WSIs must be approached differently from the storage of other medical images. DICOM Workgroup 26 is working on DICOM Supplement 145 to create a mechanism for the storage, retrieval, and reviewing of pathology WSI within the DICOM framework. To simulate the process of viewing a physical microscopic glass slide, the proposed solution involves storing large WSIs in small subsets and sending them in fast response to interactive image requests, enabling dynamic WSI viewing. It will be some time before the DICOM supplement is finalized and supported by vendors on their products.

10.3.3.4 Tile Image Organization

Because pathologists review slides by panning through at a low resolution and then zooming in to a higher resolution and adjusting the focus for a region of interest, the software for digital WSI storage and review must be designed for fast panning, zooming, and focusing on the multiple planes acquired during WSI digitization. The solution for this requirement is that large digital pathology WSI data can be stored in a sophisticated two-dimensional "tiled" form. This allows efficient image data access while avoiding the structural limitations of DICOM, which are also applicable to other imaging modalities. As shown in Figure 10.11, each tile is square or rectangular, and the entire WSI is divided and stored in many, much smaller tiled images. The data matrix size of each tile ranges from 240×240 pixels to 4096×4096 pixels, resulting in individual tile image sizes of approximately 173 kB to 50 MB without data compression. To review or process a specific region of interest from the WSI, only tile images encompassing the region need to be loaded. Having smaller tiled images means having a greater number of tiled images but less total data to be loaded each time. Therefore, the size of each tile affects image access performance.

WSI high-resolution image review and analysis with a high zoom factor requires the loading of a small number of original image tiles, while a low-resolution review and analysis with a low zoom factor requires a large number of original image tiles. The fast zooming capability required for WSI review and analysis is achieved through the precalculation and storing of images at several spatial resolutions, thus allowing the rapid formation of subregions of the slide at any spatial resolution without processing the entire slide data. A similar approach is adopted to handle

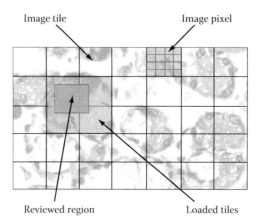

Figure 10.11 Tile image arrangement for WSI. Only the shaded image tiles that encompass the reviewed region are loaded for displaying.

thick samples in which images are acquired at multiple focal planes, termed Z-planes, and stored as separate images.

10.3.4 DICOM Standards for Digital Pathology

Anatomic pathology images are considered as visible light imaging in DICOM. There are various DICOM supplements that can be used for digital pathology. As mentioned above, DICOM Supplement 145 incorporates a solution to handle tiled WSI digital slide images so that subregions of a very large WSI can be stored, managed, and accessed individually at various spatial resolutions.

10.3.4.1 Pyramid Image Data Model for WSI

The pyramid model, based on tiled image organization, has been proposed by DICOM as the solution for storing digital pathology WSI. Although it has not yet been finalized, the proposed model divides a large WSI into many small regions stored in a tiled fashion, with each tile as an individual image of a DICOM series, so that a DICOM series can contain many images of small tiles, or fewer images of large tiles. To facilitate fast panning while reviewing the digital pathology WSI on a workstation, lower-resolution versions of the original high-resolution WSI are precalculated, and a pyramid model of image data organization used for image storage, as shown in Figure 10.12. The WSI thus consists of several images at different spatial resolutions, with the base of the pyramid being the original high-resolution digital pathology image acquired from scanning and the top of the pyramid being the lowest-resolution thumbnail image of the entire WSI. Each image in the pyramid, with different resolutions (or magnifications), may be stored as a series of tiled images for fast image retrieval during panning and zooming. Image tiles related to the shaded regions in Figures 10.11 and 10.12 need to be retrieved to compute the synthesized region of interest at the desired spatial resolution.

When images are precalculated for lower spatial resolutions to facilitate fast zooming, images located at the same "level" of the WSI pyramid are stored together, separate from other levels. Typically, the high-resolution image tiles located at the base of the WSI pyramid are stored first, followed by successively lower-resolution image tiles located at higher levels of the WSI pyramid. It is also proposed that for a specific layer of the pyramid, all tiles must have the same size, shape, image format, and compression; they should correspond to the same Z-plane, and they are not to overlap. Multiple Z-planes (e.g., focal planes) for thick samples will be represented by separate layers in the pyramid. Image tiles belonging to the same plane are stored together, and all Z-planes corresponding to the same level of the WSI pyramid are also grouped together. Therefore, there might be multiple layers with the same magnification or spatial resolution, each correlating with a different focal plane or compression algorithm. Other images that may be included in the DICOM WSI pyramid series include a low-resolution thumbnail image and ancillary images such as a slide label.

Each image in the DICOM WSI series is defined by four coordinates: X, Y, Z, and Zoom (or Magnification), where Zoom refers to the precalculated resolutions.

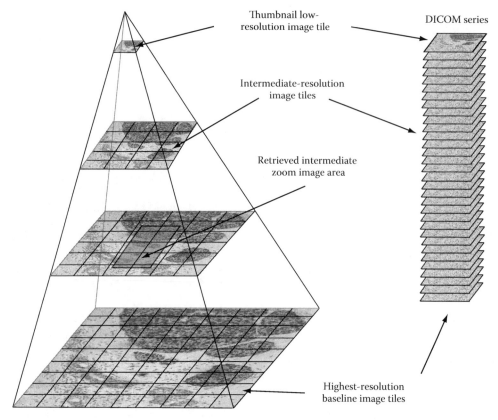

Figure 10.12 Pyramid image organization model for WSI storage and review.

10.3.4.2 Slide Image Orientation

As shown in Figure 10.13, when the slide is scanned, the reference slide orientation is defined by DICOM Supplement 15. The origin of the slide is at the corner across from the label when the label is at 12 o'clock. The Y-axis is a line including the left edge of the slide, while the X-axis is a line perpendicular to the Y-axis and includes at least one point of the specimen edge of the slide. The Z-axis is a line passing through the intersection of the X- and Y-axes perpendicular to the surface of the slide, with the origin at the top of the slide but below the specimen and any coverslip.

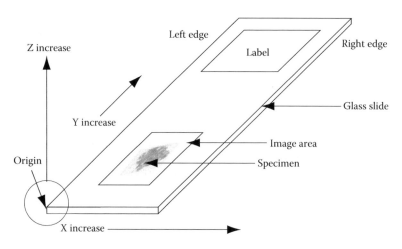

Figure 10.13 Slide reference orientation for scanning.

10.3.4.3 DICOM Modality Worklist and Modality Performed Procedure for Pathology Workflow

As in radiology, where workflow is improved by using the DICOM Modality Worklist (MWL), it is also beneficial for a pathology department to have the modality worklist, using which the different modalities can obtain identifications for pathology specimens and label them in acquired images. Using the DICOM MWL, a pathology department LIS can provide identifications for the patient, study, specimen, and describe each acquired image. Unlike a radiology department, where the imaging subject is a patient, in a pathology department, the imaging subject is a specimen collected from a patient.

10.3.4.4 Digital Anatomic Pathology Workflow and LIS Integration

In a digital pathology department, it is not enough to just be able to scan and digitize glass slides into virtual slides. To realize digital pathology's full potential for healthcare improvement, the management and viewing of digital slides must be integrated into the digital pathology department's workflow. The critical component for digital pathology workflow is the LIS, which applies identifiers to specimens and their containers, records specimens' processes, and manages all workflow processes in the department to track images and specimens. Depending on their specific implementations, the LIS systems may process digital pathology imaging workflow differently. Shown in Figure 10.14 is a pathology PACS system with (a) case management and (b) a digital slide management tool.

DICOM Supplement 122 facilitates specimen-based pathology laboratory digital imaging workflow by introducing a new pathology specimen identification system and revising information object definitions (IODs) for its use. The imaging workflow in a digital pathology department is supported by defining specimen identification attributes both in the MWL and the Modality Performed Procedure Step's (MPPS) SOP, allowing the LIS to pass appropriate information to DICOM-compliant imaging systems such as slide scanners. The DICOM-defined specimen module allows storage of specimen information within the DICOM image information object, which is needed during its interpretation and analysis in case no LIS is available.

10.3.4.5 Image Query

In a typical radiology PACS system, PACS workstations query the PACS server for the list of images for an identified patient, accession, or study. For digital pathology images, it is possible for the LIS to manage a database registry that links patients, accessions, studies, specimens, and image identifiers to the PACS for image retrieval when a query service is implemented. This can be achieved when pathology containers and specimen identifiers are added as optional keys to the image-level query.

10.3.5 IHE Anatomic Pathology

To consistently apply standards such as DICOM and HL7 in anatomic pathology, IHE Anatomic Pathology was formed. It is sponsored jointly GMSIH, the French Association for the Development of Informatics in Pathology (ADICAP), the Spanish Health Information Society (SEIS), the Spanish Society of Pathology (SEAP), and the French Society of Pathology (SFP). The purpose of the IHE Pathology Technical Framework is to enable the integration of pathology laboratories, their clinical images and data, their workflow, and their imaging equipment with their healthcare enterprises' information systems. The Technical Framework defines the pathology workflow, its functional components, and the transactions among them. Integration profiles describing these relationships are the functional units of the Framework, and each integration profile addresses specific clinical needs. The diagnostic process of a pathology department requires close communication between the pathologists and technicians from the pathology department and clinicians from the radiation oncology, radiology, surgery, and other departments. The Technical Framework will define a comprehensive approach for managing digital pathology images and associated clinical data for patient care. For subspecialties within the pathology domain, they are going to be covered by the IHE Pathology Technical Framework progressively.

10.3.5.1 Integration Profiles in the Anatomic Pathology Domain

IHE integration profiles in the Anatomic Pathology Technical Framework define real-world scenarios or specific sets of capabilities of integrated systems in pathology departments. In each integration profile, those capabilities are supported by specified sets of functional units and associated transactions. Additionally, integration profiles under other technical frameworks can

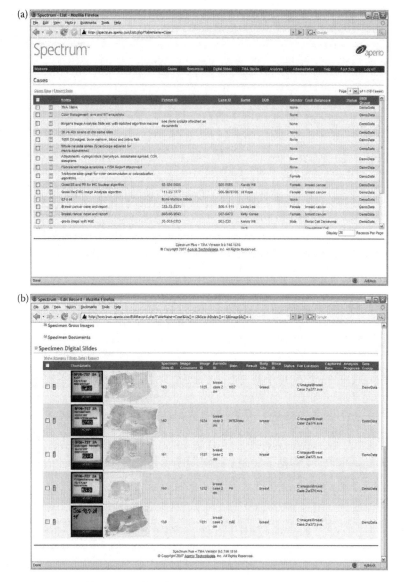

Figure 10.14 Spectrum (Aperio Technologies, Vista, CA) (a) pathology case management tool, and (b) digital slide management tool. (Courtesy of Ole Eichhorn from Aperio Technologies.)

be used in the anatomic pathology domain, or their components can be reused by anatomic pathology integration profiles.

10.3.5.2 Anatomic Pathology Workflow

The Anatomic Pathology Workflow (APW) integration profile allows reliable basic pathology data and image exchange among various systems in the pathology domain, as shown in Figure 10.15. These systems are used for test orders and for performing pathology laboratory studies, or for image production, communication, archiving, and display. This integration profile defines a specific set of actors to finish transactions for supporting various capabilities. For test ordering, the APW defines multiple transactions to ensure the consistency of ordering and of laboratory specimen management information. In terms of image management, the APW defines multiple transactions for the image acquisition worklist, image creation, and archiving. The profile also supports basic reporting workflow, which creates and stores various observations and reports.

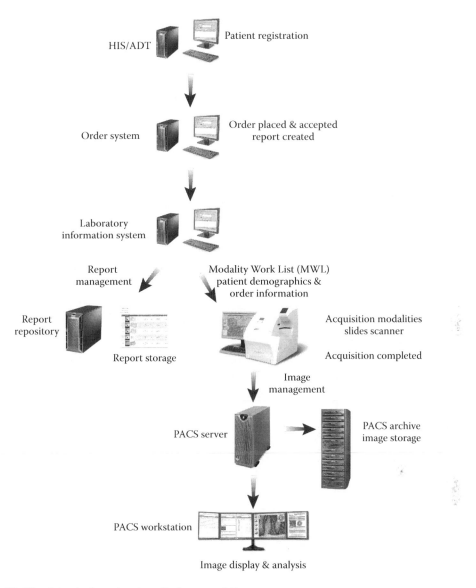

Figure 10.15 A typical anatomic pathology workflow.

Some components from the Radiology Technical Framework and Laboratory Technical Framework are used in the APW integration profile.

10.3.5.3 Anatomic Pathology Reporting to Public Health

Public health agencies such as the Centers for Disease Control and Prevention (CDC) and cancer registries routinely collect disease diagnostic data from anatomic pathology laboratories. Healthcare institutions also report their relevant data to cancer registries, which is essential for public health authorities to be able to provide cancer statistics. The data submission process is traditionally labor intensive and prone to errors. IHE Anatomic Pathology is in the process of developing the Anatomic Pathology Reporting to Public Health (ARPH) integration profile for transmitting anatomic pathology data directly to public health organizations from anatomic pathology laboratories. In this profile, appropriate actors and transactions are defined for reporting of anatomic pathology results to public health agencies. The ARPH profile is designed for international use, and it enables anatomic pathology laboratories, public health agencies, and software vendors to have a consistent and common methodology for data transmission and processing.

10.3.5.4 *Future Developments*

Digital pathology's development relies on standards that are still evolving. It is important that efforts from the IHE pathology initiative, DICOM, HL7, and the Systematized Nomenclature of Medicine (SNOMED) are coordinated to define the information model for specimens and the templates for structured reports because they are all independent and distinct organizations with overlapping goals. Additionally, further development of WSI standards is needed. DICOM Workgroup 26 is in the process of developing and finalizing standards to support whole-slide microscopic imaging, the navigation mechanism for selecting specific ROIs from the entire slide image, a multiresolution image format in the pyramid data model, a default image display order among various multiple focal planes, a default order of images and series, and new IODs for WSI series and series indexes.

In addition to the integration profiles defined in IHE Anatomic Pathology, integration profiles defined in other IHE Technical Frameworks in Chapters 6 and 12 can also be useful for the integration of pathology images and data with EHR/PHR systems. These integration profiles include KIN, ED, Enterprise User Authentication (EUA), Patient Identifier Cross-Referencing (PIX), PDQ, Patient Consistent Presentation of Images (CPI), XDS, RID, ATNA, and Personnel White Pages (PWP). With the development of all IHE technical frameworks, other integration profiles may also be applied to the pathology domain.

As pathology undergoes digitization, the increased capabilities of digital workflow are creating many new opportunities. All pathology records, digital slides, laboratory results, and clinical information will be readily available through the integration of pathology PACS and the department LIS. Applying PACS and digital workflow to pathology processes will produce the same operational efficiency improvement that has been realized in radiology.

10.4 RADIATION ONCOLOGY PACS

Radiation oncology is another medical specialty with very specific needs. Most radiation oncology departments generate their own CT images while at the same time receiving diagnostic images from radiology PACS. As a result, a radiation oncology PACS needs to handle images from both standard radiology PACS and its own radiation therapy treatment planning systems. It is challenging to integrate images from both sources for diagnostic information and treatment planning. Additionally, the radiation oncology PACS must be seamlessly integrated with the healthcare enterprise's EHR/PHR systems.

10.4.1 Treatment Management System and Treatment Delivery System

In a radiation oncology department, a treatment management system or radiation oncology information system is used for worklist management and procedure performance tracking. User terminals or workstations of a treatment management system can be located in the treatment delivery area or throughout the healthcare institution. A treatment delivery system is used to perform radiation therapy procedures according to the worklist. The treatment delivery system also updates the procedure status and stores treatment records and verification images. The workstation of a treatment delivery system is typically located in the treatment control area.

10.4.2 DICOM for Radiation Oncology

Established in 1997, DICOM Workgroup 7 has developed Supplements 11, 29, 74, 102, and 147 specifically for extensions of the DICOM standards for radiation oncology applications. DICOM Radiation Therapy (RT) is not designed for patients' diagnostic procedures with imaging modalities such as CT, MRI, ultrasound, or digital x-ray. It is designed for patients' treatment with a treatment plan, radiation dose application, the quantity of the radiation, the method of delivery, and the treatment simulation. These supplements, together with other DICOM developments, are used in IHE Radiation Oncology for system integrations.

10.4.2.1 *Basic DICOM RT Objects*

The DICOM RT-defined objects allow communication of all types of data used in radiation therapy. Currently, the widely used basic radiotherapy IODs defined in DICOM Supplements 11, 29, and 102 include:

- *RT Image IOD*: RT images are RT simulator images and portal images acquired by conventional simulators and portal imaging systems, and digitally reconstructed radiographs (DRR). They are not diagnostic radiology images such as the CT, MRI, or PET images typically stored in a radiology PACS. This IOD defines RT images' semantic content. The requirements of image transfer in general RT applications on conventional simulators, virtual simulators, and portal imaging systems have been addressed. The corresponding DICOM storage SOP for the IOD to be interchanged through the network or digital storage media is also included, together with numeric data of the beam parameters at the time images are taken.

- *RT Dose IOD*: RT dose is the characteristics of dose distribution determined through calculations on the RT treatment planning system. The IOD defines the semantic content of the RT dose in 2D or 3D dose grids, groups of dose points, a dose matrix, isodose curves, and dose volume histograms determined by the treatment planning system for total dose distribution. Also included is the storage SOP class for its exchange through network or digital storage media.

- *RT Structure Set IOD*: RT structure set is the patient-related structure, which is determined by CT scanners, virtual simulation workstations, treatment planning systems, and data obtained by other diagnostic equipment. It is also capable of being transmitted through the network or digital storage media with the appropriate DICOM storage SOP class.

- *RT Plan IOD*: RT plan is the dosimetric and geometric data arrangement for treatment courses with an external beam or brachytherapy. It defines the semantic content of RT treatment plans and includes the DICOM storage SOP class for interchange on network and storage media.

- *RT Beams Treatment Record IOD*: RT treatment record is for the documentation of treatment sessions data throughout an external beam radiation therapy course. There is also an optional treatment summary for recording the cumulative effects of a treatment course. The IOD can be interchanged through the network or portable storage media.

- *RT Brachytherapy Treatment Record IOD*: This IOD records sessions during a treatment course of brachytherapy, with an optional treatment summary describing the cumulative effects of the treatment.

- *RT Treatment Summary Record IOD*: This IOD is for recording treatment summaries documenting the cumulative state of a radiation therapy treatment.

- *RT Ion Plan IOD*: RT ion plan is used for the exchange of treatment plans generated by manual entries, virtual simulation systems, or treatment planning systems during or before the start of a course of ion therapy treatment. These plans may include ion beam definitions and fractionation information.

- *RT Ion Beams Treatment Record IOD*: This IOD is used for the interchange of ion beam treatment session reports, which are generated by treatment verification systems during the course of a treatment. It may include cumulative summary information and be used for treatment information exchange during treatment delivery.

- *RT Beams Delivery Instruction Storage Composite IOD*: This IOD has all the parameters needed to instruct a treatment delivery system to deliver an RT treatment fraction that is not defined in the referenced RT Plan IOD. It contains information about the treatment plan, the beams to be delivered according to the plan, and verification images for these beams.

- *RT Conventional Machine Verification IOD*: This contains the required attribute descriptions for an external machine parameter verifier. Before a conventional photon or electron RT treatment is delivered, the machine parameter verifier verifies the correct setup of the treatment delivery system.

- *RT Ion Machine Verification IOD*: This IOD contains descriptions of the required attributes for an external machine parameter verifier that verifies an ion RT treatment system's correct setup before treatment delivery.

Each of the above RT IODs is created with a well-defined data model, and these data models provide a standard that facilitates interoperability among various RT systems, making it feasible to interchange RT data between different systems and integrate them into the EHR/PHR systems.

Currently, basic radiotherapy DICOM objects are implemented in many products offered by vendors. Workflow-related improvements are increasingly considered important, and second-generation DICOM RT objects are being developed for workflow improvement.

10.4.2.2 DICOM Second-Generation RT Objects

Since the development of the above basic IOD objects in late 1997, the field of radiation oncology and the DICOM standards have both evolved considerably, making it necessary to develop new DICOM objects to support technological advances. New treatment techniques such as tomotherapy and robotic therapy, new imaging technologies, and new treatment strategies such as adaptive therapy and image-guided therapy are becoming widely used. Workflow management is also becoming an increasingly important key area for the application of the DICOM RT standard. The DICOM Workgroup 7 is actively developing Supplement 147 to address the needs of these new technologies and techniques with a new generation of DICOM objects and processes for RT. The second-generation RT objects are specifically designed to support modern RT equipment, while basic RT objects still can be used to support legacy systems or simpler processes. Unlike basic RT objects, in which various types of data are encoded in a single, multipurpose object with relatively unconstrained contents, in second-generation RT objects, different representations of data are encoded in explicit, separate IODs developed for particular uses. Additionally, DICOM Supplements 74 and 96 can be implemented to enhance workflow management for RT delivery, and these supplements introduce extensions for existing RT objects to accommodate new techniques.

10.4.2.3 RT Machine Verification Service Class and RT Ion Machine Verification Service Class

The DICOM RT Machine Verification Service Class and RT Ion Machine Verification Service Class define new classes of service at the application level. The services allow treatment delivery systems to instruct machine parameter verifiers to independently verify geometric and dosimetric setups on a radiation delivery system before the treatment is delivered. These verification service classes can override treatment parameters if necessary. These service classes are designed for use with conventional photon or electron therapy as well as proton or ion particle treatments. A treatment management system in the radiation oncology department typically plays the role of a machine parameter verifier when the treatment delivery system is in the external verification mode. The machine parameter verifier oversees and can stop the treatment delivery if necessary. The machine parameter verifier is usually located in the treatment delivery control area, either sharing a user interface with the treatment management system or on a separate control console.

10.4.2.4 Unified Worklist and Procedure Step Service Class for Radiation Oncology

The Unified Worklist and Procedure Step (UWPS) Service Class introduced in DICOM Supplement 96 can be used in radiation oncology for workflow improvement. It defines a new mechanism for the exchange of information regarding requested and performed medical procedures and is used in the IHE Radiation Oncology treatment delivery workflow integration profiles. The general application of the service class includes simple worklist management, worklist item creation, worklist query, and progress and results communication. The UWPS can be used along with the RT Beams Delivery Instruction Storage IOD, the RT Conventional Machine Verification Service Class, and the RT Ion Machine Verification Service Class to support a wide variety of activities related to workflow and treatment monitoring in RT. This service class can be used in dose calculation push workflow, in which users schedule dose calculation tasks to a shared dose calculation server that requires progress tracking. It can also be used in RT pull workflow, where treatment delivery systems or patient positioning systems query treatment management systems, which typically schedule multiple procedure steps. The treatment delivery system retrieves a treatment delivery unified procedure step from the treatment management system that acts as a worklist manager. The treatment delivery system then retrieves related SOPs specifying the treatment setup verification, the treatment delivery, the associated result generation, and the unified procedure step update. In the workflow, the machine parameter verification step may be performed by an external machine parameter verifier instead of by the treatment delivery system.

10.4.2.5 RT Course Introduction

One of the important current developments for second-generation DICOM RT objects is the RT Course, which is a top-level entity describing a given treatment course. It links the necessary entities in radiation oncology to prepare, execute, and review patients' RT treatment. All relevant RT objects are referenced, such as acquisition images, registrations, segmentations, physician intent, beam sets, reference and verification imaging, and output records. In addition, it contains descriptions of all information relating to the current status of treatment, treatment phases, changes caused by adaptation of the therapy plan, and overall planned treatment delivery scheme including fractionation for multiple phases of treatment. The RT Course also includes phase-specific fractionation schemes for phase prescription of beams or catheter combinations. Finally, the RT Course enables the complete archiving of patients' treatment delivery and the exchange of data needed for treatment planning or treatment steps.

10.4.3 IHE Radiation Oncology

The American Society for Therapeutic Radiology and Oncology (ASTRO) has formed a multisociety task force to undertake an initiative to promote IHE Radiation Oncology, fostering seamless connectivity and integration of RT equipment and patient health information systems. The task force includes members from the RSNA, AAPM, HIMSS, NEMA, and other organizations. The purpose of the task force is to define consistent use of existing standards such as DICOM and HL7 during systems integrations in radiation oncology.

10.4.3.1 Radiation Therapy Objects Integration Profile

The Radiation Therapy Objects (RO) integration profile is the first IHE Radiation Oncology integration profile. Its purpose is to reduce ambiguity and enable basic interoperability in the interchange of DICOM RT objects for 3D conformal and external beam RT. The exchange of these DICOM objects is necessary during the flow of DICOM images and RT treatment planning data from CT scans through dose display on workstations.

10.4.3.2 Image Registration

In a radiation oncology department, there are typically workstations performing image registration for RT. These images are acquired from multiple modalities, and the registration information is usually in proprietary formats designed for specific treatment planning systems. The Image Registration (REG) integration profile addresses the interoperability issue of exchanging registered data among various systems for subsequent processes. It specifies a standardized method of exchange, storage, processing, and display for images, RT structure sets, RT doses, and image spatial registration information. This is achieved by applying existing DICOM spatial registration objects in a constrained and clarified way to avoid any misinterpretation, so that display workstations can correctly identify relevant image sets, match data with image sets, match coordinate systems, and perform spatial translation of images. This integration profile only addresses issues of content for REG with multiple coordinate systems, not the REG workflow itself, which can be managed by the PWF integration profile in IHE Radiology. Currently, only rigid registration can be handled; deformable registration is not yet supported.

10.4.3.3 RT Treatment Delivery Integrated Positioning and Delivery Workflow

In many radiation oncology departments, workflow is manually managed, and RT treatment delivery systems use proprietary interfaces. This makes it difficult to integrate new technologies and systems into the radiation oncology department. The Integrated Positioning and Delivery Workflow integration profile is an initial effort to deal with these workflow-related issues. It defines the RT workflow, which involves a single system for the positioning of a patient and subsequent treatment delivery, as shown in Figure 10.16a. The positioning and delivery system acquires a set of 2D planar projection or 3D CT positioning images. Then spatial REG is performed using previously retrieved acquired reference images. If it is deemed necessary, the patient is repositioned. Finally, the intended RT treatment is delivered. Essentially, a positioning and delivery system combines the roles of a patient positioning system and a treatment delivery system.

10.4.3.4 RT Treatment Delivery Discrete Positioning and Delivery Workflow

The Discrete Positioning and Delivery Workflow integration profile defines the RT workflow involving the patient's positioning and subsequent treatment delivery, which are accomplished through separate systems, as shown in Figure 10.16b. The treatment management system serves as a workflow manager for RT scheduling. In a particular product implementation, the treatment management system and the archive can potentially be combined into a single system. As in the integrated positioning and delivery workflow integration profile, the patient positioning system acquires planar projection or CT positioning images or other information along with previously

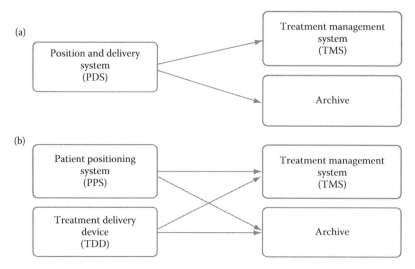

Figure 10.16 Radiotherapy treatment workflow with (a) integrated position and delivery system, and (b) discrete positioning and delivery device.

acquired reference images to perform a image spatial registration. The patient positioning system then verifies the patient's positioning prior to treatment delivery. The intermediate results are not required to be exported, and the patient positioning system's output only indicates whether the positioning is correct or canceled. If necessary, the patient's position is adjusted for treatment. Following the positioning step, the intended treatment is delivered to the already correctly positioned patient by the treatment delivery system, which delivers therapeutic radiation to the patient with internal verification.

10.4.3.5 Future Developments

The RT treatment delivery workflow integration profiles are still evolving. The above two initial integration profiles handle scheduled workflow for the patient's positioning and treatment delivery. IHE Radiation Oncology is developing new integration profiles in coordination with DICOM Workgroup 7 in the following areas:

- *Scheduled workflow for the remaining stages of the procedure in radiation oncology*: Covered treatment steps include simulation, planning, plan review, treatment review, other modes of positioning and treatment delivery, and charge capture and posting using treatment management systems or billing systems. This workflow covers most cases in radiation oncology.

- *Admission, discharge, and transfer (ADT) support*: This profile, in conjunction with existing IHE integration profiles and HL7 messaging protocol, will enable radiation oncology treatment management systems to directly receive patients' demographics from HIS.

- *Nonmanaged workflow*: Radiation oncology treatment management systems will be updated for nonmanaged workflow, such as emergency treatments, and such treatments can be documented and charged through normal processes.

- *Partially managed workflow and media archive*: This profile is for treatments involving some procedures that are not performed under managed workflow but whose output objects are introduced into the workflow through the digital data storage media. The profile also will support media archive generation.

- *Advanced RT object interoperability*: The purpose of this profile is to reduce ambiguity during the exchange of RT plan information among treatment planning, management, and delivery systems.

Together with other organizations, DICOM and IHE are constantly developing standards to deal with evolving RT treatment prescriptions and dynamic plans, and new IHE integration profiles are being developed to implement these standards. For many radiation oncology departments, the new challenge is how to enhance department workflow by leveraging existing and upcoming DICOM RT objects, the unified procedure step-based workflow, and IHE Radiation Oncology integration profiles because many existing RT technologies require the new objects, and workflow implementation can potentially lead to very significant productivity and efficiency improvement.

10.5 ENDOSCOPY AND OTHER VISIBLE LIGHT SPECIALTIES

A variety of devices can be defined as endoscopes, and they are perhaps one of the most widely used categories of visible light image-capture instruments. They are used in different medical specialties for diagnostic and therapeutic interventional procedures, such as GI endoscopy, intraoperative imaging, laparoscopy, orthopedic imaging, arthroscopy, otolaryngology, gastroenterology, gynecology, and urology. Compared to traditional open surgeries, endoscopic procedures can reduce healthcare costs by eliminating sedation, monitoring, and hospital stays. An increasing number of patients prefer less invasive procedures over invasive surgeries because of reduced pain, less trauma, fast recovery, shorter hospital stay, less likelihood of infection, reduced post-surgical complications because of minimally exposure of surgical sites to the external environment, and reduced cost. Since HIS, RIS, and radiology PACS systems have been adopted by many healthcare institutions, and their advantages for workflow improvements are well recognized and documented, they provide a model for the integration of endoscopes and other visible light imaging systems into the electronic healthcare system with digital image storage and workflow improvement. Figure 10.17a shows the procedure schedule, and Figure 10.17b the worklist of an endoscopy PACS that is integrated into a radiology PACS system. However, there are technical challenges that require to be solved during the endoscopy integration.

10.5.1 Endoscopy-Specific Requirements and Technical Challenges

Endoscopy has a complex workflow, which, like that of cardiology catheterization laboratories, is quite different from that of radiology. The endoscopy procedure, especially in an emergent situation, is often started before an order is placed. After the start of the procedure, the patient's identifier is usually used multiple times by various imaging devices and by measuring and reporting systems. An endoscopy may be diagnostic, therapeutic, or both.

Because still images and dynamic cine movies are acquired during endoscopic procedures, ergonomic and minimally procedure-interruptive controls for image access and data entry are required by physicians and technologists. When images and dynamic cine movies are captured, their anatomic locations need to be annotated immediately to avoid subsequent confusion during postprocedure review. Touch-screen and foot-pedal controls have been found to be appropriate to meet this demand, which is specific to the endoscopy department. In addition, many endoscopy procedures are performed while patients are conscious or minimally sedated, which requires that observations and data entry during the procedure be made without causing any disturbance or uneasiness to the patient. In this case, point-and-click and pull-down menu selection provides a suitable data entry option for standard category and observation selections. After a procedure is completed, interpreting physicians need to be able to select key images or cine video clips. The relevant images and videos can then be included in the report to support and illustrate the clinical findings of the procedure.

Endoscopic images are acquired and stored in formats different from those used in radiology images, which presents a challenge to their integration into PACS systems. However, because of the benefits of improved patient care and consolidated image storage, solutions need to be developed to integrate image acquisition, clinical document management, and department workflow with existing PACS systems. Endoscopy has the same data storage needs as the imaging modalities discussed in earlier sections, but integrating digital endoscopy equipment with healthcare enterprise image management systems is not solely a matter of acquiring images compliant with the DICOM Visible Light standard; improving workflow by linking hospital or department information systems with the PACS also plays an important role in integration. The challenge to integrating digital endoscopy systems is that a few of them support direct links to the hospital or department information systems for the exchange of patient demographics. This same challenge is faced by other departments whose primary focus is not imaging, because there are no department information systems and processes in place to facilitate image acquisition, workflow, and to associate all patients' demographics, metadata, images, and final reports together in the EHR/PHR systems.

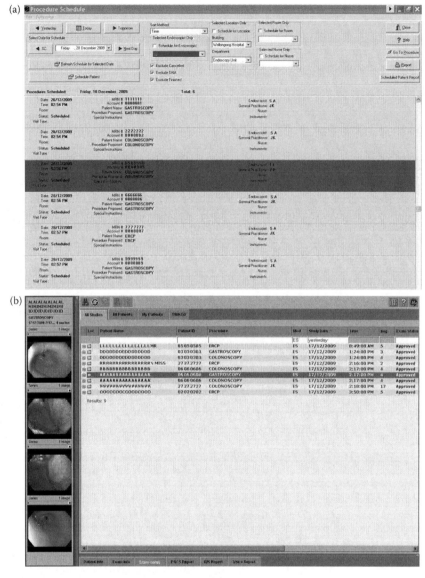

Figure 10.17 An endoscopy PACS with (a) procedure schedule and (b) worklist. (Courtesy of Steve Flinn, Wollongon Hospital, NSW, Australia.)

10.5.2 Legacy Systems

Almost all of the latest-generation endoscopy equipment are digital and support some of the DICOM standards, and so the most significant challenge to systems integration is that healthcare institutions may have acquired different pieces of endoscopic equipment over several years. Some of these legacy systems produce analog images captured on video tapes, CD, or paper. More recent endoscopic equipment may generate images in digital form; however, their image formats may not conform with DICOM standards. If the department has its own department-wide information system or shares another department's information system, such as RIS, interface modules are needed to convert endoscopic images into DICOM-compliant images for storage in a PACS. These interface modules can act as image digitizers and protocol converters. They enable the digital automation of endoscopy across the healthcare institution or enterprise by capturing still images and dynamic cine movies in DICOM formats and storing them in PACS. These interface modules typically support DICOM Secondary Capture Store, Print, Modality Worklist, Modality Performed

Procedure Step, and Storage Commitment so that digital images originating from endoscopic equipment can be treated the same as radiology images, and workflow features can also be incorporated. Through DICOM and HL7 interfaces, HL7 messages and DICOM service classes can be implemented for workflow improvement.

10.5.3 DICOM Visible Light Standard and Video Formats

DICOM Workgroup 13 has developed Supplements 15, 42, 47, 137, and 149 for visible light imaging technologies. They are extensions of the DICOM standard and are applicable to endoscopy imaging. New extensions for DICOM Visible Light standards are going to be proposed when needed to fit new clinical requirements.

Driven by the consumer video equipment market shifting from standard definition to high definition (HD) at a spatial resolution of 1902 × 1080, HD video sources are increasingly being used for video capture by imaging modalities operating in the visible light domain. Medical camera vendors are offering medical HD video recording products for endoscopy equipment. HD video and associated still-image captures are much larger in file size. They require more storage space on PACS, and they need a higher network bandwidth to be archived and shared by relevant departments and clinicians. Since video files usually remain large even after data compression, it is critical to store and transmit video files efficiently and quickly and to reduce the video file size while maintaining video quality in order to control the data storage and network infrastructure

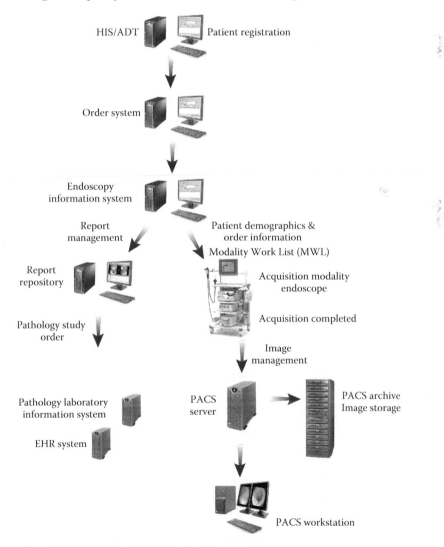

Figure 10.18 Endoscopy workflow specified by IHE-Japan.

cost. DICOM has adopted Moving Picture Experts Group (MPEG)-2 as an HD video data compression technique for HD video. The defined HD MPEG-2 DICOM extension enables vendors and users to have DICOM-compliant still and compressed motion images both in the standard and the new HD resolution. Since DICOM routinely updates its standard to accommodate new needs, MPEG-4 AVC/H.264 (MPEG-4 Part 10) for use in DICOM is currently under development. MPEG-4 is capable of very high data coding efficiency and meets the needs of small video file size and low network transmission bandwidth while maintaining HD video quality and frame rate. Using MPEG-4, different configuration options can be applied for various tradeoffs between video file size and video quality. Although MPEG-4 has a higher compression ratio than MPEG-2, its limitations are higher computational requirements than MPEG-2 and limited compatibility with existing coder and decoder products. Since MPEG-4 itself is not limited to the spatial resolution and aspect ratio of the current HD broadcast formats, future DICOM standard developments may allow different video specifications without changing the video coding standard.

10.5.4 IHE-J Endoscopy Workflow Integration Profile

Currently there is no IHE integration profile developed and adopted by the IHE international community for endoscopy. Since applications of endoscopy cross many medical specialties, it is not feasible to incorporate all their workflow together. Therefore, IHE-Japan (IHE-J) initiated the Endoscopy Workflow (ENDO) integration profile for gastrointestinal endoscopy in 2006. Although this integration profile was designed specifically for workflow in Japan, it may apply to other countries as well, because despite differences, (Figure 10.18), the concept is the same, which is to use HL7 messaging standards for linking ordering information systems and DICOM standards for imaging systems.

Radiology and endoscopy share common workflow procedures: order placement, registration, image acquisition, and report generation. Shown in Figure 10.19 is the endoscopy worklist, and Figure 10.20 is an endoscopy report with images of an endoscopy PACS that has been integrated into a radiology PACS system.

The ENDO workflow profile defines the mechanism for managing and distributing the department's workflow of upper gastrointestinal tract endoscopies. This integration profile provides a common language for endoscopic equipment manufacturers and clinicians to communicate the product integration requirements for specific capabilities in integrated systems. It allows consistent patient demographic data flow in the endoscopy domain, even in emergent cases in which patients' demographics may not be available until after the procedure has begun or for a significant amount of time after the procedure has been completed. It also defines procedural

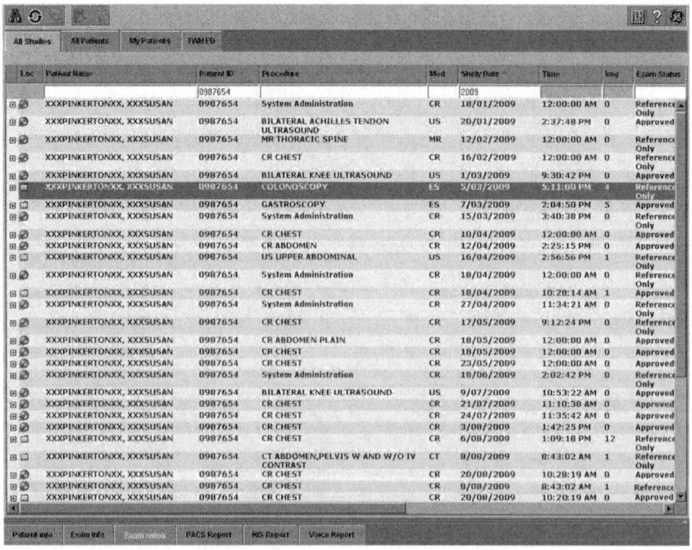

Figure 10.19 The worklist of an endoscopy PACS which is integrated into a radiology PACS. (Courtesy of Steve Flinn, Wollongon Hospital, NSW, Australia.)

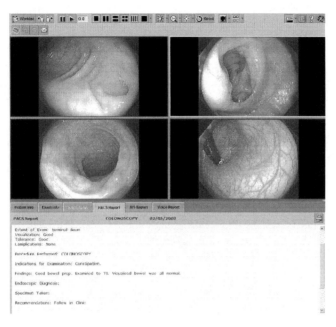

Figure 10.20 An endoscopy report with associated images displayed on an endoscopy PACS. (Courtesy of Steve Flinn, Wollongon Hospital, NSW, Australia.)

data coordination among various imaging, analysis, and data archival systems, so that the data are ready to be used in subsequent steps of the endoscopy workflow. If the workflow includes both diagnostic and therapeutic procedures, the profile also coordinates each step's completion status. The integration profile incorporates common components from other IHE technical frameworks' integration profiles such as SWF and PIR to coordinate an endoscopy department's specific data and workflow requirements, as shown in Figure 10.21.

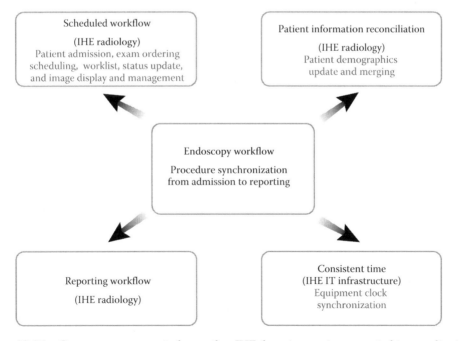

Figure 10.21 Common components from other IHE domains are incorporated to coordinate an endoscopy department's specific data and workflow requirements as by IHE-J.

Endoscopy will continue to see increased deployment in various medical specialties. While some imaging technologies may redefine the role of diagnostic endoscopy, it will remain an important tool in the hands of specialists practicing in different sectors of the medical field.

10.5.5 Enterprise Integration with other PACS and EHR/PHR

It is well established that the multidisciplinary approach to patient care can improve outcomes. When endoscopic procedures are integrated into PACS and the department information system, visible light images and dynamic cine movies from endoscope cameras can be digitally captured and directly transferred to the PACS system, while scheduling and patient demographic information can be exchanged between the department information system and endoscopy modalities using the DICOM modality worklist. If structured reporting capabilities are incorporated, reports can be completed even as procedures are performed, making images and reports available only minutes after procedures are complete. Endoscopic images on PACS also allows clinicians to review them alongside radiology images such as CT, MRI, and other diagnostic modalities also in the PACS, at any time, anywhere within the healthcare enterprise. The ability to compare endoscopic images side by side with radiology data sets and pathology digital slides will promote better coordination and collaboration among various medical specialties. The integration of enterprise imaging with EHR/PHR will extend this coordination throughout the patient care continuum.

BIBLIOGRAPHY

Avrin, D. PACS and the IR suite. *Radiological Society of North America (RSNA) 2007 Annual Meeting.* Chicago, IL.

Bakalar, R. S. 2008. Clinical image storage archives & distribution: Enterprise solutions for medical imaging. *Pathology Visions Conference 2008.* www.pathologyvisions.com/us2008slides/Clinical/Monday/08_at_4_00_Bakalar_medical_imaging_storage.ppt. Accessed January 15, 2010 San Diego, CA.

Beckwith, B., C. D. Le Bozec, and M. Garcia-Rojo, et al. 2008. Pathology in DICOM: Progress from working group 26 and IHE. *DICOM International Conference & Seminar.* Chengdu, China.

Chang, P. J. Image workflow beyond radiology. *The Society for Imaging Informatics in Medicine 2008 Annual Meeting.* Seattle, WA.

Chang, P. J. Advanced integration strategies toward the multimedia EHR. *The Society for Imaging Informatics in Medicine 2008 Annual Meeting.* Seattle, WA.

Chang, P. J. What you need to know about pathology. *The Society for Imaging Informatics in Medicine 2009 Annual Meeting.*

Curren, B. 2008. IHE-RO 2008 managed workflow profiles. www.ihe.net/Participation/upload/ro3_ihewkshop07_workflow_curran.pdf. Accessed January 15, 2010.

CVIS fuels integrated, enterprise cardiology. *Cardiology PACS & CVIS.* Supplement to *Cardiovascular Business*, Nov/Dec 2008. www.cardiovascularbusiness.com/index.php?option=com_articles&view=article&id=15051:cvis-fuels-integrated-enterprise-cardiology. Accessed January 15, 2010.

Daniel, C., W. Scharber, and F. Macary. 2009. IHE anatomic pathology technical framework supplement: Anatomic pathology reporting to public health (ARPH). Draft for public comment.

Desjardins, B. IT challenges in cardiology. *The Society for Imaging Informatics in Medicine 2009 Annual Meeting.*

DICOM Standards Committee, Working Group 7 Radiotherapy Extensions. 1997. DICOM supplement 11: Radiotherapy objects.

DICOM Standards Committee, Working Group 13. 1999. DICOM supplement 15: Visible light image for endoscopy, microscopy, and photography.

DICOM Standards Committee, Working Group 1 Cardiovascular Information. 2003. DICOM supplement 66: Catheterization lab structured reports.

DICOM Standards Committee. Working Group 1 Cardiovascular Information. 2003. DICOM supplement 72: Echocardiography procedure reports.

DICOM Standards Committee, Working Group 1 Cardiovascular Information. 2004. DICOM supplement 76: Quantitative arteriography and ventriculography structured reports.

DICOM Standards Committee, Working Group 1 Cardiovascular Information. 2004. DICOM supplement 77: Intravascular ultrasound (IVUS) structured reporting.

DICOM Standards Committee, Working Group 1 Cardiovascular Information. 2005. DICOM supplement 97: CT/MR cardiovascular analysis report.

DICOM Standards Committee, Working Group 1. 2008. DICOM supplement 128: Cardiac stress testing structured reports.

DICOM Standards Committee, Working Group 6. 2007. DICOM supplement 96: Unified worklist and procedure step.

DICOM Standards Committee, Working Group 7 Radiotherapy Extensions. 1999. DICOM supplement 29: Radiotherapy treatment records and radiotherapy media extensions.

DICOM Standards Committee, Working Group 7 Radiotherapy Objects. 2006. DICOM supplement 102: Radiotherapy extensions for ion therapy.

DICOM Standards Committee, Working Group 7 Radiation Therapy. 2007. DICOM supplement 74: Utilization of worklist in radiotherapy treatment delivery.

DICOM Standards Committee, Working Group 7 Radiation Therapy. 2009. DICOM supplement 147: Second generation radiotherapy.

DICOM Standards Committee, Working Group 13 Visible Light. 2004. DICOM supplement 47: Visible light video.

DICOM Standards Committee, Working Group 26, Pathology. 2008. DICOM supplement 122: Specimen module and revised pathology SOP classes.

DICOM Standards Committee, Working Group 26, Pathology. 2008. DICOM supplement 145: Whole slide microscopic image IOD and SOP classes.

Eichhorn, O. Unlocking the value of digital pathology. *Healthcare Information and Management Systems Society 2009 Annual Meeting.*

Flinn, S. Images in scope. www.archi.net.au/e-library/performance/e-health/images-scope.

Foord, K. PACS outside radiology. www.pacsgroup.org.uk/forum/messages/437/K_Foord_15_35-11828.ppt. Accessed January 15, 2010.

Gilbertson, J. Pathology informatics: Introduction & the approaching convergence with radiology informatics. *The Society for Imaging Informatics in Medicine 2009 Annual Meeting.*

Greenleaf, T. 2006. Cardiology PACS: Ready for integration? www.imagingeconomics.com/issues/articles/2006-03_10.asp?SkipInterstitial=TRUE. Accessed January 15, 2010.

Gundurao, A. 2008. Implementing a unified DICOM broker for cardiology—An experience. *DICOM International Conference & Seminar.*

Halin, N. J. Workflow issues in the interventional suite. *The Society for Imaging Informatics in Medicine 2008 Annual Meeting.*

HealthImaging. 2008. Digital endoscopy: Challenges for enterprise imaging. www.healthimaging. com/index.php?option=com_articles&view=article&id=15210:digital-endoscopy-challenges-for-enterprise-imaging. Accessed January 15, 2010.

HIMSS. 2003. *Applying the IHE Technical Framework to Cardiology Rev. 3.0.* www.himss.org/content/ files/Applying%20the%20IHE%20Technical%20Framework%20to%20Cardiology%20v3.pdf. Accessed January 15, 2010.

Huang, H. K. 2003. Enterprise PACS and image distribution. *Comp. Med. Imaging & Graphics* 27 (2–3): 241–253.

IHE-Japan. 2006. IHE gastrointestinal endoscopy technical framework (upper gastrointestinal tract): integration profiles.

IHE Japan Endoscopy Committee and DICOM WG13. 2009. Update, IHE-J endoscopy committee activity.

Integrating the Healthcare Enterprise. 2006. Cardiac catheterization workflow and evidence documents integration profiles (white paper).

Integrating the Healthcare Enterprise. 2006. Cross-enterprise document sharing integration profile—Cardiology use cases (white paper).

Integrating the Healthcare Enterprise. 2006. Echocardiography workflow and evidence documents integration profiles (white paper).

Integrating the Healthcare Enterprise. 2006. *IHE Cardiology Technical Framework Year 2: 2005–2006, Volume I, Integration Profiles* (white paper).

Integrating the Healthcare Enterprise. 2006. Integration profiles for the electrophysiology lab (white paper).

Integrating the Healthcare Enterprise. 2006. Integration profiles for stress testing (white paper).

Integrating the Healthcare Enterprise. 2006. Nuclear medicine integration profile—cardiology option.

Integrating the Healthcare Enterprise. 2006. Purchasing using IHE cardiology (white paper).

Integrating the Healthcare Enterprise. 2006. Retrieve ECG for display integration profile (white paper).

Integrating the Healthcare Enterprise. 2006. Urgent implantable rhythm control device identification (PDQ-IRCD) profile (white paper).

Integrating the Healthcare Enterprise. 2007. IHE cardiology: Cardiology data handling white paper.

Integrating the Healthcare Enterprise. 2007. *IHE Radiation Oncology Technical Framework Volumes 1–2.*

Integrating the Healthcare Enterprise. 2007. *IHE-Radiation Oncology Technical Framework Volumes 1–2*, Draft for Trial Implementation.

Integrating the Healthcare Enterprise. 2007. IHE cardiology technical framework supplement 2007, evidence documents profile cardiology options: stress testing CT/MR angiography.

Integrating the Healthcare Enterprise. 2007. *IHE Radiation Oncology Technical Framework Volumes 1–2: Managed Delivery Workflow Addenda for Version 2.0*, Draft for Public Comment.

Integrating the Healthcare Enterprise. 2008. *IHE Anatomic Pathology Technical Framework Volume 1 (PAT TF-1) Profiles.*

Integrating the Healthcare Enterprise. 2008. *IHE-Radiation Oncology Technical Framework Volumes 1–2, Managed Delivery Workflow Addenda for Version 2.0*, Draft for Public Comment.

Integrating the Healthcare Enterprise. 2009. *IHE Radiation Oncology Technical Framework Volume 1: Integration Profiles, Supplement Proposal for Advanced RT Objects Interoperability (re-planning and plan management)*, Draft for Public Comment.

Mello-Thoms, C. The Future of Radiology and Pathology Informatics. *The Society for Imaging Informatics in Medicine 2009 Annual Meeting.*

Paris, J. S. and J. de Dios Cardia. 2007. Introducing SESCAM's Serendipia Project: Scope and Technologies. *Pathology Visions—Digital Pathology Solutions Conference.* www.conganat.org/digital/index.htm. Accessed January 15, 2010.

Reilly, A. J. DICOM radiotherapy. www.oncphys.ed.ac.uk/downloads/confs/DICOMRT.pdf. Accessed January 15, 2010.

Rojo, M. G., C. Peces, and J. Sacristan, et al. Healthcare information standards (IHE, DICOM, HL7) in the management and integration of virtual slides in pathology. www.conganat.org/digital/index.htm. Accessed January 15, 2010.

Simon, D. A. 2005. Radiation therapy in oncology. *DICOM Conference.* medical.nema.org/Dicom/DCW-2005/PROGRAM_DICOM_2005_International_Conference.pdf. Accessed January 15, 2010.

Sippel Schmidt, T. M. 2006. Cardiology PACS: The new deal. www.imagingeconomics.com/issues/articles/MI_2006-07_03.asp. Accessed January 15, 2010.

Smith, E. M. APIII 2005. Pathology PACS: Lessons learned from radiology. apiii.upmc.edu/breakout/archive/2005/Smith05/Smith05%20part3.ppt. Accessed January 15, 2010.

Wendt, G. J. Image management beyond radiology. *The Society for Imaging Informatics in Medicine 2008 Annual Meeting.*

Winfield, C. Radiation therapy in oncology. medical.nema.org/dicom/Conf-2005/Day-1_Seminar/B30_Winfield_Oncology_v2.pdf. Accessed January 15, 2010.

Yokoi, H. IHE-J Endoscopy in clinical practice. medical.nema.org/Dicom/minutes/WG-13/2009/2009-04-22/IHE%20activity090422.ppt. Accessed January 15, 2010.

Yokoi, H. IHE-J Endoscopy: The concept of Japanese activities. IHE-Japan Endoscopy with DICOM WG13. medical.nema.org/Dicom/minutes/WG-13/2007/2007-05-21/IHEendoscopy_withDICOM_WG13.ppt. Accessed January 15, 2010.

11 Electronic Health Record and Personal Health Record Implementation

11.1 INTRODUCTION

Healthcare institutions and medical practices have historically relied on paper medical records. Today, many medical records are in digital form and can potentially be shared and exchanged electronically more quickly and efficiently than in the past. The electronic collection, aggregation, and reporting of healthcare information has been a central theme in the effort to improve the quality, safety, efficacy, and cost of healthcare. EMR, EHRs, and PHR systems have been bringing significant changes to the healthcare industry as traditional paper-based medical records are gradually phased out. In this chapter, existing standards for EMR, EHR, and PHR, the functional model underlying these systems, and implementation guidelines for these systems are discussed.

11.1.1 Electronic Medical Record and Electronic Health Record

Frequently, the terms EMR and EHR are used interchangeably in the press, in government, and in the healthcare industry. Although they both contribute to improved patient care, quality, and efficiency, EMR and EHR are two different concepts. An EMR system is generally defined as the data repositories that contain healthcare information of patients within a given healthcare institution or enterprise. Thus, an EMR contains the aggregated clinical data and information from a variety of sources including clinical laboratory tests, diagnosis, pharmacy records, patient registration forms, radiology reports, treatments, surgical procedures, clinic and inpatient notes, preventive care services, emergency department records, and billing records from a single healthcare institution or enterprise that owns the EMR. Furthermore, an EMR system includes clinical applications that can act on the data contained within the repository, such as a clinical decision support system, a computerized provider order system (CPOE), a controlled medical vocabulary, and a results reporting system. In general terms, an EMR is clinician focused in that it quickly provides clinicians and administrators with the information they need to facilitate patient health information exchange and thereby streamline healthcare delivery.

An EHR extends the notion of an EMR to include the concept of cross-enterprise data sharing. Thus, an EHR typically contains subsets of EMR data from several healthcare institutions and enterprises; the information contributed to the EHR may vary depending on the individual institutions' policies. An EHR can only be assembled if the participating institutions or enterprises all have EMR systems in place and are able to exchange data using standardized data transmission coding and messaging formats. An EHR system for recording, retrieving, and manipulating information should possess the following capabilities:

- Longitudinal collection of patients' electronic healthcare information

- Results management

- Order entry and management

- Immediate electronic access to an individual's or a population's information

- Electronic communication and connectivity

- Provision of knowledge and decision support to enhance patient care quality, safety, and efficiency

- Administrative process support

- Healthcare delivery process support

- Population health reporting and management

Since a patient's clinical healthcare information stored in an EHR system is usually generated by multiple healthcare institutions or enterprises at various times, it is likely to include multiple sets of data regarding patient demographics, medical history, medications, medical problems, allergies, vital signs, radiology reports, and laboratory reports. When reviewed over time, a patient's EHR provides a longitudinal history of the patient's health and of the healthcare services the patient has received. After each healthcare service is delivered, the appropriate data are collected and stored, and all downstream users can utilize the information to avoid redundant or inconsistent data collection. In addition to improving patient safety and continuity of patient care, EHRs also

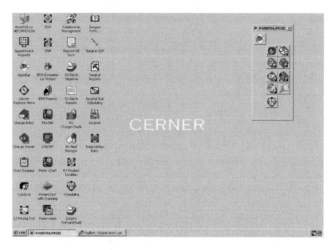

Figure 11.1 A workstation with various scheduling, RIS, EHR, billing, and other clinical information systems of a large hospital.

contribute to public health improvement and healthcare cost control. Figure 11.1 shows a screen shot of the opening page of an enterprise EHR system with many different functionalities.

From the above description, it is obvious that an EMR is different from an EHR. An EHR can include a patient's complete healthcare history across multiple healthcare institutions, enterprises, and geographies. It requires compiling information from various EMRs and having cross-enterprise data exchange mechanisms in place; true EHR will be available only when EMR systems are widely adopted by healthcare institutions and enterprises. However, since the term "EHR" is increasingly gaining popularity, throughout this book "EHR" is used to encompass both terms.

11.1.2 Personal Health Record

A PHR is similar to an EHR, but as the name implies, a PHR is more patient centric. Like an EHR, a PHR is a comprehensive record of a patient's health history comprising both clinical and administrative information collected from various healthcare institutions or enterprises at different times. But unlike an EHR system, a PHR system provides the patient with immediate electronic access to manage one's own record, which may include personal demographics, insurance coverage, healthcare provider information, health advice, public health data, data input and/or accessed by the patient, medication information, and personal healthcare planning and management. A PHR will also include a health condition summary along with records of problems, symptoms, allergies, laboratory and radiology test results, and immunizations record. Since a PHR is mostly controlled by the patient, an institution- or enterprise-based PHR system requires security and privacy features to ensure proper user authentication and access control.

Because a PHR system may be electronically connected to the information systems of several healthcare institutions and enterprises, they should all use standard clinical nomenclature, coding, and data exchange for consistency and interoperability. The effective use of PHR should contribute to improved healthcare by facilitating both the patient's self-management and communication between the patient and healthcare providers.

With the high cost of healthcare, increasingly more healthcare providers and payers are engaging patients with personal health awareness, care management, and education programs, anticipating cost control through these measures. The success of these initiatives could depend largely on the implementation and proper application of robust PHR systems.

11.2 HL7 EHR MODEL

HL7 has developed a standardized EHR system functional model to describe the functions that EHR systems may have and an interoperability model to ensure interoperability among EHR systems for data interchange.

11.2.1 HL7 EHR Functional Model

The system functional model outlined in the *HL7 EHR System Functional Model* consists of a functional outline, function components, function profiles, and conformance criteria. In the functional model, there are healthcare-specific application functions for electronic healthcare information management, as well as common IT infrastructure to enable EHR related interoperability and integration capabilities. The EHR system functional model also supports requirements from various research activities to ensure patients' privacy, confidentiality, and security when their data are used in analyses. The functional model does not endorse specific technologies to be used for practical implementation, nor does it specify the EHR content to be included. In addition, the functional model does not define specifications for conformance testing.

11.2.1.1 Functional Outline

The HL7 EHR system functional outline categorizes all functions of the EHR functional model into three main sections and 13 subsections according to the functions' purposes. The three main sections are direct care, supportive, and information infrastructure. The subsections and their relationships to the main sections are described in Figure 11.2. Each function is described in user-oriented language so it can be clearly and easily understood. The EHR system functional model provides:

- Standard descriptions of the user-defined benefits of an EHR system

- Common understanding for planning and evaluation of EHR system functions by vendors and users

- A framework for next-level standards for information portability within and between EHR systems

- A standards-based methodology for international adoption to meet various countries' unique requirements

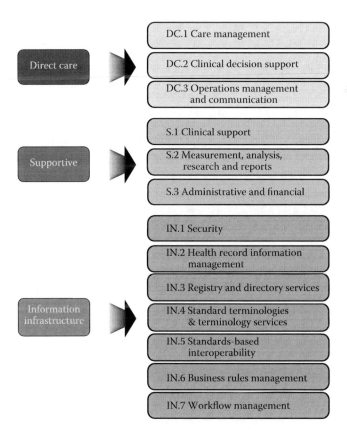

Figure 11.2 HL7 EHR functional model has 3 main sections and 13 subsections according to the functions' purposes.

11.2.1.2 Function Components

The functional outline has 13 subsections. Within these subsections are more than 160 individual functions that could possibly be used in an EHR system. A particular EHR system would use some, but not necessarily all, of these functions. Each function has its own unique function name, function statement, conformance criteria, and description. Illustrated in Table 11.1 are titles for each column of the EHR system function table, which lists all of the functions in rows with their elements and components described in 7 columns.

Although the EHR system functional model includes all possible EHR system functions, it is not practical to expect any single EHR system to be equipped with all of them.

11.2.1.3 Functional Profile

A functional profile is a collection of selected functions from the entire function list used to perform a specific task. The EHR system functional model defines how functional profiles are created and requested with HL7. Within a functional profile, each function can be assigned a different implementation priority of "essential now," "essential future," or "optional." The functions listed in the functional model are only usable when a functional profile is created. Therefore, instead of conforming to the functional model in its entirety, a particular EHR system conforms to a functional profile. A functional profile is created by compiling a list of the subset of applicable functions needed to fulfill a set of specific requirements. Buyers of EHR systems can generate a functional profile based on what are desired from such systems, while developers or vendors of EHR systems can use functional profiles to describe the capabilities of their specific products.

11.2.1.4 Conformance Criteria

Each function within the EHR system functional model has a set of conformance criteria that constitute the basis for the implementation of the function. Within a functional profile, all of the functions have conformance criteria, which are taken directly or adapted from the functional model. In addition, *HL7 EHR System Functional Model* has a conformance clause chapter, which describes the minimum requirements that must be met in order to conform to the EHR system functional model. The conformance clause describes critical concepts for the interpretation of the functional model and its implementation at a high level, so that the functional profile, the functional model conformance criteria, and the mandatory and optional requirements can be fully understood during the actual implementation of the EHR system functional model. A conformance clause enables EHR system equipment vendors (as the functional model implementers) and buyers (as users) to effectively communicate what they mean by the terms "conforming profile" and "conforming EHR system" and their associated requirements. In addition, third-party testing and certification organizations such as the Certification Commission for Healthcare Information Technology (CCHIT) can use the conformance clause as the foundation for establishing their testing criteria.

The HL7 EHR system functional model allows EHR functions to be described in a standardized way for various medical specialties, regions, or countries. However, this functional model itself is not a clinical messaging, implementation, or conformance specification criteria list. It is intended only to be used as a standardized language tool for discussions about EHR system functions between healthcare institutions, vendors, government agencies, and other involved organizations.

Table 11.1: Titles for Each Column of the HL7 EHR System Functional Model

ID	Identifies unique function using a naming convention with a prefix followed by a number. The prefix indicates which of the three main sections of the functional outline the function belongs to: DC for direct care, S for supportive, and IN for information infrastructure. The numbers show the function's place in the functional outline.
Type	Identifies the item as a header (H) or a function (F).
Name	Names the function
Statement/Description	Briefly states the function's purpose
See Also	Indicates relationships with other functions
Conformance Criteria	Clarifies function conformance
Row#	Indicates the item's location within the entire function table

11.2.2 EHR Interoperability

Typically, an independent EHR system receives data from standard messages sent by other clinical information systems belonging to the same healthcare institution or enterprise, which controls the policies for all its information systems. The desired data are incorporated into the data repository of the EHR system, and the message is deleted. Since the EHR is not linked to external clinical information systems, there is no need to specify the authoring system. However, when an EHR system is linked to a variety of external clinical information systems for data exchange, it becomes important to identify the authors, contexts, alterations, and transformations of exchanged clinical messages, so that each EHR is a valid clinical record that can be understood by the clinicians involved in patients' care. In this scenario, it is critical for the EHR system and its records to be interoperable with various healthcare information systems and other EHR systems both within and outside the healthcare enterprise.

The term "interoperability" has different meanings in different situations. For general business and telecommunication organizations, interoperability is usually defined as successful information exchange among systems. In the clinical healthcare environment, it refers to converting information from one information system to another, such as between a RIS, or a LIS, and the institutional EHR system or patient PHR. Patient clinical information exchange is related to different interoperability issues depending on whether it occurs among different healthcare institutions belonging to the same healthcare enterprise, among the various healthcare enterprises of a regional healthcare information organization (RHIO), or at a national level. Additionally, using patient information from EHRs to generate longitudinal PHRs, which enable patients to use the entire healthcare system over time for their well-being, requires an even higher level of interoperability. In the following sections, different levels of interoperability are discussed so that EHR buyers can better understand the various related standards, hardware and software capabilities, and system functionalities required for the degree of sophistication they seek in their EHR system's functionality.

11.2.2.1 Technical Interoperability

From the viewpoints of data format, syntax, and physical connectivity, technical interoperability is described as the very basic, hardware level of interoperability. Technical interoperability refers to the hardware, data transmission, data access, and security management that enable two different physical systems to be connected and exchange data successfully. The core of technical interoperability is the link across networks and application software. Technical interoperability is the foundation of the other two, higher levels of interoperability (semantic and process, discussed below). This is also the location of the security systems that allow the secure storage and exchange of patient health information. The complexity of information links ranges from the simple successful exchange of messages over the network, as defined by HL7 V2.x messaging standards, to the intermediate level of message exchange in which the data elements in a message are arranged in a defined order but are not assigned meaning specifications and interpretations. At this level, data mapping, the process in which the data in a message are related to other data, is done in a predefined, mutually agreed form, as defined by the HL7 RIM. The emphasis of technical interoperability is to exchange bits and bytes of data, not their meaning. Technical interoperability ensures the proper transmission and reception of data to be used by the receiving system, but it does not involve the software applications that process data into semantic equivalents.

11.2.2.2 Semantic Interoperability

Since healthcare has its own unique vocabulary, technical interoperability alone is not sufficient for efficient, effective, and safe delivery of healthcare information. Semantic interoperability is a higher level of interoperability than technical interoperability, which means it can provide clear, consistent meaning to communication. Using application software, semantic interoperability optimizes the exchanged data so that they can be processed and understood by the receiving system. Semantic interoperability involves several levels of information processing accessibility, including:

- Data that are meaningful to users at both the sending and receiving ends but are not machine readable and therefore require the receiving user to manually incorporate these data into information systems.

- Free text without defined structure that is readable by the receiving application software.

- Classification systems such as the Common Procedural Terminology (CPT) and International Classification of Diseases (e.g., ICD-9, ICD-10), for unified clinical diagnostic and therapeutic procedure descriptions.

- Standardized nomenclature within structured messages, such as the SNOMED.

In the HL7 RIM, some data elements can be semantically characterized. Each of these elements has specifically defined structured fields, so that the same semantic meaning of the HL7 message can be established with different terminologies. Each field in the RIM may point to a clinical coding vocabulary, such as the Logical Observation Identifiers Names and Codes (LOINC). Within LOINC, for example, each concept contains semantic axes such as body systems and test timing. With complete semantic interoperability, disease and medical interventions can be characterized, public health information can be summarized and alerts concerning epidemics can be passed on, free text can be extracted for computable information, terms from various clinical terminologies and classification systems can be mapped or translated, and communication among healthcare providers at the national or international levels can be improved. In addition, with this high level of software semantic interoperability, the error-prone, redundant human manual processes of data entry and analysis can be minimized. High levels of semantic interoperability should be thoughtfully designed, fully tested, and carefully implemented to avoid the intrusion of misleading information and misguided policies into patient healthcare delivery processes.

11.2.2.3 Process Interoperability

Process interoperability, an emerging concept sometimes referred to as "workflow" or "social" interoperability, is required for the successful implementation of an information system in a real work environment system implementation. Process interoperability is related to process and workflow management, workflow and data presentation with advice, and resource allocation for planning and protocol fulfillment. Process interoperability is also closely related to workflow management or system engineering because it involves the design and implementation of human work processes as well as human interaction with computer information systems and applications systems. It defines a way of providing information for use in a real-life setting, and it coordinates workflow. Process interoperability has aspects of prescribed processes in which clinical healthcare organizations utilize engineering techniques for routine or emergent processes and for disaster management.

For optimized application of any healthcare IT system, proper integration of the system with the designated workflow is essential. Otherwise, the deployed system's performance may not meet expectations or realize its performance potential. In a healthcare institution, process interoperability optimizes process coordination for healthcare delivery teams so that documents critical to patient care, such as previous diagnostic and therapeutic reports as well as safety and quality reminders, can improve human involvement during patient treatment, which could result in improved patient care.

Through the use of technical, semantic, and process interoperability, together with an analysis of the impact associated with the necessary workflow and process changes caused by a newly implemented information system, patient healthcare delivery can be done in the most efficient, effective, and safe manner possible.

11.2.3 HL7 EHR Interoperability Model

The HL7 EHR interoperability model is complementary to the EHR system functional model. The EHR system functional model is focused on EHR systems and defines their functional characteristics and conformance criteria, while the EHR interoperability model is focused on the individual EHRs and defines the records' interoperability characteristics and conformance criteria. The EHR interoperability model also specifies a methodology for tracking an interoperable EHR's status from its origin to the point where it is used. During the information exchange, the record may go through processes of verification, modification, amendment attachment, or translation from one language to another, or from one clinical vocabulary to another (e.g., from LOINC to SNOMED). Therefore, the objectives of the EHR interoperability model are:

- Establishing a common frame of reference for the EHR industry

- Establishing interoperability characteristics for records to complement the HL7 EHR system functional model's focus on the functional characteristics of systems

- Establishing conformance criteria for validation of EHRs

- Establishing a framework for EHRs to be used as legal documents

- Specifying an EHR's flow and life cycle, including its creation, retention, security, access, use, and exchange

- Specifying the role of EHRs as healthcare delivery documentation integrated into clinical workflow and practice

- Specifying the rationale and meaning of EHR interoperability characteristics

- Establishing specifications for EHRs with a neutral approach for technology, vendors, and products

- Facilitating unique conformance profiles for specific healthcare settings, realms (countries or regions), products, implementations, and uses

- Ensuring EHR exchange with data integrity and persistent semantic meanings

It is important for the EHR product manufacturing industry to have a common understanding of the EHR interoperability definition. The HL7 EHR interoperability model provides a definition that has industry consensus, as well as a framework for analyzing and planning interoperable record requirements, so that patients' EHRs are complete, accurate, persistent, and traceable by an audit trail. Such records serve as clinical, business, and legal documents. This model uses a series of assertions to generate EHR interoperability characteristics, which can be measured and tested when they are implemented. The interoperability model also establishes a function and requirement reference list for interoperable EHRs, allowing them to be validated at their source system, during transmission, or at the receiving system using the model's conformance criteria. However, the interoperability model itself is not a specification list for the implementation of any single EHR product; it only provides requirements and characteristics for interoperable EHR records. Similar to the EHR functional model, the EHR interoperability model may also be specifically profiled for various medical specialties, practice settings, regions or countries, implementations, and uses.

11.2.3.1 EHR Interoperability Model Components

The EHR interoperability model is published as a long table with rows representing discrete EHR interoperability definitions and characteristics, and columns describing EHR interoperability characteristics and applicability. These assertions and EHR characteristics serve as the foundation for EHR interoperability. Each characteristic clarifies its own requirement for the EHR to be interoperable, and it can be tested. Listed below are the names of the HL7 interoperability model columns, which are derivatives of EHR assertions, or characteristics, with brief descriptions when necessary:

- *ID*: EHR interoperability assertion or characteristic identifier

- *EHR Interoperability Assertion/Characteristic*: EHR assertion or characteristic statement

- *Elaboration*: Additional definition or description

- Attribute Class (Testable for EHR Interoperability Characteristic)

- *Example*: EHR interoperability or characteristic example

- *Use Case Example*: EHR interoperability or characteristic use case example

- *Legal Record Requirement*: National or regional requirement for being a legal record; "Yes" or "No"

- *Source EHR System/Application*: EHR interoperability characteristic conformance criteria

- *Source EHR System/Application/Outbound at Point of Record Transmittal*: Outbound, at point of health record transmittal, EHR interoperability characteristic conformance criteria

- *Health Record Interchange Standard*: Conformance criteria for industry standard health record exchange

- *Intermediary Application*: EHR interoperability characteristic conformance criteria

- *Receiving EHR System/Application*: EHR interoperability characteristic conformance criteria

- *Patient*: Subject of care

- *Act Participant*: Record author, or originator

- *Record Recipient*

- *Standards Reference*: Relevant industry standards

- *HL7 EHR System Functional Model Reference*: Reference to relevant functions of the HL7 EHR system Functional Model

- *Normative Clause*: Normative language example

- *Progression*: EHR interoperability characteristic implementation strategy

- *HL7 CDA r2 Reference*: References to HL7 Clinical Document Architecture Release 2

- *Testability Criteria*: Details of HL7 CDA r2 testable criteria

11.2.3.2 EHR Interoperability Model Profiles

Four categories of EHR interoperability profiles can be developed from the EHR interoperability model: implementation, product, care setting, and realm. Although different EHR profiles may be developed within this framework, the legal requirements profile and the EHR implementation profile must be applicable to any implementation for it to meet the basic interoperable EHR requirements.

11.2.3.3 EHR Record Validation and Conformance Test

The EHR interoperability model provides the foundation for EHR conformance testing in the areas of identifier, organization, coding, content, persistence evidence, and authentication during both data storage and transmission.

In summary, the HL7 EHR interoperability model provides basic guidance and benchmarks for current and emerging standards to facilitate interoperable EHR exchange.

11.2.4 Certification Commission for Healthcare Information Technology

The CCHIT itself is not another healthcare information technology standard. It is a private nonprofit organization formed to certify EHR systems using a set of minimum functional, interoperability, and security criteria.

Although it is agreed by many healthcare institutions and providers that patient medical records should be moved into electronic form using EHRs, there are many different views held by EHR system manufacturers as to what functionalities such a system must possess and how security during information exchange should be ensured, and the market is filled with several hundred EHR products. The role of CCHIT is to define EHR systems' functionality, means of communication, and patient information protection using established industry standards such as HL7 and IHE integration profiles.

CCHIT also provides optional certifications for cardiovascular, child health, emergency department, and behavioral health EHR systems, which have unique specialty requirements. Currently, CCHIT uses a list of detailed product capability criteria to evaluate each EHR system submitted for review. The list includes about 500 criteria for the functionality, interoperability, and security of EHR systems. These criteria include basic functionalities such as management of patient clinical records, histories, and clinical chart notes, and more advanced features such as integration of laboratory information and pharmacy information systems. For a large healthcare institution or enterprise, these capabilities are important for the integration of various existing healthcare information systems with the EHR system for workflow efficiency and patient care improvement. However, a small clinic or an independent physician's office might not use many of the advanced features. An EHR system can be certified by CCHIT only when all listed criteria for a specific type of EHR are met, even though some of the criteria may not be used by a specific healthcare institution or enterprise.

One benefit of choosing CCHIT-certified products is that CCHIT is recognized by the U.S. federal government as the certification body for EHR systems. There is a payment incentive program from the U.S. Centers for Medicare and Medicaid Services (CMS), and CCHIT-certified

EHR products are exempt from antikickback laws, so a physician's procurement of EHR products may be subsidized by local healthcare institutions or healthcare enterprises. With the CCHIT certification, healthcare institutions have an array of EHR products that meet every functional requirement of the CCHIT criteria.

The following considerations can help purchasers decide whether to buy CCHIT-certified EHR products:

- Evaluate and determine workflow-related and required capabilities from CCHIT's criteria list

- Evaluate and understand specialty requirements not covered by CCHIT

- Understand that some good EHR products may meet all of the purchaser's functionality needs but may not be certified because of the nature of the certification process

- Evaluate vendor reputation, system software user interface, and service support, which are not reviewed by CCHIT

Another consideration for selecting CCHIT-certified products is that the cost of such systems may be higher than that of similar noncertified products, because for each EHR product submitted for certification, the manufacturer must incur certification costs during the initial review and the 3-year certification period, and pay an annual maintenance fee.

11.3 HL7 PHR SYSTEM MODEL

The HL7 PHR system functional model is a very broad model that defines various profiles for healthcare industry stakeholders and PHR system users. The PHR system functional model is divided into three main sections: personal health, supportive, and information infrastructure. These sections are then subdivided into 14 subsections, as shown in Figure 11.3, each comprising a number of individual functions. In each function, a behavior of the PHR system is described in plain language so that all the key stakeholders can understand the meaning. Although the PHR system functional model lists all reasonably expected functions found in all PHR systems, it is not anticipated that any single PHR system would include all these functions. PHR system functional profiles are to be developed to include a limited selection of functions from the PHR functional model for various specific intended uses. And although the PHR system functional model defines possible functions for retrieval and distribution of clinical documents, event notes, and claim attachments, the functional model itself does not provide specifications for messaging, implementation, records, conformance, and requirements for a particular PHR system.

11.3.1 Personal Health Functions

All PHR functions belonging to the personal health main section have the prefix of "PH," followed by a series of numerical number. They are used for patient self- and provider-based care information and feature management, generating a summary record of a patient's care over time. These functions give direction to help patients maintain their own health and demographic information so that their PHRs are accurate and up to date with a complete history of their healthcare activities, including personal wellness and disease prevention activities, and clinic or physician office visits. These functions are intended to actively engage patients in their own healthcare and give them better access to educational and other resources for their personal health condition monitoring.

11.3.2 Supportive Functions

Supportive PHR functions with prefix of "S" pertain to administrative activities related to patient healthcare. These functions input data to medical research systems and promote public health for overall healthcare quality improvement.

11.3.3 Information Infrastructure Functions

Information infrastructure PHR functions with prefix of "IN" are common functions supporting the functions belonging to the personal health and supportive sections. Information infrastructure functions ensure the privacy and security of patients' information as mandated by various government authorities, promote interoperability among PHR systems or between PHR and EHR systems, and make PHR systems user friendly. The information infrastructure functions described by the PHR system functional model may be carried out by software applications for health record management and information technology supporting infrastructure.

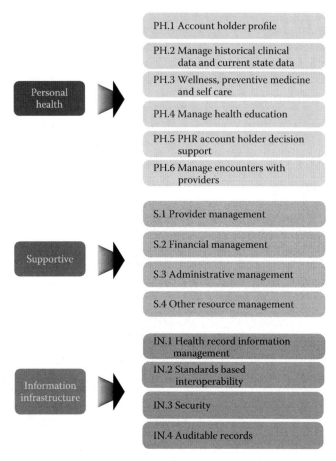

Figure 11.3 HL7 PHR functional model has 3 main sections and 14 subsections according to the functions' purposes.

11.3.4 Components of PHR Functions

In the HL7 PHR system functional model, all functions, and the set of elements and components for describing them, are organized in a long table format, with the functions listed in rows and the columns representing the elements and components as in Table 11.2.

PHR functions can be applied to provide:

- A common understanding of PHR system functions for vendors and users during the development and procurement of PHR systems

Table 11.2: Titles for Each Column of the HL7 PHR System Functional Model

ID	Identifies a unique function using a naming convention with a prefix followed by a number. The prefixes are PH for personal health functions, S for supportive functions, and IN for information infrastructure functions.
Type	Identifies the item as a header (H) or a function (F)
Name	Names the function
Statement/Description	Briefly states the function's purpose, often with examples for illustration. Some functions are for clinical use, while others are for business or other uses.
See also in EHR-S FM	Indicates relationships between PHR system functions and those of the EHR system functional model (EHR-S FM)
See also in PHR-S FM	Indicates relationships with other functions in the PHR system functional model (PHR-S FM)
Conformance Criteria	Clarifies conformance clarification
Row#	Indicates the item's location within the entire function table

- A framework for next-level standards of information portability within and between PHR systems

- A standards-based methodology for international adoption to meet various countries' unique requirements

- Integrity of clinical data

11.3.5 PHR System Functional Profiles

A PHR system functional profile is a collection of functions from all the three main sections to support a PHR system use for a specific goal, user group, required level of interoperability, region, or country. With a functional profile, functions from the entire list of the PHR system functional model can be practically used and managed. It is not expected that any single PHR system would possess all the functions listed. Just like an EHR system, a specific PHR system conforms to a functional profile instead of to the entire PHR system functional model. The PHR system functional model also defines the way to create and register functional profiles with HL7. When a functional profile is created, each function included in the profile is assigned a priority of "essential now," "essential future," or "optional." It should be noted that PHR functions allow standards-based interoperability; however, the functions do not provide implementation details. Actual implementations of PHR system functions need to be carried out in the context of the entire PHR system functional model so that universal requirements such as privacy, security, and audit functionalities are incorporated.

11.3.6 Conformance Criteria

The PHR system functional model defines the minimum requirements that must be met in order to claim conformance in a conformance clause, which describes the functional profiles and their implementations at a high level. Important concepts of functional profiles, conformance criteria, and mandatory and optional requirements are all clarified in the conformance clause. In addition, the conformance clause serves as an effective communication tool between vendors (PHR system functional model implementers) and buyers (PHR system users), and it provides the foundation for testing and certification done by third-party organizations such as CCHIT.

11.3.7 Types of PHR Systems

Currently, PHR implementation is in its early stage. There are basically three types of PHRs: provider-linked PHRs, payer-linked PHRs, and Web-based hybrid provider- and payer-linked PHRs.

11.3.7.1 Provider-Linked PHR Systems

A provider-linked PHR system is different from other types of PHR systems in that it is connected to provider-controlled EHR systems for the consumer to view his or her stored medical records, because the PHR is capable of providing communication between the consumer and healthcare providers by means of secure e-mail exchanges, e-prescribing, electronic requests for prescription refill, and patient appointment scheduling. Patients can correct patient-recognizable inaccurate information about themselves in the healthcare providers' EHR systems, thus improving the data quality of EHR systems and potentially improving patient safety. Patient self-entered data, medical devices data, and administrative data from other sources are also stored on provider-linked PHR systems, each marked with its origin. Provider-linked PHR systems can determine how their patient clinical data can be accessed by the PHR and EHR systems of other healthcare enterprises with which they are interoperable.

11.3.7.2 Payer-Linked PHR Systems

Payer-linked PHR systems are initiated by health insurance companies, which are sometimes called "care management" or "care coordination." Through such PHR systems, health insurance carriers support and engage patients and other consumers in their own healthcare so that these consumers can better understand their overall health conditions and involve themselves in continual healthcare during times of both health and illness. This relatively new, proactive approach changes the focus from treating illness to overall enhancing patient and consumer population well-being; the anticipated results are cost control and improved health outcomes.

In these insurer-based PHR systems, insurers play active roles while patients and consumers are engaged.

In addition to patient and consumer self-entered data such as demographics, histories, and lists of allergies, clinical data concerning medication, laboratory tests, lists of providers and their contact information, and diagnostic and therapeutic procedures can be consolidated from submitted claims. These data may come from multiple healthcare providers, which may belong to different healthcare enterprises that do not have health information exchange, or they can be interoperable with EHR systems from healthcare institutions and enterprises.

11.3.7.3 *Web-Based, Hybrid Provider- and Payer-Linked PHR Systems*

Because the unique characteristics of any PHR system are patient- or consumer-centric, simple and secure patient access is a critical component of the PHR system. The wide availability of Internet access has led to Web-based PHR systems becoming the model for PHR data access, allowing patients to conveniently review, manage, and share their healthcare information and improve their health management. The core of these Web-based systems can be either provider-linked, payer-linked, or hybrid provider- and payer-linked PHR systems. Related to but different from either provider-linked or payer-linked PHR systems, a hybrid provider- and payer-linked PHR system integrates and supports the use of clinical and administrative data from both healthcare providers and insurers. Multiple independent healthcare enterprises or provider organizations may supply clinical data for the patient. A provider-linked PHR system with provisions to incorporate information from various payers and external PHR systems can also be categorized as a hybrid. For any hybrid PHR system, it is critical to tag all information sources and to integrate the data related to a single patient in a seamless manner so that the patient and healthcare providers can use the data effectively.

11.4 STANDARDS USED IN EHR AND PHR

In addition to EHR and PHR models discussed in the previous section, there are several sets of standards, vocabulary, and code systems used in EHR and PHR systems. In this section, some frequently used ones are discussed: IHE document sharing related integration profiles for system interoperability integration, SNOMED-Clinical Terms (SNOMED-CT), CPT codes, ICD, and LOINC for standardized vocabulary and coding, continuity of care record (CCR), and continuity of care document (CCD) for document formats.

11.4.1 IHE IT Infrastructure for Health Information Exchange

Efficient and secure access to comprehensive EHRs is necessary for optimal patient care. IHE is the process for achieving standards-based interoperability. It provides a common framework using established standards for seamless information exchange among healthcare institutions and regional and national healthcare information networks. IHE improves patient care by reducing medical errors, improving workflow efficiency, and lowering systems interfacing costs through greater interoperability and data sharing. IHE helps widen the adoption of information standards and EHR systems. Each year, more manufacturers implement IHE in their systems. IHE systems provide interoperability across various clinical care settings for clinical documents while maintaining the confidentiality, authenticity, integrity, and consistency of the patient care data in EHRs. Although details about IHE are discussed in Chapter 6, IHE integration profiles used in healthcare information exchange and EHR systems are introduced in the following sections.

11.4.1.1 *Patient Identifier Cross-Referencing*

Patient Identifier Cross-referencing (PIX) enables the matching of patient identifiers from multiple issuing institutions or enterprises. Information about the same patient from different sources can then be correlated by an information system and used by healthcare professionals.

11.4.1.2 *Patient Demographics Query*

PDQ is an integration profile for a list of patients to be queried from a central patient information server by multiple distributed applications. The query is based on search criteria defined by the user so that applications can directly retrieve the patients' demographic and visit-related information.

11.4.1.3 Cross-Enterprise Document Sharing

The IHE XDS integration profile is designed to be used with a wide variety of healthcare business policies and rules for sharing clinical documents based on common standards and policies, without defining institution-specific business policies and rules. It allows multiple institutions within an XDS affinity domain to cooperate in healthcare delivery. XDS affinity domains are groups of independent healthcare enterprises such as national EHR systems, RHIO, clinical specialty networks, private insurance provider-supported communities, or government-sponsored facilities. Inside each XDS affinity domain, common policies and business rules are defined for patient identification, patient clinical information access control, patient consent documentation, and the organization and presentation of patient clinical information based on its structure, format, and content. The enterprises within an affinity domain collaborate to use a common policy set and infrastructure to share their own patient care clinical information and EHRs with each other. A patient's longitudinal healthcare information and records are available from distributed healthcare document repositories and registries belonging to different healthcare enterprises in the XDS affinity domain. The document repositories are responsible for secure and reliable document storage, management, and fulfilling document retrieval requests, while the document registries are responsible for storing and managing information about these documents. Using document registries, clinical documents requested by physicians and healthcare professionals for patient care can be easily searched and retrieved regardless of which document repository these documents are physically stored in.

An XDS document is a basic component unit for a document repository, and is registered as an entry in the document registry. The XDS integration profile treats an XDS document as the smallest unit of information within a repository; therefore, an XDS document cannot be partially accessed. An XDS document consists of information about the clinical services received and observations made with regard to a patient, and can be read when an appropriate application is used. In addition to textual information, other standard formats for document structure, encoding, and content definitions can be used. These standard documents include HL7 clinical CDA documents, images in the DICOM format, and structured and vocabulary-coded information such as DICOM structured reports. However, common policies about specific document structure, format, and content need to be agreed upon by all members of the XDS affinity domain to ensure interoperability.

XDS is not the only integration profile designed to meet all cross-enterprise patient healthcare document exchange needs. Other integration profiles such as ATNA, PIX, Cross-Enterprise User authentication (XUA), and RID may complement XDS during document exchange across enterprise boundaries.

11.4.1.4 Cross-Enterprise Document Media Interchange

Cross-Enterprise Document Media Interchange (XDM) allows patients' healthcare documents to be exchanged between healthcare providers using common storage media and common file and directory structures. For healthcare institutions and enterprises that either do not have XDS in place or choose not to use it, XDM is a viable alternative for exchange of patient healthcare information across enterprise boundaries, although it cannot address all cross-enterprise patient EHR communication needs. The use of XDM integration profile requires information to be manually imported, sorted, and reconciled at the receiving end. For example, patients may carry their EHR from one healthcare enterprise on common standard portable storage media during their visit to another healthcare enterprise. The benefits and risks should be carefully evaluated if the storage medium containing a patient's EHR is re-writable and is under the patient's control. In addition to portable physical storage media such as CD-R and flash memory drives, XDM is applicable to e-mail attachments used to exchange of medical documents in standard ZIP file compression format.

11.4.1.5 Cross-Enterprise Document Sharing for Imaging

The XDS-I integration profile provides specifications that allow imaging information to be shared among healthcare enterprises. This imaging information includes DICOM images, presentation states, evidence documents, reports, and key images associated with reports. The imaging information is stored in repositories and registered in central registries, which can be queried by users for imaging information retrieval. XDS-I relies heavily on the XDS integration profile and is the essential imaging component of a regional or national EHR system.

11.4.1.6 Cross-Enterprise Sharing of Scanned Documents

There are many patient clinical history documents in various paper, film, electronic, and scanned output formats for archive and exchange. Paper clinical documents include computer printouts, handwritten or typed documents, and chart notes. These legacy document formats were designed for general use, not for the healthcare environment; hence, there is no unified means to associate these documents with patient identification, demographics, and clinical orders. In order to effectively manage medical records in electronic form, users sometimes must correlate legacy patient healthcare documents with structured healthcare metadata so that they can be archived and accessed together electronically. Cross Enterprise Sharing of Scanned Documents (XDS-SD) is an integration profile used to correlate structured healthcare metadata with general-purpose nonhealthcare-specific scanned document formats so that patient healthcare information can be retrieved from the management system in which it was originally archived while record integrity is maintained. The integration profile defines patient identity, demographics, scanner operator identity, scanning technique, and time information in the structured HL7 CDA header while storing patient clinical information in PDF format, which preserves the quality of the text, graphics, and line drawings of original clinical documents better than image-based compression formats such as JPEG.

11.4.1.7 Patient-Synchronized Applications

The Patient-Synchronized Applications (PSA) integration profile allows a user to run independent and unlinked applications for a single patient on a workstation without having to select the same patient in multiple applications. The PSA profile can be used to synchronize the patient's identification codes for these applications. The goal of patient-synchronized applications is to improve patient care by reducing possible medical errors caused by viewing the wrong data. The integration profile leverages the HL7 CCOW standard, which has been discussed in Chapter 6, to achieve context management.

11.4.1.8 Retrieve Information for Display

RID allows physicians and other healthcare professionals to quickly and easily retrieve patient clinical information that is not in their current desktop computer application but is critical for better patient care or improved workflow. RID provides access to key information about a patient, such as report summaries and lists of allergies and current medications. These documents are in the standard presentation formats: HL7 CDA, PDF, or JPEG. The RID integration profile has no mechanism for access control or security measures for patient healthcare information exchange; therefore, other appropriate security-related integration profiles need to be implemented to maintain the necessary level of security. When the implementation of the RID integration profile is associated with the Enterprise User Authentication (EAU) and PIX integration profiles, it can access information systems from multiple healthcare institutions belonging to the same healthcare enterprise.

IHE can make EHR adoption easier by improving the exchange of information among healthcare information systems and devices. The end result is that both healthcare institutions and patients can efficiently and securely create, manage, and access EHRs, which ultimately contributes to improved patient clinical care.

11.4.2 Clinical Vocabulary and Classification Systems

Clinical vocabularies, data sets, terminologies, and classification systems enable standardized clinical data collection, processing, storage, and retrieval for both healthcare delivery and administrative functions. Several standard clinical vocabulary and classifications used in EHR and PHR are discussed in the following sections.

11.4.2.1 Systematized Nomenclature of Medicine: Clinical Terms

In a healthcare environment, clinicians and healthcare institutions have traditionally used several terms that bear the same meaning; however, to a computer, these different terms are considered to have different meanings. In order for clinical information to be exchanged consistently among many information systems, healthcare institutions, or healthcare enterprises, the healthcare information infrastructure requires a unified medical terminology system. SNOMED-CT is a comprehensive clinical terminology originally created by the College of

American Pathologists (CAP); since 2007, it has be owned, maintained, and distributed by the International Health Terminology Standards Development Organization (IHTSDO), a nonprofit organization in Denmark. The CAP continues to support operations of SNOMED-CT under contract to the IHTSDO and provides SNOMED-related products and services as a licensee of the terminology.

SNOMED-CT is a systematically organized compositional concept system with a collection of medical terms that can be processed by computers. Designed to maintain content as a dynamic resource, SNOMED-CT covers most clinical information areas, including microorganisms, diseases, clinical findings, pharmaceuticals, and diagnostic and therapeutic procedures. The SNOMED-CT clinical vocabulary is available in many languages and covers most aspects of clinical medicine, with more than 344,000 concepts.

The application of SNOMED-CT includes EHR, CPOE clinical data retrieval across medical specialties and between healthcare institutions, and laboratory reporting. It provides consistent medical terminology across all healthcare domains, allows precise recording of clinical information and content management, has an inherent structure, and is now a developing international standard.

11.4.2.2 Current Procedural Terminology Code

The CPT code is maintained and copyrighted by the American Medical Association (AMA). The code describes various medical, surgical, and diagnostic services and is designed to allow the uniform communication of medical procedures and services among physicians, medical coders, patients, accreditation organizations, and payers for administrative, financial, and analytical purposes.

11.4.2.3 International Classification of Diseases

Published by the WHO, the ICD is an international diagnostic classification standard used in clinical practice, healthcare management, and for epidemiological purposes. It provides codes to classify diseases along with a wide variety of signs, symptoms, abnormal findings, complaints, social circumstances, and external causes of injury or disease recorded on health records, death certificates, and many other types of vital records. Every health condition can be assigned to a unique category and given a code. The classifications also include analysis of the general health situations of various populations as well as the incidence and prevalence of diseases and other health problems, together with their relation to other social and economic variables. The code set is revised periodically and is currently in its 10th edition, ICD-10, which includes more than 155,000 codes and permits tracking of many new diagnosis and procedures. ICD-10 is a significant expansion over the 17,000 codes available in the previous edition, ICD-9. ICD-9 is still widely used in the United States, but the Centers for Medicare and Medicaid Services within the U.S. Department of Health and Human Services will start to use ICD-10 in 2013.

11.4.2.4 Logical Observation Identifiers Names and Codes

Using a universal code system allows healthcare institutions to exchange and compare clinical data nationally or internationally. LOINC is a database and standard used to categorize clinical laboratory observation and identification. It was created and is maintained by the Regenstrief Institute, a U.S.-based nonprofit organization, to meet the demand for an electronic database for clinical terminology. The laboratory portion of the LOINC database includes codes for specialty observation reports in clinical chemistry, hematology, serology, microbiology, cytology, surgical pathology, fertility laboratories, and blood banks. In addition to these clinical laboratory code names, the LOINC database has been expanded to include codes for patient clinical care data, nursing interventions, and outcome classifications. These clinical portions of the LOINC database include measurements of blood pressure, heart rate, respiratory rate, cardiac output, body temperature, body weight and height; results of ECG, echocardiography, obstetric and urologic sonography, gastrointestinal endoscopy; and other reports including those related to dental care, ventilator management, emergency services, radiology, history and physical examinations, discharge summary, operative note report, and claim attachments.

LOINC identifies each test or clinical observation with a distinct six-part name in the form of "xxxxx-x," where each "x" is a numerical number. There are more than 41,000 observation terms in the database for universal access and identification. The six-part name is subdivided into six data fields for classification:

- Measurement, evaluation, or observation component of potassium, hemoglobin, hepatitis C antigen, or another clinical laboratory value

- Measurement property of length, area, volume, mass, time stamp, mass concentration, enzyme activity, catalytic rate, or a similar value

- Observation or measurement time span

- Observation context or specimen type

- Measurement scale

- Measurement or observation methods and procedures

When clinical laboratory test results in LOINC are sent electronically to healthcare information systems used by healthcare institutions, they are accepted, understood, and filed properly with other HL7 messages that use LOINC. The benefits of using a universal standard include:

- Improved communication within integrated healthcare enterprises

- Improved patient EHR systems

- Automatic reportable disease epidemic detection and control transfer to public health authorities

- Improved billing and payment information exchange

- Improved healthcare delivery quality through error reduction

Since LOINC uses universal identifiers and names for medical terms related to EHR, it can be used with healthcare messaging standards such as HL7 to facilitate the electronic collection and exchange of clinical results from laboratory tests, clinical observations, and outcome management among heterogeneous clinical information systems. LOINC is endorsed by the American Clinical Laboratory Association and the CAP. There is an increasing international interest in the adoption of LOINC, and its terms and documents have already been translated into German, simplified Chinese, Spanish, and Estonian.

11.4.3 Continuity of Care Record

The CCR was developed jointly by the American Society for Testing and Materials (ASTM) International, HIMSS, the American Academy of Family Physicians (AAFP), the American Academy of Pediatrics (AAP), and the Massachusetts Medical Society (MMS), and was released in 2006. The goal of a CCR is to summarize, store, and pass patients' health and administrative information electronically with little or no human involvement in the information and data interchange, because frequently, healthcare information systems are not interoperable. CCRs allow electronic communication of patients' most relevant clinical information from various healthcare institutions and enterprises. CCRs used in PHR systems can also benefit patients. The CCR standard is open, without any fees associated with its use. It is sponsored by more than a dozen medical associations and organizations, and has been adopted by many (RHIO) networks.

11.4.3.1 CCR Content

A CCR does not include the patient's entire medical history. Instead, it provides the most relevant clinical health information about a patient in a summarized form contained in a single digital document so it can be used effectively during the patient's next visit with a different healthcare provider. A CCR contains clinical information such as laboratory test results, procedure reports, visit documents, and discharge notes. A CCR can contain up to 17 sections:

- Patient demographics

- Insurance information

- Advance directives

- Problems/diagnosis

- Allergies and alerts

- Medication list
- Immunizations
- Family history
- Social history
- Vital signs
- Results
- Procedures
- Encounters
- Medical equipment
- Plan of care
- Healthcare providers
- Functional status

Each of these sections consists of coded data and free or structured text from various contributing sources; each data element is marked clearly to indicate when and where it was acquired. A CCR document does not have to contain all these sections. Its highly structured format makes the CCR both human readable and machine readable.

11.4.3.2 CCR Structure

In a CCR, extensible markup language (XML) is used to represent data. XML is a general-purpose markup language and is an international IT standard. The primary purposes of XML are to permit the easy exchange of structured data across different information systems and to allow users to define their own data elements. XML is widely used for data exchange interoperability in other business sectors. XML embeds structured data in the document, which not only allows its display and presentation by various common software applications, but also enables the data to be read, processed, and interchanged by computers without manual involvement.

A CCR is not a traditional medical document, nor is it intended to replace any traditional medical documents. It has important functionalities that combine clinical data and information from multiple sources in a summarized form. A CCR can easily be incorporated into EHR and PHR systems. Depending on the capabilities of the recipient systems, a CCR can be exchanged in direct transactions, Web documents, e-mail document attachments, and hardcopy.

11.4.4 Continuity of Care Document

The CCD is also an XML-based standard that defines the encoding, structure, and semantics of a patient's clinical summary document for interchange. It was released in 2007 as a collaborative effort between ASTM International and HL7 for interoperability in clinical data interchange. Combining the benefits of the CDA from HL7 and the clinical data representation of CCR from ASTM, the CCD integrates 2 separate but complementary standards. The result is the mapping and representation of CCR data within the CDA structure. A CDA document consists of a mandatory textual part designed for human interpretation of the document content and an optional structured part for data to be processed by computer application software. The structured part is based on the HL7's RIM, and it provides a framework for using concepts defined in coding systems such as LOINC and SNOMED. (More details about CDA can be found in Chapter 6.) Thus, a CCD document allows a healthcare institution to aggregate all of a patient's pertinent demographic, clinical, and administrative information covering one or more visits by the patient. The summarized information can then be forwarded to other healthcare providers for continuity of care for the patient, which results in improved patient care and increased operational efficiency.

11.5 GUIDE FOR EHR IMPLEMENTATIONS

Since the U.S. government has called for the adoption of EHR and PHR systems, many healthcare institutions in the country are in different phases of implementing these systems. While healthcare institutions and enterprises are ready to implement EHR and PHR, few have implemented one that is fully integrated into all areas of the institution or enterprise. In previous sections, various models and standards used in EHR systems were discussed. This section addresses some key practical issues pertinent to the successful implementation of these systems in healthcare institutions and enterprises of small or large size.

Since EHR implementation brings significant changes, it demands support from all levels of the healthcare institution or enterprise for success. EHR implementation is a complex task; it includes planning, software product and vendor selection, computer hardware selection, staff training, workflow and practice procedure adjustments, service, and support. Each of them relates to the final success of the implementation and transition from paper-based medical record to the electronic-based EHR. As with any large project, the success of implementing an EHR system relies on good planning. It is necessary to analyze not only EHR products and vendors, but also workflow changes, implementation logistics, budget, and how the EHR system fits into the unique operation of the healthcare institution. The importance of arranging resources to manage both implementation and system support cannot be underestimated. Other factors to be considered include the staff's proficiency in computer use, and communicating with staff about the impact of the new EHR system for more productive and efficient workflow.

User involvement in the planning and product and vendor selection process depends on the size and type of the healthcare institution, and whether or not it is associated with a large healthcare enterprise. During the planning phase, a team should be formed and a leader selected to coordinate the implementation. For a large healthcare institution, a leader from the administration and a representative physician typically are appointed to lead the project. The leaders are responsible for planning, organizing a team, taking decisions, testing, workflow analysis and change, staff training, maintenance, and future upgrades. They should be enthusiastic about the change and about the use of new technology. When there is frustration during initial implementation, they can provide encouragement to staff. They are key players not only during the planning phase but also after the installation is complete, because the operation of the EHR system needs to be monitored and evaluated so that progress can be reviewed and operation data can be analyzed, after which appropriate workflow changes and more EHR system customization can be implemented. Depending on the size of the healthcare institution, one of the leaders may also be the contact person for discussions with vendors, although sometimes a contact person is appointed from the team. The contact person's role is to ensure that vendors have a clear understanding of the institution's needs and prepare a contract that will generate the desired outcome. Since both the vendor and the healthcare institution are equally important for the implementation's success, the healthcare institution should allocate adequate financial and IT resources for the project during and after the implementation.

After the implementation team is organized, workflow should be carefully analyzed and redesigned to get the maximum benefit from the EHR system; it should not be a simple automated version that uses paper-based medical records. In the workflow analysis, some identified improvements can be immediately implemented without the EHR system, while others will require testing after the EHR system is in place. Immediately after the system's installation, there will be a learning curve and possibly reduced patient throughput while employees get used to the new system. This adjustment period should be kept as short as possible. Since implementing an EHR system is a continually evolving process, workflow may be adjusted after system installation as users become familiar with the new EHR system. After the initial implementation phase, a good EHR product appropriate for the healthcare institution or enterprise and properly implemented will result in improved workflow.

11.5.1 Financial and Needs Analysis

Evaluating the impact of EHR systems implementation on cost and business operations will depend on the nature of the institution itself. Before decisions are made about the product and vendor, the budget for the system must be determined, along with the overall desired workflow change, expected achievement from the new EHR implementation, and anticipated impact on clinicians, staff, and patients. Potential business cost reductions may be in the areas of personnel, medical transcription, billing, storage space, and chart management because of increased productivity. The ultimate benefit of EHR adoption, improved patient care, is difficult to quantify. It should be remembered that the cost savings from an EHR system are offset by the costs incurred by the maintenance, usage, service support, and upgrades for using the new system. For medium or large healthcare institutions, there is a need to hire technical staff for user training and system administration. Additional indirect cost includes staff training, infrastructure (e.g., wired or wireless networks), interfacing with existing medical information systems, and hardware and software upgrades.

11.5.2 Product Selection

Choosing the right EHR product and vendor that match the needs of the healthcare institution or enterprise is critical for a successful implementation. Otherwise, the system may not be used as effectively as expected. Easy use of the EHR system is important; the more intuitive the system is, the shorter the learning curve will be, and the easier it is for all users.

For a small healthcare institution, decisions can be made by a few key individuals, and the implementation is less complicated than for a large or multispecialty medical practice. However, decisions for large enterprises need to be made strategically by an EHR system implementation team with an overall view of the institution, because the system should meet the needs of all areas, and consensus on the best product for all may mean compromises for some. For both small and large healthcare institutions, communicating with all users of the system and others affected during the process may prevent unnecessary problems from arising; no user area should be left out.

One of the major benefits of a fully functional EHR system is improved efficiency and workflow. Different medical specialties have different needs, and within the same specialty each institution has its own unique ways of operating with different levels of technical proficiency. Since there are a few hundred EHR products on the market, selecting the right EHR and vendor to match the needs of a particular healthcare institution or enterprise may not be a trivial task. Some EHR systems have a complete suite of applications, which can be used to completely replace current practice management systems with a single patient database and easy access to various applications. Other EHR systems incorporate only a few applications, in which case interfacing with other medical information systems or applications is needed. Selecting an EHR system that is part of the institution or enterprise's existing vendor's suite of products can ease enterprise integration and user training.

Buyers of EHR systems should be reminded that features that allow these systems to interface with other medical information systems may be an additional product, may need to be developed, or may not be available at all. Interfacing is very important for integration. It is not difficult to imagine that the benefits of a new EHR system will not be fully realized if independent, disparate medical information systems do not communicate with each other. Therefore, interfacing issues should be identified in the early stages of product and vendor selection.

Some EHR systems can be uniquely customized to fit the healthcare institution without too much difficulty or expense, while other systems are not flexible. To use these noncustomized EHR systems, the workflow and practice procedure of the healthcare institution may need to be changed, not for efficiency improvement but because of the inflexibility of the EHR system, in which case the healthcare institution might be better off choosing a different EHR system. Customization issues should be discussed with potential vendors in the early phase of product and vendor selection, because some vendors are more willing to customize their products than others.

Vendors usually provide minimum hardware requirements for servers, desktop workstations, and portable computers or devices. It should be noted that portable computers and devices are convenient to use; however, they may require a wireless network infrastructure covering the entire medical practice, which may not yet be in place. Other considerations for portable computers and devices are limited battery capacity and physical security, which is related to sensitive patient healthcare information.

Other issues to be considered include the payment schedule, CCHIT certification, existing medical practice management systems and their interfacing with the new system, acceptance testing, training, and service support. Topics relating to the network infrastructure required for EHR system implementation can be found in Chapter 4.

11.5.2.1 Application Service Provider and Client/Server EHR

There are two types of EHR system designs: application service provider (ASP)-based systems and client/server-based systems, each with advantages and disadvantages. For ASPs, an EHR vendor provides both EHR server hardware and software. Using an ASP-based EHR service requires a monthly or annual fee. The physical EHR server can reside either at the vendor's location or in the healthcare institution.

A Web-based ASP design uses the Internet to access servers and EHR software that are remotely located and maintained by an EHR vendor. The server has all the security features and is HIPAA compliant. Since it is managed and maintained by the vendor, there are no technical issues or future upgrades to be handled by the healthcare institution except for maintaining office computers and a reliable and secure Internet connection. The initial cost for the implementation is lower than that of client/server-based systems and can be easily planned and budgeted as an operational cost instead of a capital expenditure because the vendor fees are based on services used rather than equipment. Additionally, the hardware requirements for institutional computers are lower, because most computations are performed on the EHR system's Web server instead of local EHR workstations or devices. Wherever there is Internet connection, the EHR system can be accessed at any time. If the healthcare institution has multiple locations, Web-based EHR can be a viable option. However, over the years, the accumulated cost of Web-based ASP EHR systems can be higher than that of a client/server-based system. Other disadvantages include software that cannot be customized, vendor control of data, slower data exchange than in client/server systems, and dependence on Internet connection quality and speed. Since the daily business operation of the healthcare institution relies on the EHR system service provided by the vendor, the vendor's business stability should be checked periodically and provisions made for EHR data backup.

A client/server EHR system uses a different approach. An EHR server is installed in the healthcare institution. Many client EHR workstations or devices with installed EHR software connect to the server through either a wired or wireless local network. Compared to a Web-based ASP system, a client/server system enables faster response times and overall operational speed, easier software management, better scalability, and higher reliability. If the healthcare institution requires high performance and high speed and does not have multiple physical locations, a client/server-based system may be the right choice. Client/server EHR systems are also more easily integrated with other devices and information systems than Web-based ASP systems, and the healthcare institution has complete control over its own data. However, along with these benefits, client/server-based EHR systems have their disadvantages. Since the server hardware, software, and higher-quality workstations or devices must be purchased up-front, the initial capital expenditure is high. Compared to Web-based ASP EHR systems, remote access to client/server-based EHR systems is limited in functionality and the process is more complicated. The locally hosted server requires the healthcare institution to be responsible for arranging network infrastructure, data security, data integrity, data backup, frequent server and workstation upgrades, and disaster recovery measures, which often means additional IT service costs. Most EHR systems currently used by healthcare institutions and enterprises are the client/server type. Either design may fit the specific needs of a healthcare institution. Unless the new EHR system replaces all existing medical information systems in use, the need to integrate them should be considered and communicated with the vendor. When either type of EHR system is selected, system downtime and disaster recovery should be planned. When the healthcare institution depends on computerized information systems for its routine operation, these issues need to be considered; the question is not whether these problems will happen, but when. The downtime plan should be tested regularly to verify its functionality.

11.5.2.2 EHR System Hardware Selection

Hardware and software costs for EHR systems are usually quoted separately. Software systems are developed and sold by vendors, while hardware can be purchased separately with the required specifications provided by EHR system vendors. For a client/server-based system, hardware includes one server and many client computers. Since the server is the most important piece of hardware, its reliability and performance must be much higher than those of a consumer-grade personal computer. Server products are readily available from vendors, like the PACS servers discussed in Chapter 2. Client computers are usually consumer-grade workstations or devices. There are two main types, desktop workstations and mobile computers and devices. Desktop workstations are usually connected to the server through a wired network and placed at key locations throughout the healthcare institution for easy access during patients' treatments. Mobile computers can be placed on carts and moved around the healthcare institution, with either wired or wireless network connections to the server (see Figure 11.4). Placement on carts also improves the physical security of mobile computers. Other portable wireless devices can also be used to access EHR systems if they are specified and supported by the vendor, either for Web-based or client/server-based systems. Usually a combination of fixed and mobile computers and devices is used to meet an institution's needs.

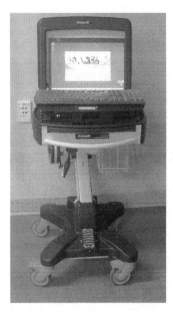

Figure 11.4   An EHR laptop computer placed on a mobile cart for mobility and security.

11.5.2.3 Vendor Evaluation

Like EHR product features, the vendor also needs to be evaluated according to its size, composition, and history; product history; number of years in business; number of products offered; number of installation sites; reputation for training and service support after the sale; stability; and willingness to work with the healthcare institution in all aspects of installation, integration, and maintenance. How well the vendor responds to needs of the healthcare institution prior to EHR system implementation is a good indicator of the quality of service support it will provide after the system is installed. The IT industry in general is evolving quickly, and the EHR industry is no exception. Today's vendor could be gone tomorrow, leaving the EHR system without service support.

11.5.3 Request for Proposal and Request for Information

Healthcare institutions and enterprises typically issue a RFP or a request for information (RFI) from vendors as part of the product and vendor selection process. An RFI is a simplified version of an RFP and is typically used if the buyer lacks the resources to prepare detailed specifications. The quote issued by the vendor in response should provide separate costs for the EHR software, hardware (if provided by the vendor), installation, software application training, and service; the terms of the service should also be specified. In the RFP or RFI, the buyer must clearly communicate with the vendor about the details of the healthcare institution or enterprise, the total number of licensed users of the EHR system, interfacing of the EHR system with other information systems currently in use, and customization of the software for the specific needs of the healthcare institution. Since a low-priced EHR product can become expensive when useful features, necessary training, and services are added, a product should be selected based on how its features will be used by the healthcare institution and its overall value, not just its price.

11.5.4 Site Visit and Contract Negotiation

When EHR products and vendors are narrowed down to a few, the implementation team should visit each vendor's demonstration site. Including a mixed group of key staff members along with the implementation team in the site visit can help ensure a clear understanding of all aspects of the EHR system. During the site visit, the group should have a list of questions ready and pay attention to the entire workflow, especially when it pertains to their individual areas of expertise, and compare it with that of their healthcare institution. Although general features of the EHR system are usually well understood and will have been demonstrated by the vendor's sales representatives, it is still necessary to confirm which features and functionalities are available and which are still in development. However, features alone do not mean much without proper logical orders to fit the workflow of the healthcare institution. Ultimately, workflow improvement, without disruption, is the most significant advantage of adopting an EHR system. If it is possible, talking to various users of the system without the vendor sales representative around can elicit very useful information about the product and vendor support. Asking questions such as what the staff like about the system, what features are unsatisfactory, which features are needed but missing from the product, how service requests are handled, how the product is supported by the vendor, and whether the same EHR system would be selected if the decision were being made again, and eliciting answers without the vendor's representative present will usually reveal the real situation about the product and service support from the vendor.

Gaining an understanding of what will occur after contract signing is also important. The easily accessible, friendly sales representative may not be reachable anymore; therefore, all commitments made by the vendor should be well documented on the contract. Before signing the contract, it should be thoroughly reviewed in detail from the legal, technical, and medical perspectives. The contract negotiation should include the total cost of ownership instead of just the initial hardware and software costs. It is not uncommon for the combined implementation fees, training fees, and annual service and maintenance costs to exceed those of the hardware and software. EHR system customization to fit the needs of the healthcare institution and upgrades are also best discussed during contract negotiation. If the vendor is able to provide customization for its EHR system, it should be documented in the contract. Software upgrades are usually released every 6–12 months, some have significant feature improvements and problem fixes, while others do not. It is necessary to confirm whether these upgrades can be skipped or whether they are mandatory in order for the EHR system to remain supported by the vendor. The healthcare institution can also request that the system be maintained at the current software release level without additional cost for a negotiable period of time following the initial installation. Other key terms of the contract include service support escalation procedures and penalties for not fulfilling the contract terms.

11.5.5 System Installation and System Go-Live

Since most EHR systems are capable of providing customized templates, forms, and workflow for the healthcare institution, it is crucial to test the customized features before installation and going live in order to confirm that the system is the right choice for the healthcare institution. The degree of customization for templates varies depending on both the EHR system and healthcare institution or enterprise. A small healthcare institution may allow clinicians to have their own templates, while in a large healthcare enterprise, a dedicated staff is required to develop and maintain templates used by the various specialties and clinicians. In the rare case that a healthcare institution is unsatisfied with the performance of the EHR system during testing, it is much easier to return the software at this time than after the entire system is installed. A vendor will also be more motivated to customize its system to the healthcare institution or enterprise's satisfaction before the installation, when the project is not yet complete.

An EHR system is usually installed during a weekend. Testing the system in a real operational environment before go-live is essential for successful implementation, because systems are rarely completely operational immediately after installation. For both testing and training purposes, most EHR systems provide a separate training environment, as shown in Figure 11.5. The testing and training environment replicates all features and applications of the real version of the EHR system. Test items should include customized features; workflow; site-specific templates with all drop-down menus; any interfaces to other medical information systems such as pharmacies, laboratories, and imaging services, under a realistic maximum load; data verification; and a comparison between the EHR system and the paper environment.

After the testing and installation are complete, the system go-live can be planned. Various user accounts should be created according to the users' job functions and roles within the healthcare institution. Only a very limited number of individuals should be authorized and have access to change systems configurations and manage user accounts. Both on-site and affiliated sites' clinical areas and staff affected by the EHR system implementation should be notified of the go-live so that they are well prepared and know whom to contact if there are problems. Vendor application and technical support, plus in-house IT support, should be arranged for the go-live date so that they are readily available. Fewer than usual patient visits can be scheduled for the days or weeks immediately following the go-live date to give the staff additional time to learn to use the new system. After the implementation, resources for system

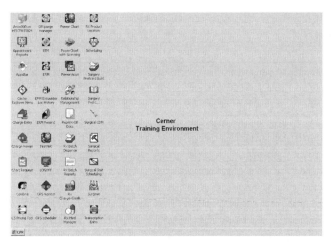

Figure 11.5   A training environment for RIS, EHR, and other clinical information systems.

maintenance should be allocated as planned, and any problems found during use must be documented and communicated to the vendor for appropriate solutions. Initially, support from the vendor should be available at any time, with on-site coverage during the first few weeks of the go-live. On an average, one support person is needed for every 5–7 EHR system users. A few weeks after the go-live, but before the last payment, users should be surveyed to determine if problems with system configurations and response times have been found. Regular staff meetings can help identify issues related to training, application, and workflow and to convey solutions to them in a timely manner. Ongoing support from the vendor should be easily accessible and responsive. Depending on the size of the institution or enterprise, the scale of internal support resources ranges from a single individual responsible for all staff training, system configuration, service support, and maintenance to a dedicated team whose responsibilities include help desk support, training, technical support, and custom software application development and management.

11.5.6 Paper Medical Record Conversion and Abstraction to EHR

In the first few weeks or months after the go-live, integrating data from other clinical information systems, including paper records, will be necessary to make the EHR system useful from the beginning. Converting prior paper medical records to the new EHR system is not a trivial task. The conversion strategy depends largely on the total numbers of patients, currently active patients, and medical records.

There are two approaches for the conversion of prior paper medical records. In the first approach, the paper records are either scanned or manually abstracted into the new EHR system for patients who are going to be seen in the near future. The most important information abstracted to EHRs includes allergy lists, immunization records, medication records, clinical health problem lists, recent test results, and family histories. These records are marked appropriately so that they can be easily recognized as having been converted into the new EHR system. New data that are usually entered into the EHR system after the go-live include current medication lists, referrals, and encounter notes. After the system go-live, the day before a patient's appointment, the patient's EHR is checked to make sure it is already in the system, or updated if the records were converted weeks or months before the EHR system's go-live date. During a patient's visit, doctors can double-check the records' consistency. Over time, all active patients' paper medical records will be converted using this approach. Because of the labor and costs associated with paper medical record conversion and abstracting, however, many healthcare institutions or enterprises choose the second approach.

In the second approach, clinicians still use paper medical records during patient visits after the EHR system goes live. The process is a gradual transition from paper to the EHR system. This approach also benefits training and learning. During or after a patient's first visit using the EHR, key data selected from the prior paper medical record are entered into the EHR. After a period of time, when frequently used patient data are already in the new system and all staff members are familiar with it, the complete switch to the new EHR system is relatively easy and smooth. It is also necessary to have a policy and procedure in place to manage paper medical records received from external sources after the go-live. These paper medical records can be scanned or abstracted into the EHR system. In addition to charts, EHR modules such as CPOE, electronic medication administration records (eMAR), electronic prescribing, electronic laboratory orders, and electronic charge posting can be adopted and integrated with the EHR system.

11.5.7 Physician and Staff Training

Training is a very important component of successful EHR implementation. The healthcare institution or enterprise should pay special attention to training during contract negotiation because the vendor may or may not include enough training in its standard product offer. Staff members' technical proficiencies in computer use can determine the amount of training needed from the vendor, and sufficient training can maximize the benefits of the EHR system.

The training schedule should be planned well before the actual training starts. During training, routine staff meetings should be held to discuss any issues that arise. Conducting staff training in stages can provide staff members with enough time to thoroughly learn different features and functions of the new system. If needed, training may be divided into separate sessions for general computer use and EHR system applications. The physicians' training schedule is different from that of the staff. Since most physicians are busy, training should be at their convenience, typically early in the morning or at the end of the day.

Combining training provided by the vendor with that provided by expert staff users frequently generates the best training outcomes. An expert staff user is an individual with a good understanding of computers who is trained extensively by the vendor in the EHR system applications. Because constant communication between the healthcare institution and the vendor is the key to success during the entire implementation process, often the expert staff user is also the main contact person to communicate with the vendor when there are application and service issues. This expert also provides assistance to other staff members during and beyond the EHR system's installation because there will always be new clinicians and staff members who will need to be trained. Since EHR system implementation is a continuous process, continuous staff training is also necessary to help the staff adjust to software upgrades.

11.5.8 Enterprise Integration

Information in the previous sections about EHR systems implementation applies to healthcare institutions of all sizes. For large healthcare institutions or enterprises, in addition to the previously discussed considerations, there is the challenge of enterprise integration. Large healthcare institutions and enterprises usually have many healthcare information systems already acquired from different vendors at various times, and many of these systems do not communicate and exchange information with other systems. Enterprise integration is the process of linking healthcare information systems to achieve interoperability, so that they can interchange healthcare data for improved patient care and operational efficiency. Enterprise integration can be achieved either by acquiring a new, completely integrated system from a single vendor, which most healthcare institutions are not able to do because of the expense, or by building interfaces among the existing information systems. An integrated application suite usually has its own included applications that are interoperable because they share a common database, a search engine, and protocols. Using interfaces, software applications from different healthcare information systems can be functionally linked. Interfaces may have various levels of automation. They initiate data exchange among disparate systems. Generally speaking, if there is no communication between information system applications or application suites, interfaces are required. In a healthcare enterprise, critical issues concerning the integration of disparate systems include:

- *Medical record identification*: to recognize and link various medical records for the same patient located on different systems, even if the patient has different identifiers in each.

- *Data reliability*: to ensure consistent data quality after they are collected, stored, processed, and presented by various healthcare information systems within the institution or enterprise.

- *Data synchronization*: to ensure data stored on different healthcare information systems are consistent and synchronized.

- *Business operation rules management*: to ensure business operational and financial information is interchanged among departments and institutions within the healthcare enterprise.

- *Clinician collaboration*: to enable clinicians and staff to effectively collaborate for improved patient care through various information systems accessible from anywhere at any time.

- *Security and privacy*: to ensure healthcare information security and privacy when multiple healthcare information systems are integrated for data exchange.

In order to achieve enterprise-wide integration of healthcare information systems, it is necessary to have the following three essential components:

- *Master person/patient index*: The master person/patient index (MPI) assigns a unique identifier to each individual patient, even though the patient may have different names or medical identifiers at the various healthcare institutions belonging to the same healthcare enterprise.

- *Context management with single sign-on*: This allows a single sign-on to access patient information and data located in multiple information systems. HL7 CCOW discussed in Chapter 6 was developed for this purpose.

- *Data warehouse*: The data warehouse serves as a central data repository, from which patient healthcare information can be accessed across the healthcare enterprise.

The integration of healthcare information systems throughout the enterprise can be achieved with the combined effort of healthcare IT departments and vendors. Enterprise integration also requires the use of common standards and protocols; it is a constant process because new information systems, applications, and application upgrades are continually added to the existing system.

11.6 HEALTH INFORMATION EXCHANGE AT REGIONAL AND NATIONAL LEVELS

In addition to internally exchanged patient healthcare data, there is an increasing need to share these data and information externally as healthcare institutions and enterprises in the United States move toward the adoption of EHR and PHR systems. Two important developments are going to impact healthcare data exchange at the regional and national levels: RHIO and the National Health Information Network (NHIN). Patient care can be greatly improved and efficiency increased if healthcare information is easily interchanged at the regional and national levels. The health information interchange depends on establishing RHIOs throughout the nation.

11.6.1 Regional Health Information Organization and Health Information Exchange

A Regional Health Information Organization (RIHO) is a group of regional healthcare stakeholders coming together to exchange electronic healthcare data about patients for the purpose of improving the quality, efficiency, and safety of patient care while reducing costs. The stakeholders of a RHIO may include healthcare institutions and enterprises, private medical practices, health insurance carriers, billing organizations, and any other individuals or organizations involved in healthcare. A RHIO's region may be established on the basis of its geographic location or community (e.g., state, multiple states, a county, multiple counties, or a city), or on the basis of nongeographic factors (e.g., medical specialties). Stakeholders in an RHIO collaborate through a secure electronic exchange called a health information exchange (HIE), which defines the initiatives of data exchange between two or more healthcare enterprises or stakeholders using common technologies and standards. This healthcare information includes clinical, administrative, and financial data. Since there is an overlap between the concepts of RHIO and HIE, sometimes the terms are used interchangeably.

11.6.1.1 RHIO Centralized Model

There are two RHIO models. The centralized RHIO model refers to a centralized data repository, from which clinicians within the RHIO can perform patient healthcare information transactions. Patient information is aggregated into a shared and centralized repository that contains health information from various stakeholders of the RHIO. When a data transaction request is received from a stakeholder, the central data repository authenticates the requesting stakeholder, responds to the request, and records the transaction for audit and reporting. This model requires interoperability among disparate clinical information systems used by various healthcare institutions and enterprises within the RHIO, so that commonly agreed standards are used for consistent inter-pretation of data. The central data repository may contain clinical healthcare, administrative, and insurance information. Challenges to this model include patient identification, sensitive clinical information protection, and concerns about the ownership of the data located in a common data repository.

11.6.1.2 RHIO Federated Model

The RHIO federated model coordinates partial and controlled exchange of electronic healthcare information among autonomous databases within a RHIO. The federated model consists of distributed databases for all data that are physically stored behind the firewalls of contributing healthcare institutions and enterprises. These data may be converted to a standard format to facilitate searching with cross enterprise document registry for data access and retrieval, as shown in Figure 11.6. The advantages of this model are that it allows health information to be shared without the need for a central data repository, and it does not require a unified national patient identifier. Additionally, contributing healthcare institutions and enterprises still retain ownership and control of the data. This model is especially useful for medical image sharing across enter-prise boundaries, because radiology image data sets from PACS can be very large; a centralized image repository may not be technically or financially feasible. The most significant limitation for this model is the lack of standardized data models and code systems, thus limiting medical reporting and clinical decision support capabilities.

11.6.2 RHIO Business Sustaining Model

The potential of cost savings is one of the frequently speculated benefits of an RHIO. Cost reduction can be either in the form of direct savings, such as that realized by reducing the number of unnecessary and uncompensated costly emergency room visits by the uninsured or underinsured, or indirect, difficult-to-quantify savings, such as that resulting from efficiency improvement. Most RHIOs in the United States are initially supported by major grant funding from various sources. However, an important issue faced by any RHIO is sustaining it in terms of finances and human resources. Typical RHIO business models include providing services such as clinical messaging, health data collection and aggregation, patient demographics, medical records, and insurance data access, and the revenues may be generated by membership and

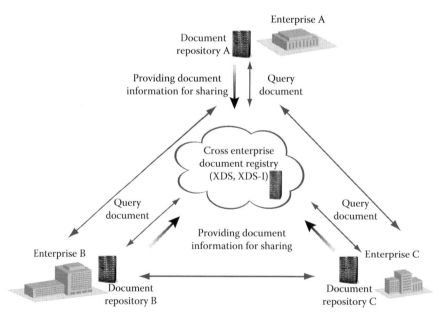

Figure 11.6 Enterprise A, B, and C belong to the same XDS affinity domain. They maintain their own document repository A, B, and C. This is a federated model for sharing images and clinical documents among healthcare enterprises. The cross enterprise document registry is responsible for storing and managing information about these documents.

subscription fees or by service-based fees. Additional potential revenue-generating models for a RHIO may be in the areas of public health data reporting and aggregation, consumer PHR services, pharmaceutical clinical trials, and healthcare insurance carriers. Regardless of grant support and business models, currently a few RHIOs in the United States are financially sustainable without government support.

11.6.3 U.S. National Health Information Network

The NHIN is a U.S. government-sponsored initiative to improve the effectiveness, efficiency, and overall quality of health and healthcare in the country. It is currently being developed to provide a secure, nationwide, interoperable health information infrastructure that will connect providers, consumers, and others involved in supporting health and healthcare. This critical part of the national health IT agenda will enable health information to follow the consumer so that it can support clinical healthcare delivery and general health improvement. The goals of the NHIN are to:

- Develop nationwide standards-based secure data exchange capability

- Improve coordination of information among healthcare providers

- Ensure that appropriate healthcare information is available anytime and anywhere

- Ensure consumers' health information security and confidentiality

- Give consumers the capability to access, control, and manage their PHR

- Reduce medical errors and support evidence-based healthcare delivery

- Reduce healthcare costs

The core capabilities of the NHIN for nationwide information exchange across a diverse range of healthcare institutions and enterprises include:

- Ability to find and retrieve healthcare information within and between HIEs and healthcare organizations

- Ability to deliver a summarized patient record to support patient care and to support the patient's health

- Ability to support consumers' preferences regarding the exchange of their information, including the ability to choose not to participate in the NHIN

- Ability to support secure information exchange

- Ability to support a common trust agreement that establishes the obligations and assurances to which all NHIN participants agree

- Ability to match patients to their data without a national patient identifier

- Ability to support harmonized standards developed by voluntary consensus standard bodies for the exchange of health information among all such entities and networks

The NHIN requires an infrastructure of interconnected, interoperable healthcare information systems from all healthcare industry sectors, so that a patient's complete EHR, assembled from various healthcare institutions and enterprises, is available anytime and anywhere the patient seeks healthcare. The foundation of the NHIN is collaborative RHIOs in both the public and private healthcare sectors, with the guiding principles and requirements specified by the federal government. Major challenges to the formation of the NHIN include lack of funding, lack of financial incentives for sustainable operations, concerns of healthcare institutions or enterprises about data and information ownership, and the general public's concerns about personal privacy.

11.6.4 The Federal Health Information Technology Strategic Plan

The U.S. Department of Health and Human Services Office of the National Coordinator (ONC) for Health Information Technology has issued the Federal Health Information Technology Strategic Plan: 2008–2012; the goal of this plan is for most Americans to have access to EHRs by 2014. This plan sets forth a number of goals, objectives, and strategies that, for the first time, properly coordinates all federal efforts in health IT. It provides guidance for the advancement of healthcare IT throughout the federal government.

The U.S. government, together with state and local governments and the private sector, has established the critical processes required to move the nation toward an interoperable health IT architecture. This architecture will be supported by federal efforts to guide electronic health information exchange in the following four areas:

- *Privacy and security*: Ensure public confidence in the privacy and security of the HIE.

- *Interoperability*: Ensure a reliable and secure HIE by using consistent, specific data and technical standards.

- *Adoption*: Ensure development of standards and policies that will enable widespread adoption and ongoing use of health IT.

- *Collaborative governance*: Ensure that collaborative governance occurs across the public and private sectors.

Although many essential components are already in place to realize the goals of the strategic plan, there is still a long way to go.

11.6.5 Global EHR Implementation Status

Although healthcare costs constitute a large portion of government budgets for industrialized nations worldwide, healthcare IT's share of the total spending lags behind the percentage spent on IT in other business sectors, and as a result, the technological capabilities of health-care IT lag behind those of other business sectors. The HIMSS conducted an international survey on the status of EHR system implementation in 2008. The survey found that EHR adoption is at different stages depending on funding sources, governance models, standards utilization, barriers to adoption, healthcare delivery system, healthcare IT strategies, overall EHR adoption approach, technical standards, administrative policies, and regulatory and legal issues.

11.6.5.1 *Governance and Funding*

Governance and funding support for a national healthcare IT initiative usually comes from the government, the private healthcare sector, or both. In most countries, government initiatives and funding are the driving forces for the adoption of a national interoperable healthcare IT system. In countries with a single healthcare payer, the EHR adoption is based on a national approach. In other countries, including the United States, the support comes from both the government and the private healthcare industry. In this case, although the approach begins with the building blocks of local interoperable healthcare IT networks such as RHIOs, incentives and national leadership for strategic planning from the government remain crucial for the successful implementation of a national healthcare IT network (such as the NHIN).

11.6.5.2 *Challenges to Global Healthcare IT Implementation*

Challenges preventing the adoption of EHRs in various regions of the world typically occur in the following areas:

- Cost is still the universal major issue.

- Lack of healthcare IT standards leads to healthcare IT interoperability issues at the local and national levels.

- Lack of communication within and between countries causes noninteroperable healthcare IT at the national and regional levels.

- Lack of support from clinical users, for fear of technology and change, continues to hamper the adoption of EHRs. This issue will be resolved when more education, improved functionality, user-friendly operations, standards for interoperability, and government support are available.

Although these challenges remain, there is no doubt that healthcare information systems for healthcare information acquisition, storage, and exchange can improve patient care and operational efficiency. Because of the tremendous benefits of EHR systems, their use will be widespread in the future.

BIBLIOGRAPHY

Access. 2008. EMR white paper, www.accesseforms.com. Accessed January 15, 2010.

Anderson, H. 2007. Realizing "a network of networks": Experiences from regional involvement in NHIN phase 1 and preparing for NHIN Phase 2. *IHE Educational Workshop.* www.ihe.net/Participation/upload/3b_ihe_wkshp07_nchica_anderson.pdf. Accessed January 15, 2010. Oak Brook, IL.

Arnold, S., W. W. Wieners, and M. J. Ball, et al. 2008. Electronic health records: A global perspective. HIMSS Enterprise Systems Steering Committee and the Global Enterprise Task Force.

Boone, K. W. 2006. Patient controlled health records, standards and technical track. www.pchri.org/2006/presentations/pchri2006_boone.pdf. Accessed January 15, 2010.

CAP SNOMED Terminology Solutions. 2008. SNOMED clinical terms and SNOMED terminology solutions. www.capsts.org. Accessed January 15, 2010.

Certification Commission for Healthcare Information Technology. 2008. *Physician's Guide to Certification for 08 EHRs.* 2008. www.cchit.org. Accessed January 15, 2010.

Cohen, M. R. 2007. Tips for preparing to acquire a CIS: Presentation to HIMSS CIS task force. www.mrccg.com. Accessed January 15, 2010.

Dougherty, M. 2008. HL7 EHR-S records management & evidentiary support functional profile. www.himss.org/content/files/HL7EHR-SFunctional.pdf. Accessed January 15, 2010.

Eilenfield, V. and J. Carraher. 2007. Lessons learned in implementing a global electronic health record. www.himss.org/content/files/lessonslearned_imp_ehr.pdf. Accessed January 15, 2010.

EMR Experts. 2009. How to purchase an EMR system. www.emrexperts.com/emr-system/index.php. Accessed January 15, 2010.

Fitzmaurice, J. M. 2006. Standards and interoperability: the DNA of the EHR. www.himss.org/contenttitles/094.pdf. Accessed June 5, 2010.

Fischetti, L., D. Mon, and J. Ritter, et al. 2007. HL7 EHR System Functional Model 2007. www.hl7.org/documentcenter/public/standards/EHR_Functional_Model/R1/EHR_Functional_Model_R1_final.zip. Accessed January 15, 2010.

Fornes, D. Should CCHIT influence your EHR selection? www.softwareadvice.com/articles/medical/should-cchit-influence-your-ehr-selection/. Accessed January 15, 2010.

Garets, D. and M. Davis. 2006. Electronic medical records vs. electronic health records: Yes, there is a difference. HIMSS Analytics White Paper.

Gauvin, A. 2008. Enabling interpretation and consultation with a diagnostic imaging repository (DIR). *The Society for Imaging Informatics in Medicine 2008 Annual Meeting.* Seattle, WA.

Gibbons, P., N. Arzt, and S. Burke-Beebe, et al. 2007. Coming to terms: Scoping interoperability for health care. Health Level Seven EHR Technical Committee- Interoperability Work Group.

HIMSS. 2009. Health information exchanges: Similarities and differences. HIMSS HIE Common practices survey results white paper.

HIMSS Analytics. 2010. EMR adoption model. www.himssanalytics.org/hc_providers/emr_adoption.asp. Accessed January 15, 2010.

HIMSS Enterprise Information Systems Steering Committee and the HIMSS Enterprise Integration Task Force. 2007. Enterprise integration: Defining the landscape.

HIMSS Enterprise Integration Task Force. 2008. The importance of enterprise integration.

HIMSS Healthcare Information Exchange National/International Technology Guide White Paper. 2009.

Igras, E. Achieving diagnostic imaging data sharing. *IHE Workshop 2007*. www.ihe.net/Participation/upload/3d_ihe_wkshp07_chinfoway_igras.pdf. Accessed January 15, 2010.

IHE. 2007. Cross-enterprise user authentication (XUA). IHE IT infrastructure technical framework white paper.

IHE. 2007. *Integration Profiles. IHE Technical Framework Vol. I*.

IHE. 2007. *National Extensions. IHE Technical Framework Vol. IV*.

IHE. 2007. *Transactions. IHE Technical Framework Vol. II*.

IHE. 2008. Cross-community information exchange including federation of XDS affinity domains. IHE IT infrastructure (ITI) technical framework white paper. 2008.

IHE. 2008. Cross-enterprise document sharing (XDS) patient identity merge. IHE IT infrastructure technical framework supplement. 2007–2008.

IHE. 2008. *IHE Patient Care Coordination (PCC) Technical Framework* Volume I.

IHE. 2008. Subscribe to patient data (SPD). IHE patient care device technical framework supplement. 2007–2008.

IHE. 2008. Template for XDS affinity domain development planning. IHE IT infrastructure technical committee white paper.

IHE. 2009. A service-oriented architecture (SOA) view of IHE Profiles. IHE IT infrastructure white paper.

IHE. 2009. Care management (CM). IHE patient care coordination (PCC) technical framework supplement. 2008–2009.

IHE. 2009. Patient identifier cross-reference HL7 V3 (PIXV3) and patient demographic query HL7 V3 (PDQV3). IHE IT infrastructure technical framework supplement. 2008–2009.

IHE. 2009. Performance measurement data element structured for EHR extraction. IHE quality, research & public health technical committee white paper. 2008–2009.

IHE. 2009. Publish/subscribe infrastructure for XDS.b. IHE IT infrastructure (ITI) technical committee white paper. 2008–2009.

IHE. 2009. Query for existing data (QED). IHE patient care coordination (PCC) technical framework supplement. 2008–2009.

James, L., D. T. Mon, and P. Van Dyke, et al. 2008. Payer-linked personal health record system (PHR-S) functional profile: Draft. HL7 EHR workgroup & HL7 personal health record work group.

Koff, D. A. 2008a. Transforming healthcare delivery through the implementation of electronic health records in Canada. *The Society for Imaging Informatics in Medicine 2008 Annual Meeting*. Seattle, WA.

Koff, D. A. 2008b. The Canadian electronic health records strategy for better care delivery. *DICOM International Conference & Seminar*. Chengdu, China.

Majurski, B. 2007. Cross-enterprise document sharing (XDS). *IHE Workshop*. www.ihe.net/Participation/upload/iti6_ihewkshp07_xds_majurski.pdf. Accessed January 15, 2010.

McDonald, C., S. Huff, and K. Mercer, et al. 2010. *Logical Observation Identifiers Names and Codes (LOINC) Users' Guide*. www.loinc.org. Accessed January 15, 2010.

Mendelson, D. S. 2008. Achieving interoperability and adoption of the electronic health record. www.ihe.net/Connectathon/upload/mendelson_ihe_na_connectathon_conference_2008_01_29.pdf. Accessed January 15, 2010.

Miller, J. 2005. Implementing the electronic health record: Cases studies and strategies for success. Health Information and Management System Society (HIMSS).

Mon, D. T., J. Ritter, and C. Spears, et al. 2008. *HL7 EHR Work Group, Personal Health Record System Functional Model, Release 1, Draft Standard for Trial Use.*

Noumeir, R. 2008. Delivering images from the EHR to the interpretation process. *The Society for Imaging Informatics in Medicine 2008 Annual Meeting*.

Nusbaum, M. 2007. An overview of IHE IT infrastructure profiles and their relationship with other domain. *IHE Educational Workshop 2007*. www.ihe.net/Participation/upload/iti1_ihewkshp07_overview_nusbaum.pdf. Accessed January 15, 2010.

Nuance Communications. 2008. Speech recognition: Accelerating the adoption of electronic medical records. www.nuance.com. Accessed January 15, 2010.

Peytchev, V. 2007. PIX/PDQ—Today and tomorrow. *IHE Workshop 2007*. www.ihe.net/Participation/upload/iti14_ihewkshp07_pix_pdq_peytchev.pdf. Accessed January 15, 2010.

Public Health Data Standards Consortium. 2007. Building a roadmap for health information systems interoperability for public health (public health use of electronic health record data) white paper. www.ihe.net/Technical_Framework/upload/IHE_PHDSC_Public_Health_White_Paper_2007_10_11.pdf. Accessed January 15, 2010.

Rosenfeld, A., S. Koss, and S. Siler. 2007. Privacy, security, and the regional health information organization. www.allhealth.org/briefingmaterials/chcf-rhioprivacy-1091.pdf. Accessed June 5, 2010.

Schanner, A. 2008. National electronic health record in Austria based on IHE: Challenges & opportunities for DICOM in telemedicine. *DICOM International Conference & Seminar*.

Seifert, P. 2009. Cross-enterprise document sharing for imaging (XDS-I.b) integration profile. IHE radiology technical framework supplement.

Seifert, P. 2009. Cross-enterprise document sharing for imaging (XDS-I.b) integration profile, trial implementation. IHE radiology technical framework supplement.

Thielst, C. B. and L. E. Jones. 2007. Guide to establishing a regional health information organization. Healthcare Information and Management Systems Society (HIMSS).

Trabin, T., L. Geis, and A. Khosravi. 2008. Health authority-based personal health record system (PHR-S) functional profile: Draft. Health Authority-Based PHR-S Profile Working Group.

Weisfeiler, L. 2008. Combining JPIP with WADO within and XDS-I framework for efficient, standard-compliant streaming of EHR imagery. *DICOM International Conference & Seminar*.

medical.nema.org/dicom/conf-2008/Day_2/D2_2_1110_Weisfeiler_en.ppt. Accessed January 15, 2010.

Zaroukian, M. H. 2007. Achieving interoperability through implementation of standards. *IHE North America Connectathon and Conference.* www.ihe.net/Connectathon/upload/IHE_overview_ zaroukian_final_2007_01_16.pdf. Accessed January 15, 2010.

12 Quality Control, Security, and Downtime Management

For healthcare institutions and enterprises using PACS, EHR, and other clinical information systems, it is necessary to have quality control, disaster recovery, data backup, emergency operation, and test plans in place to meet government and regulatory requirements, and more importantly to ensure continuity of operation under emergency situations. These plans constitute a critical component of the overall enterprise healthcare information management plan. It is most likely that there are existing policies and procedures at the healthcare institution for security, privacy, and asset management. Additional policies and procedures that are unique to PACS, EHR, and other health information systems need to be developed and implemented, with reference to other existing policies and procedures.

12.1 PACS PATIENT FILE RECONCILIATION AND QUALITY CONTROL

Although less frequent than in the past, occasional information errors still occur, such as mismatched patient data file due to misspelling of patient demographics and mislocated studies. It is often necessary to correct these type of errors manually. These tasks are usually performed by the designated PACS QC/QA coordinator or the PACS administrator. For most PACS, modifying existing examination or patient information requires special administrative privileges and it is often done on a QA/QC workstation by the PACS administrator. When the PACS administrator logs on to this workstation under the special privileged QA/QC account, various QA/QC tasks can be carried out; for instance a patient's file reconciliation, discussed in Chapter 2, can be done when the "QC" function is selected.

Patient data reconciliation should be performed regularly or as needed on the QA/QC workstation. When unmatched studies appear, such as images located in the wrong study, or studies belonging to the wrong patient, they need to be corrected. If the imaging series of a patient is accidentally placed in a different patient's folder, these imaging series should be moved to the correct folder. Sometimes, there is a need to merge various studies into one study, or to relocate studies, or delete an imaging series.

In case there is any error in examination attributes or patient demographic information, they can also be corrected, as shown in Figure 2.10. Besides, corrections can be made during audits of the PACS and the long-term usage statistics in addition to using QA/QC tools. With these QA/QC tools, the PACS administrator can perform functions such as auditing the time and duration of a user log in and log out, the examinations that a specific user accessed, and the particular workstation at which these events occurred.

These are very basic PACS QA/QC procedures. However, these functions are also very restrictive, and should only be available to the PACS administrator or the QA/QC technologist to ensure the integrity of patient data. If there are any database mismatches that cannot be corrected, service support from the vendor may be necessary.

Another important aspect of PACS quality control is the QC of image display monitors (i.e., computer monitors used for primary diagnosis). The main objective of display monitor QC is to ensure optimal setting, to provide accurate image display for clinicians, to identify problems before they become clinically significant, and to make sure that the images displayed have the highest achievable quality. The performance standards of PACS image display monitors and associated quality control procedures have been discussed in detail in Chapter 3.

12.2 DOWNTIME PLANNING

Every healthcare institution or enterprise may suffer system downtime. Any interruption, interference, or system downtime resulting in inaccessibility of PACS or EHR applications and data may cause additional risk, complexity, cost, and delay. There is no question about whether there will be system failures, only a question of when. Although some unplanned downtimes are caused by natural events or disasters, most of them are caused by human errors, hardware or software failures, and security violations that are unpredictable and difficult to prevent. These downtimes cause much stress to staff and users because they impact availability of mission-critical clinical applications for patient care. Nevertheless, through proper planning, monitoring with problem detection, response with communication and issue tracking, and documentation of solutions for future reference, both scheduled and unscheduled downtime can be effectively managed to minimize their impact on the clinical operation of the institution or enterprise. With the right technical solutions and appropriate policies and procedures, a healthcare institution can be protected to mitigate the negative effects of emergencies and disasters.

12.2.1 Fault Tolerance

Fault tolerance is the ability of systems to continue operations when an unexpected component failure occurs. The quality of operation may be reduced depending on the severity of the failure. Natural disasters such as earthquakes, hurricanes, fires, power outage, and other catastrophes can bring down crucial servers, routers, switches, other devices, and networks. With the increasing use of electronic clinical information systems such as PACS and EHR, the disruption of networks and servers causes more problems than in a paper-based clinical environment. This requires that the network and servers not only have high speed, but also have high reliability, because they are critical to healthcare institutions and enterprises. The fault tolerance of PACS servers with redundancy and failover has been discussed in Chapter 2, and RAID in Chapter 5. Since redundancy can significantly reduce the risk of failures adversely impacting clinical operations, both internal and external communication carriers' networks that link various institutions of the enterprise should be carefully planned and designed to maximize redundancy. In theory, every server and network device, including the switches in a communication closet and the network cables, can be duplicated to achieve redundancy, and ideally the entire network infrastructure can be completely built with redundancy without any single point of failure. Unfortunately, complete redundancy is not financially feasible or practical in many cases, because it requires two or more sets of every component. However, through careful planning and design, network fault tolerance can be achieved with nonredundant components thus bypassing failed components, while keeping the rest of the network minimally affected, reducing the risk of complete network failure.

When network services are provided by carriers to link multiple institutions of the enterprise, and if there are more than one network carriers in the region, the redundant network should be provided by a different carrier. Communication cables should be physically placed in different trenches and connected to separate central facilities, so that the redundant cables are not subject to the same accidents, such as damage to the cable trench or a systems failure of the same communication carrier's central facility. Sometimes different communication carriers may use the same physical communication cable, cable trench, or central facility, because the carrier marketing and billing for the communication service may not be the one that actually provides the service. Institutions or enterprises located in rural areas or regions with only one communication carrier may find this very difficult, if not impossible, to implement.

For servers and workstations, uninterruptible power supply (UPS) units should be utilized with remote alerting mechanisms for servers and critical systems, so they can keep systems up and running when power outage occurs. In addition, many healthcare institutions and enterprises are equipped with emergency backup power generators. They can be utilized to ensure continuous power supply to servers and critical network devices and workstations, because the UPS will eventually fail if the power outage lasts for an extended period of time. For most institutions, the likelihood of an occasional power outage caused by meteorological activities and network outage by human mistakes such as digging, cutting, or unplugging communication cables is higher than a devastating disaster. If it is possible, an institution should avoid placing all vital systems and network devices in a single location.

12.2.2 Unscheduled Downtime

Almost all institutions and enterprises have unscheduled PACS and EHR systems downtime. Unscheduled downtime caused by hardware failures, natural disasters and human errors has the most negative impact on the clinical operation of an institution or enterprise. During an unscheduled downtime, it is vital that the healthcare institution reacts quickly according to a plan, identifies whether the problem originates from the servers, workstations, network devices, server overload, or network overload, and controls the downtime and its negative impact to the minimum. Unscheduled downtime should be planned for by:

- Server performance proactive monitoring and active notification tools for network connectivity, hardware errors, hard disk and CPU usage, hard disk failures, and other aspects of PACS or EHR.

- Technical monitoring for quick warning and notification in the event of a system problem, and avoiding unnecessary or missed warnings.

- Redundant architecture, such as redundant networks for critical systems or components.

- Vendor collaboration in terms of server function and status, monitoring, and management capabilities.

- Downtime communication to key stakeholders such as key personel in PACS administration, IT, PACS/EHR vendors and communication service providers.

- Alternative communication route if the primary is not available.

- Documented downtime policies and procedures easily accessible to all staff affected by such outages.

- Policies and procedures of escalation to involve appropriate resources for solutions.

- Vendor service contracts that include details of downtime responsibilities, service at occurrence of system failure, guaranteed service terms, and penalties for excessive downtime.

- Practicing and testing downtime procedures periodically.

- Documenting downtime events; identifying issues, their causes, business impact, and solutions; taking action to prevent and improve response to similar problems the next time; and facilitating knowledge transfer when staff changes.

Unscheduled downtime planning not only copes with system recovery, but also deals with business operation during the outage because it takes time to completely restore the proper functioning of the system. Therefore, it is crucial to understand the business impact on the clinical environment caused by such downtime, to prepare downtime policies and procedures for different levels of PACS or EHR failures (i.e., system is partially available, or system is down but the IT infrastructure is still functional, or system and the IT infrastructure are both down), to collaborate and communicate with vendors to reduce the likelihood of system failure, and to minimize the impact caused by such downtime. With a well-designed unscheduled downtime plan, mission-critical operations and services of healthcare institutions and enterprises can still maintain essential functions, resume full functionality as soon as possible, minimize interruptions to the business operation, thus ensuring continuity of patient care.

12.2.3 Scheduled Downtime

Scheduled downtime for systems upgrades and routine backups is much more manageable than unplanned downtime. Scheduled downtime of clinical information systems should be planned during off-peak clinical business hours such as evenings and weekends, and appropriate operational procedures should be communicated to all clinical areas and staffs before the outage:

- Tasks scheduled prior to the downtime should be completed before the onset of the scheduled system downtime.

- Complete as many tasks as possible by the outage time.

- All activities throughout downtime need to be documented on paper.

- All paper-documented activities need to be entered when the system comes back up.

- Activities entered to the system should be checked for accuracy.

It is possible that scheduled downtime for systems upgrade can sometimes turn into unscheduled downtime, because some features of the system, or the entire system can unexpectedly fail to function after the upgrade because of unknown hardware or software problems. Therefore, it is necessary to have a backup plan to restore the system to the status before the upgrade in case this happens.

Both plans for scheduled and unscheduled downtime should be documented on paper and easily accessible by all staff and users, so that they are still available even in the worst case that all network, computers, and workstations are down.

12.3 HEALTH INSURANCE PORTABILITY AND ACCOUNTABILITY ACT

HIPAA enacted in 1996 by the US Congress is well known to the healthcare industry. The purpose of HIPAA is to provide insurance portability and administrative simplification for the healthcare industry, so that personal healthcare information can be kept private, data and transactions can be standardized, and processes can be streamlined to reduce the overall healthcare cost.

HIPAA Title II, also known as the Administrative Simplification provision, is about establishing national standards for electronic healthcare transactions and national identifiers for all federal and state healthcare providers, private health insurance plans, and healthcare clearing houses defined as "covered entities" by HIPAA and the U.S. Department of Health and Human Services (HHS).

These covered entities involved in healthcare information exchange are governed by five electronic data interchange Administration Simplification rules: the Privacy Rule, the Transactions and Cod Sets Rule, the Security Rule, the Unique Identifiers Rule, and the Enforcements Rule. HIPAA mandates that all covered entities must ensure the integrity, confidentiality, and availability of protected patient electronic information they collect, use, maintain, or transmit. Since security and privacy of health data are addressed in this provision for improving effective and efficient electronic data exchange in the U.S. healthcare system, it is most relevant to PACS and EHR implementations. It defines standards of safeguards for administration processes, physical and technical infrastructure, rights of individuals, and obligations for healthcare organizations in terms of electronic protected health information (EPHI). The principles of HIPAA have been adopted by many other nations into their healthcare industry regulations as well. In this chapter, only the introductions of the most important security-and privacy-related issues of HIPAA are discussed. To be compliant with HIPAA, readers should seek legal advice and assistance from professionals who are specialized in HIPAA rules.

12.3.1 Privacy Rule

The HIPAA Privacy Rule establishes a national standard for the collection, maintenance, access, use, and disclosure of individually identifiable health information, and sets limitations on the use and release of health records. Any information that is collected and stored relating to an individual's health and that identifies the patient is confidential, and is considered as protected health information (PHI). PHI is about individuals' health status, medical history, provision of healthcare, payment information for healthcare, medical record, billing information, and any personal information provided by the patient including name, address, date of birth, phone number, insurance number, social security number, and can be in the form of paper, verbal, white boards, photos, medical container labels, or electronic media. The Privacy Rule requires all personally identifiable information in any form to be adequately safeguarded when it is collected and used in patient care. The Privacy Rule also gives individuals the right to request correction of inaccurate PHI.

Keeping PHI confidential is about giving good patient care while ensuring patients' trust in the care providers to use their information properly, accessing it only as needed, sharing it only with those who have the need to know it in the process of delivering care. When healthcare institutions keep PHI private, they gain patient trust, and in turn, it leads to a patient's willingness to share important clinical information for better care. Since the threat of identity theft has become a real concern for patients, following the Privacy Rule has become more important than ever before. Sensitive PHI may only be disclosed to facilitate treatment, payment, healthcare operation, or if there is authorization from the individuals. In such circumstances, it is required that only the necessary information is disclosed, and individuals are notified of the use of their PHI. Further, covered entities need to keep track of all PHI disclosures, document privacy policies and procedures, take reasonable procedures to ensure the confidentiality of communication with individuals, appoint privacy officials and contact persons to be responsible for receiving complaints, and provide training for all members of the institution or enterprise in privacy procedures concerning PHI.

For the staff at a healthcare institution, it is never appropriate to access patients' PHI for reasons other than to perform their job. For those requiring PHI access to fulfill their role, only the minimum information needed should be accessed and used. For some caregivers or other workforce members, PHI should never be used. The Privacy Rule also gives individuals the right to file a complaint with the HHS if they believe that the Privacy Rule is not being upheld. Those who are convicted of the Privacy Rule violation are held responsible through civil and criminal penalties under HIPAA.

12.3.2 Security Rule

The HIPAA Security Rule is complementary to the Privacy Rule in that the Privacy Rule deals with all PHI in paper, electronic, and other forms, while the Security Rule pertains to electronic PHI (EPHI) only, which applies to all PACS, EHR systems and records. The relation between privacy and security is that to achieve the goal and fulfill the requirements of the Privacy Rule, many Security Rule safeguard requirements must be implemented, and institutions or enterprises must effectively integrate security and privacy management strategies.

EPHI is any PHI that is created, maintained, transmitted, or received electronically through the Internet, networks between healthcare institutions, CD, DVD, magnetic tapes, flash storage media, or any other digital storage media. Generally, it does not apply to conventional paper fax systems, although it applies to computer-based fax or automated voice systems. The purpose of the HIPAA Security Rule is to keep confidentiality, integrity, and availability of EPHI. "Confidentiality" of EPHI is to ensure that data stored or transmitted is only accessible to authorized individuals or processes, and unauthorized persons or processes are not able to access or intercept EPHI. "Integrity" of EPHI means data are free from unauthorized alteration or destruction by unauthorized individuals or processes, and "availability" of EPHI is that when it is requested by authorized users, EPHI data, software applications, and systems are available for access and proper use. Both "confidentiality" and "integrity" require security measures to prevent the EPHI from being tampered with, or if such tampering occurs, it ought to be detected. To achieve these goals by covered entities, the Security Rule requires protection of EPHI against reasonably anticipated security threats and disclosure of EPHI that violates the Privacy Rule, compliance with the Security Rule by employees, and reviewing security policies and procedures periodically.

The HIPAA Security Rule recognizes that institutions and enterprises are all different. It does not impose a single set of security measures for all covered entities. On the other hand, it is designed to be technology neutral without specifying what technologies are to be used. This allows flexibility and for institutions to have the discretion to select security measures primarily based on their unique technical infrastructure, hardware/software, and security capabilities, so that the regulation can be reasonably and appropriately implemented. The Security Rule is scalable as well, allowing institutions or enterprises to make decisions depending on their size, nature, complexity, and capability of their operations, the cost of particular security measures, and the likelihood of risks and their impact to the EPHI.

In addition to privacy, protection the Security Rule also lays down security rules of EPHI that is stored, accessed, displayed, or transmitted electronically; it is comprehensive, covering administrative and technical issues. A policy for HIPAA Security Rule compliance must include risk analysis, risk management, sanction policy, information system activity review, and provide practical guidance to workforce members and the healthcare institution or enterprise regarding acceptable conduct and expectation in the following areas:

- Rules across the institution or enterprise

- Risk determination and management

- Mandatory security measures

- Expectation between workforce members and the institution or enterprise

- Clarity of meaning, information should not be subject to interpretation

- Being accessible on paper form to those impacted, available even when the entire computer system is down

- Periodic review and updates to reflect changes in clinical operations, technologies, risks, and regulations

Compliance with the HIPAA Security Rule is not just about the technologies used, nor specific processes implemented, but a combination of both. The Security Rule specifies 3 types of safeguards required for compliance: Administrative Safeguard, Physical Safeguard, and Technical Safeguard. It further clarifies various security standards. Some standard specifications are "required," which means they must be implemented by all institutions or enterprises regardless of their risk analysis results, and these standards are mandatorily to be adopted and administered according to the Security Rule; other specifications that are "addressable" must be considered on the basis of the risk analysis, and are flexible and can be implemented according to unique situations of the institution or enterprise. The nature of clinical operations in the institution or enterprise determines the risks and the security policy, and the implementation of the policy decides processes and technologies needed to achieve security for EPHI.

12.3.2.1 Administrative Safeguards
The HIPAA Security Rule is designed for implementing appropriate and cost-effective security standards by health plans, healthcare clearing houses, and healthcare institutions. The Security Rule's Administrative Safeguards are nontechnical measures imposed by healthcare institutions

or enterprises on employees concerning their conduct and procedures to comply with HIPAA. They are policies of general statement of intent, documented procedures for the implementation of required processes, and administrative actions to manage the selection, development, implementation, and maintenance of security measures to protect patients' EPHI. Healthcare institutions or enterprises are thus required to implement the Administrative Safeguards to establish reasonable and appropriate security policies and procedures by:

- Designating a security official for the development and implementation of required written policies and procedures, with the roles and responsibilities known to all staff and business associates.

- Managing policies and procedures for documented security control processes.

- Setting up procedures for access authorization and modification.

- Identifying employees who need access to EPHI to finish their job, and limiting such access only to these employees in procedures.

- Creating contingency plans, including plans for data backup, disaster recovery, and emergency mode operations to ensure the availability of EPHI during emergencies.

- Mitigating harmful effects of known security incidents, and documenting the outcomes.

- Establishing policies of scope, frequency, and procedures of routine and event-based internal audits.

- Identifying and responding to suspected or known security incidents identified during routine operation and audits.

- Undertaking ongoing training programs on EPHI handling for employees.

- Ensuring HIPAA compliance by a third-party contractor during business outsourcing.

- Evaluating technical and nontechnical compliance to the HIPAA Security Rule periodically.

The Administrative Safeguards contain the required implementation specifications of risk analysis, risk management, sanction policy, and information activity review, which are defined to prevent, detect, contain, and correct security violations. The security policies and procedures developed after risk evaluation is an essential component of the Administrative Safeguards. With the changing vulnerabilities, risks, and threats, the policies and procedures need to be updated periodically, because the protection of EPHI is mandated by government regulations, and should be coordinated with overall general security policies and procedures of the healthcare institution or enterprise.

12.3.2.2 Physical Safeguards

HIPAA Security Rule's Physical Safeguards are physical measures, policies and procedures to protect physical healthcare information systems, related facilities, and equipment against natural and environmental hazards such as earthquakes, tornadoes, hurricane, fire, and flood, as well as human unauthorized intrusion, hazards, and threats. Within the Physical Safeguards there are 4 standards, including facility access controls, workstation use, workstation security, and device and media controls. Each standard has implementation specifications, which are either mandatory as required by HIPAA or addressable, so that the following procedures can be implemented:

- Appropriate EPHI facility or data center site selection

- Access control of facility security procedures, access records, visitor entry log and escort

- Proper workstation use and placement

- Security and use of workstations that store or access EPHI

- Controlling and monitoring equipment access

- Accessing to hardware and software by authorized personnel only

- Device and media control

- Controlling hardware and software removal and disposal

- Off-site computer backups

- Third-party contractor's physical access responsibility training

- Equipping fire prevention, detection, and extinguishing devices where EPHI is located

- Appropriate humidity, flood, heating, ventilation, air conditioning, and power control and backup

The Physical Safeguards should also be implemented in areas that require strengthening through risk assessment, and incidental or unauthorized access to EPHI should be prevented, such as unnecessarily visible or public accessible computer monitors and keyboards exposing patients EPHI to nearby casual observers or to potential tampering. While the physical facilities and buildings of healthcare institutions or enterprises where health information systems containing EPHI are located should be physically safeguarded; so also should electronic information systems, which because of the extensive use of laptop computers, cell phones, or portable devices, can be easily accessed. Medical images and EHR are susceptible to access through the Web or e-mail, and the concept of a regulated physical facility may potentially extend to anywhere these mobile devices are used to access EPHI. Therefore, healthcare institutions and enterprises need to develop and implement reasonable and appropriate physical security policies and procedures for the use of such devices, and physical security of these devices should be checked, and updated if necessary.

12.3.2.3 Technical Safeguards

Technical Safeguards within the HIPAA Security Rule are the technologies, policies, and procedures that protect EPHI and control access to computer systems containing EPHI. To prevent EPHI from being intercepted by any unauthorized recipient during electronic exchange over a network, EPHI should be protected by:

- Utilizing data encryption technologies for exchange over open networks such as the Internet

- Optional data encryption use for EPHI exchange over closed network

- Ensuring data integrity with no unauthorized data alteration by using technologies such as message authentication and digital signature

- Receiving system identity authentication for data exchange

- HIPAA practice documentation of policies and procedures for government compliance check

- Documentation for network component configurations and access records

- Documentation of risk analysis and management programs for preventing nonhealthcare-related EPHI use

The purpose of the Technical Safeguards is to use commercially available, administratively and financially reasonable, and appropriate mechanisms to achieve technical safeguards, and commercially available technologies at reasonable cost should be sufficient for such purposes.

12.3.3 Other HIPAA Administration Simplification Rules

In addition to the Security Rule and Privacy Rule, there are three other HIPAA Administration Simplification rules. The Transaction and Code Sets Rule is aimed at improving electronic data interchange, so that healthcare institutions may exchange medical, patient, and billing information more efficiently. The Unique Identifier Rule requires hospitals, physicians, health insurance companies, billing organizations, and large health plans to use only the National Provider Identifier (NPI) to complete electronic transactions. The NPI has 10 alphanumeric digits, with the last digit being a checksum number, to replace all other identifiers used by health plans, Medicare, Medicaid, and other government healthcare programs. However, the state license number and the tax identification number of a provider remain the same. The Enforcement Rule is the last rule for HIPAA enforcement. It sets civil monetary penalties for HIPAA rule violation, and establishes procedures for HIPAA violation investigation and hearing.

12.3.4 Policies and Procedures Evaluation

There is no single set of policies and procedures that fits all "covered entities"; they need to be customized to meet specific needs based on risk analysis. Since the clinical environment, operations, technology, and management all change over time, healthcare institutions and enterprises should review security and privacy policies periodically to ensure that procedures are current to handle issues of vulnerability, risks, and threats to information systems that contain patients' EPHI. Issues to be evaluated include the dependence of clinical operations on clinical information systems, the dependence of ancillary and administrative services, the impact of one application failure on other applications throughout the process of patient care, the time it takes to restore data from backup, and the impact on patient care if the system is down for various lengths of time. In addition to regular evaluation of policies and procedures (such as on an annual base), their implementation in real-life should also be checked, to assess if they comply with all security and privacy requirements and if they are sufficiently effective.

The security and privacy evaluations can be carried out by either internal personnel or outside assessors. During assessments, physical facilities should be inspected, information system outputs reviewed, and processes tested to determine the effectiveness of administrative, physical, and technical security controls. Once these assessments are conducted, results can be analyzed and information gathered to determine if there are any existing vulnerabilities, risks, or threats. Through the analysis, the healthcare institution or enterprise can identify weaknesses in the policies and procedures, and develop a plan for remedial action. It should be remembered that when PACS or EHR systems are replaced the security level should be maintained and verified, not degraded because of such changes. Additional evaluations may be necessary if there are any emerging new threats and risks, new regulatory requirements, or if the clinical operation environment of the healthcare institution or enterprise changes significantly.

12.4 CONTINGENCY PLAN AND DISASTER RECOVERY

Since HIPAA defines "availability" of EPHI as data or information being accessible and usable upon request, contingency plans, policies, and procedures must be in place for responding to emergencies or other human-induced or natural disasters, such as vandalism, fire, flood, and system failures that may damage information and data. The primary purpose of a contingency plan is to ensure the confidentiality, integrity, and availability of EPHI through any reasonably predictable disasters or emergencies that can be identified by risk assessment, to respond effectively to disasters as soon as possible, and to develop response procedures for detection and reporting of EPHI-related security incidents. A contingency plan can overlap with the general downtime planning, which has broader coverage as it includes scheduled downtime planning. A comprehensive contingency plan involves data backup, disaster recovery, emergency mode operation, and testing and revision procedures. Although it is difficult to prepare for all possible emergency events that have not occurred, there are measures that can be taken to mitigate risks to the security of EPHI, and minimize system failures and interruptions to the clinical operation of healthcare institutions or enterprises. To prepare for disasters, critical clinical operations and services and their reliance upon EPHI should be identified. Then business continuity plans to keep such services operational during or after a disaster should be developed. The disaster recovery plan should be technically feasible and financially practical and reduce negative impact to the minimum. Staff should have been trained to understand each step of the contingency plan so that when a disaster occurs, they can use it to handle the emergency.

12.4.1 Disaster Recovery

Disaster recovery often relates to issues that are not produced by nature. In reality, the majority of disasters in information systems are caused by human error and machine failures, such as a system's technical failure, network interruption, utility power outage, fire, and theft. The term "disaster recovery" is sometimes used interchangeably with "business continuity," which means to keep a minimum level of business operational to meet critical requirements during system outages or disasters. Disaster recovery for healthcare institutions or enterprise is to restore or recreate information systems and clinical operations to the state before the occurrence of a disaster. It includes the process of implementing policies and procedures to restore clinical operations that are critical to the institution or enterprise, and accessing business processes and resources, such as data and operational infrastructures during and after the disaster. Disaster recovery for healthcare information systems, including PACS and EHR, is required by the HIPAA regulation.

A basic and simple form of disaster recovery for system failures and errors is redundancy, which means multiple, interchangeable, backup components can take over the function of the failed components and resume operations. Purchasing a disaster recovery solution is a necessary condition to meet the needs of an emergency situation. Cost is a key element and, depending on the design of the disaster recovery solution, can vary greatly. Institutions or enterprises might include redundant network connection and multiple data backup solutions, balancing needs and expenses. According to HIPAA, the healthcare institution or enterprise has the discretion to select reasonable and appropriate commercially available solutions. It should also be noted that disaster recovery is not just about the technology; the nontechnical policies and procedures are essential for successful disaster recovery. A disaster recovery system or solution without proper disaster recovery planning cannot function as it is expected. Even though disasters cannot be totally avoided or eliminated, a properly designed disaster recovery strategy can minimize the catastrophic impact on the clinical operation of the healthcare institution or enterprise at reasonable cost.

12.4.2 Disaster Recovery Planning

The importance of providing care to patients during a catastrophic event is well known. Healthcare institutions or enterprises should anticipate that a disaster will eventually occur, and disaster preparedness should not be treated as a future problem that can be left for the next day. Disaster recovery planning addresses catastrophic events and brings affected systems back online, although restoring all systems may take a long time. Institutions and enterprises need appropriate disaster recovery plans beyond system failure repair to continue the clinical operations of the institution. The disaster recovery plan should establish, document, and implement policies and procedures for the restoration of any data that are lost in any disasters.

Healthcare institutions may have disaster recovery plans for the entire institution or enterprise. However, they may not cover the specific needs of each individual department. Since system failure, power and network outages can lead to critical operations and information systems being interrupted, the disaster recovery plan should provide fail over solutions for different scenarios. The plan should include detailed procedures and processes for restoring clinical operations; it should identify the information systems and clinical operations affected, specify the responsibility of each staff member, and lay out the circumstances in which the plan should be invoked. Additionally, the plan should be on paper, periodically tested, revised, and easily available to staff when there is a disaster. It should determine:

- The triggering events (system failures, natural catastrophic events, human attacks, etc.) for disaster recovery

- Internal communications with staff, users, and vendor

- External communication with the media and the general public

- The process and procedures to restore data and systems

Disaster recovery or continuity of business during disasters should be based on priorities as well. The goal of business continuity is to keep the clinical operation of departments and the institution or enterprise going. The transition from daily routine to emergency operation should be seamless. For PACS and EHR systems, during an emergency the clinical operations will still need to access medical images and EHR data as other clinical information systems, such as the HIS, may experience outage. The most important and basic capability of PACS is to display images just acquired from imaging modality scanners during disasters, even though it may rely on using optical disks as the transmission media when the network is completely down. The minimum requirement is to allow manual operation of the department at reduced efficiency. The next priority is to access prior images acquired recently, followed by those acquired much earlier. System outages that last several hours or a few days are treated differently.

12.4.3 Data Backup Plan

HIPAA requires healthcare institutions or enterprises to have data backup plans that can be used to restore data that are lost, damaged, or corrupted. A data backup plan establishes and implements procedures for the creation and maintenance of duplicate copies of EPHI. When the primary EPHI is not available, this duplicate copy can be retrieved. Data backup is a critical component of any disaster recovery plan, and its policies and procedures should work seamlessly with those established for disaster recovery.

Institutions must establish policies and procedures in terms of backup and retrieval of PACS and EHR data. The implemented procedure should indicate the backup frequency—constant, twice daily, once a day, once an hour—the backup rotation schedule with reusable media, and the type of backups (full, incremental, differential, data only, or including all applications and operating systems). The frequency of backup determines the maximum allowable data loss, because the amount of data that is not backed up is subject to potential loss if there is any serious incident. The data backup plan should indicate the backup technologies, for example, tapes, optical disks, network attached storage devices, SAN devices, or third-party data storage or backup service. Additionally, one duplicate copy of data may not be sufficient, especially when the same technology (such as tapes or optical disks) are used for both the primary and the duplicate copies, because they may fail around the same time and are subject to similar deterioration or failure mechanisms. The institution should also prepare for data restoration when the original technology used becomes obsolete.

The data backup policy should designate staff who will be responsible for the data backup, testing, validating, and restoring operations. The policy should also specify the maximum time required to restore the backed up data if necessary. These data backup media containing EPHI should be protected with the same security measures as any other EPHI. They should be stored in a separate physical location not subject to the same disaster (flood, fire, earthquake, or hurricane) if it occurs, and maintained at the specific temperature and humidity as required by manufacturers to ensure that data stored are not lost, destructed, and corrupted, but are ready to be recovered.

12.4.4 Emergency Mode Operation

The emergency mode operation plan documents procedures to enable continuation of critical clinical operations and protection of EPHI when the institution or enterprise is operating in emergency mode. This is very similar to and overlaps with business continuity and disaster recovery. HIPAA requires each healthcare institution or enterprise as a covered entity to establish and implement its own emergency mode operation plan. Based on the results of a risk and threat analysis, an institution or enterprise should identify information systems and clinical operations that are mission critical, and confirm system redundancy and redundant network services that link institutions of the enterprise. The emergency mode operation plan should identify those clinical operations that must be continued during disasters, and those that need not. In addition, responsibilities should be assigned to designated key personnel, so that when an emergency occurs, there is no confusion about individuals' responsibilities.

12.4.5 Testing and Revision of Contingency Plan

HIPAA recommends implementing procedures for periodic testing and revision of contingency downtime plans. To prepare for unscheduled downtime, it is imperative that healthcare institutions not only have contingency plans in place, but also periodically test them to check their viability, simulate various emergency scenarios, go through the plans to test their effectiveness, and ensure clinical operation during disasters. Since these necessary tests mean scheduled downtime, it requires administrative support, and can be conducted during weekends or evening time to minimize impact to clinical workflow. Healthcare institutions can use their own discretion to define and document the scope and frequency of such tests, and revise them accordingly. The test of an unscheduled downtime can be either in the form of reviewing all procedures with involved staff, or simulating the emergency downtime situation by shutting down the information system. The former test is convenient for the staff, while the latter test is more close to the real downtime scenario. Although the replicated downtime operation more accurately reflects the true performance of the contingency plan in case of a disaster, it may not be accepted by clinical staff because of the inconvenience during the test, or it may not be feasible because of the cost associated with such simulations. In either case, healthcare institutions can learn from these tests, train staff, accumulate experience, receive feedbacks from all staff involved in the tests, and revise appropriate policies to best fit the clinical operation of the healthcare institution. When there is a disaster, the contingency plan should be used to effectively communicate with all involved users and vendors, and start the disaster recovery process. When the disaster is over and the systems are restored, the contingency plan should be revised according to experience and lessons learnt through the entire process.

12.5 SECURITY MANAGEMENT PROCESS

The security management process implements policies and procedures to prevent, detect, contain, and correct security violations. It requires high-level administration and management support, the development and enforcement of appropriate policies and procedures, and periodic evaluation and update of such policies and procedures. The security management process includes risk analysis, risk management, sanction policy, and information system activity review.

12.5.1 Risk Analysis

Every healthcare institution faces various types of risks and threats to EPHI, including fire, vandalism, loss of utilities, cyber crimes, acts of terrorism, and IT system failures. These risks require the administration to determine the major risks to their institution and their solutions. Risk analysis is to list all possible scenarios that could lead to adverse events, detailing especially which assets are impacted, the kind of impact, the likelihood of the threats occurring, the extent of damages, and the mitigation of the risks. The goal of risk analysis is to reduce or eliminate the risks identified and to determine how to manage them through process controls and use of technologies. Healthcare institutions must conduct an accurate and thorough assessment of potential risks to the confidentiality, integrity, and availability of the EPHI they hold, and its vulnerability. The HIPAA Security Rule does not specify the exact procedures of a risk analysis, so the institution or enterprise has the discretion to determine the contents of a risk analysis.

A typical risk analysis involves identification of potential risks, vulnerability of assets such as workstations and mobile devices, asset valuation, the likelihood of threats, and the degree of damages when the assets are compromised. Risks are identified based on the likelihood of threats and the possible impact to EPHI. These should be documented for risk management, so that high risks and low risks can be managed appropriately. It should be realized that there are risks that cannot be identified during analysis, and there are risks that are likely to remain even though identified.

12.5.1.1 Identification of Vulnerability

Identification of vulnerability is to identify the weakness of assets relevant to EPHI when faced with specific threats that can jeopardize the confidentiality, integrity, and availability of assets. The vulnerability can originate from insufficient or inappropriate safeguards. When vulnerabilities to specific threats are identified, the strengths and weakness of safeguards should be determined and appropriate action taken to strengthen them. In evaluating possible system vulnerabilities to general threats, it is important to recognize the potential negative impact of a successful attack. The impact may be caused by unauthorized disclosure or alteration of patient health data, compromised patient safety, or interruption to system operation. The severity of a security incident can be ranked with regard to loss of confidentiality, integrity, and availability of patients EPHI. In general, the more number of systems or patient data affected, the more severe the impact.

12.5.1.2 Identification of Threats

Threats are potential dangers from events or persons that exploit vulnerabilities. Threat identification determines the threats that can potentially be detrimental to patient EPHI or related assets. Threats may be caused by either intentional or unintentional human conduct, or by natural disasters. The consequence and likelihood of threat can range from severe and likely to moderate and unlikely. Threats that can cause significant damage and are reasonably anticipated should be prepared for the most.

12.5.1.3 Asset Identification and Valuation

Asset identification and valuation requires listing and making an inventory of assets that store or access EPHI. Servers, PACS workstations, laptop computers, and mobile devices used for PACS or EHR are assets belonging to this category, while other assets not relevant to patients EPHI need not be included.

12.5.2 Risk Mitigation and Management

The results of a risk analysis allow appropriate risk mitigation measures to be defined. The goal of risk mitigation and management is to implement security measures that are sufficient to reduce risks and vulnerabilities to a reasonable and appropriate level. Risks can be mitigated through technology usage, extensive training of workforce, or policy and procedure implementation. Risk management should include:

- Identifying security requirements
- Reviewing risk analysis results

- Listing information systems, devices, and their intended applications, with details sufficient to identify and implement mitigation of threat

- Prioritizing risks and their mitigation

- Implementing risk mitigation

- Identifying residual risks

- Regulation change review

- Institution or enterprise structure and clinical operation change review

- New technology implementation and associated vulnerability analysis

Additionally, security requirements should be identified in terms of regulatory requirements, security best practices in the healthcare industry, documented security and privacy policies of the institution or enterprise, and lessons learnt from past experiences. It is not anticipated that all risks can be eliminated. The purpose of risk mitigation and management is to minimize and manage the risks to a reasonable and acceptable level based on the risk analysis. If new hardware, software, or service products are available to reduce risks, the institution or enterprise should determine if the additional expenses are worth the added benefits. It should be noted that there are security products that address some of the vulnerabilities or weaknesses of information systems, but they also can interrupt other services or systems by creating new issues.

12.5.3 Facility Security and Access Control

According to HIPAA, healthcare institutions and enterprises should implement access control policies and procedures to limit physical access to the electronic information systems and the facilities in which they are housed, prevent unauthorized individuals and software programs from accessing patients EPHI, while permitting access only to those persons or software programs that have been granted access rights. This requires institutions to impose physical access control to locations where patients EPHI is stored, or where information systems containing EPHI operate. Facility access control can be in the form of general security measures such as security guards, surveillance cameras and monitors, unauthorized intrusion detection and alarms, doors and locks, and antiterrorism safeguards.

All information systems should be located in physically protected or actively monitored areas. Imaging modality scanners typically meet this requirement because of operational and patient privacy and safety concerns. Clinical information systems, data archives, and network equipment are usually locked in communication closets and protected areas. Shown in Figure 12.1 is the data center security and access control room of a healthcare enterprise, where the EPHI of the enterprise is located. Additionally, a fire extinguishing system for the safety of the data center is utilized as shown in Figure 12.2. In addition to the physical security of the network, unsupervised systems should be protected against network access. It should be noted that security technologies such as encryption, firewall, and VPN discussed in Chapter 4 cannot replace physical security; however, combined together they form an overall security protection system.

It should be emphasized that one key to access control is that each user is assigned a unique identification, and staff should be granted the minimum access right necessary and sufficient for them to complete tasks required by their job. Users must not share their ID and passwords for accessing EPHI, and all activities of the user are logged, so that any activity can be traced to only one individual. However, during an emergency, ordinary access controls may be bypassed, and care should be taken that such emergency access procedures are not used for unauthorized access under normal operating conditions.

A new challenge to facility access control is the potential wide use of portable electronic devices for EPHI access and storage, and they could be used in private homes, vehicles, and public buildings and locations. Although these places are not traditionally defined as "facilities," the healthcare institution or enterprise is still responsible for the safeguard of EPHI.

12.5.4 Workforce Security

Healthcare institutions or enterprises should have policies and procedures in place to ensure that all the members of its workforce have appropriate level of access to EPHI, and to prevent those

Figure 12.1 Aurora Healthcare data center security control room. Various security and monitoring systems are utilized to ensure the safety of the data center.

who should not from obtaining access to EPHI, because EPHI is vulnerable to threats from internal workforce members. The EPHI access authorization process should be well documented, indicating the responsibilities of the individual who can authorize and supervise staff's EPHI access. Staff without the need to access patients EPHI should not be granted the rights to such access, and any unauthorized attempt should be denied. Authorized personnel should only access the data at the level (such as review, create, modify, and delete EPHI privileges) sufficient for them to do their job, and their data access activities should be supervised. Before staff whose job

Figure 12.2 Aurora Healthcare data center where data storage systems and servers are located. Special facility equipment such as the fire extinguishing system (indicated by the arrow) must be installed to ensure the security of the facility.

responsibilities requiring access to EPHI are hired, their background, qualification, skills, and experiences should be confirmed so that they are trustworthy and competent. Procedures for terminating access to EPHI should be in place as well when recruiting, because immediately after an employee is terminated, the information systems holding patients EPHI is vulnerable to misuse or abuse by the departing employee.

12.5.5 Security Requirements on Contractors and Business Associates

If healthcare institutions or enterprises have business relations with other contractors and business associates that need to create, receive, maintain, or transmit patients EPHI, it is required by HIPAA that these business associates give formal assurance to securely safeguard the EPHI. The assurance can be in the form of security provisions in business contracts requiring business associates to implement administrative, physical, and technical safeguards to protect the confidentiality, integrity, and availability of patients' EPHI. In case the business associate violates security requirements specified in the contract, or if the business associate is not a "covered entity" governed by HIPAA, the violation can only be addressed through contract termination or reporting the business associate to the US Department HHS. If the business associate is also a "covered entity" as defined by HIPAA, it is regulated by HIPAA and overseen by HHS. The business associate is thus deemed noncompliant with HIPAA and is subjected to enforcement actions by HHS. To avoid any possible confusion, if vendors have access to patients' EPHI, the responsibilities and specific safeguards that vendors must implement should be specified in the service contracts. In addition, vendors must notify and report any security-related incidents to the healthcare institutions. Since PACS and EHR vendors or data storage service providers are typically business associates providing IT services, healthcare institutions and enterprises must specify in contracts that vendors must comply with all necessary and relevant HIPAA security requirements, standards, and specifications.

12.5.6 Security and Privacy Considerations for Remote Services

Remote services to PACS, EHR, and other clinical information systems are the delivery of hardware or software maintenance and services from locations beyond the physical premise of healthcare institutions through connections provided by network or telephone carriers. The benefits of remote services are obvious for both institutions and vendors, and is widely used by vendors. Remote servicing increases system availability, productivity of equipment, effectiveness of services, performance of systems, as well as reduces the service response time and system maintenance and service costs. However, there are increased concerns about data security and privacy issues when information systems containing sensitive health information is connected through a network for remote services, and the information systems are subject to the risk of unauthorized access and disclosure of sensitive PHI. Therefore, it is critical to protect the information. To ensure compliance with national and local regulations and the security and privacy of PHI, appropriate technical measures, administrative policies and procedures must be in place. The contract with the remote service provider should contain detailed information with regard to the following.

- Remote services provided from locations with security controls in place

- Remote services provided by staff properly trained to and bound by such policies

- Sensitive health information accessed or processed in compliance with contract terms during remote service

- Maintaining confidentiality of sensitive PHI

- Explicit terms about remote service activity audit logs, including date and time of the remote services, unique identifier for the service staff performing the service, unique identifier for the accessed system, and service activity record without patient identifying information

- Monitoring of remote service activities related to processes, controls, and sensitive health and personal data

- Limit the storage of necessary sensitive data at remote service locations

- Verification of meeting regulatory requirements before the initiation of remote services

The complexity of clinical information systems and network infrastructure requires institutions or enterprises to collaborate with vendors, understand and clarify the responsibilities of each party, comply with national and regional regulations which continue to evolve, and achieve the goal of sensitive information protection while providing cost-effective service.

12.5.7 Sanction Policy

The sanction or discipline policy applies appropriate sanctions against the staff of healthcare institutions, enterprises, or their contractors who fail to comply with the security policies and procedures. The purpose of the sanction policy is to hold the staff of the institution or business contractors accountable for their conduct. In case there are violations of documented security or privacy policies, there are procedures to follow for taking action. A written sanction policy should have procedures to investigate, report, resolve security violations, and have appropriate sanction types for different levels of violation. The sanction policy should be carried out consistently and appropriately as violations occur, with the outcome to be documented. The procedures must ensure that appropriate disciplinary action is taken against individuals who do not comply with the security or privacy policies. In employment agreements or service contracts there should be provisions for sanctions, which should include termination of employment or contract, or criminal prosecution. Additionally, all staff and contractors should be educated to fully understand security policies, the consequence of violation and the implication for abuse or misuse, so that potential violation of security policies is prevented.

12.5.8 Security Incident Procedures and Responses

Security incidents are attempted or actual unauthorized access, use, disclosure, modification, or destruction of information or interference with the system in a clinical information system. Although it is not expected that all security incidents can be defended against, it is necessary to plan for the incidents that are likely to occur. It is not unusual that security threats are not considered seriously until they have caused damage or loss. Planning for security incidents is not only required by HIPAA, but is also a critical component contributing to the survival of information systems and clinical operations of healthcare institutions or enterprises when it is properly implemented. In the long run, it is less costly to implement such critical security incidents and contingency plans than to recover from disasters without such plans in place. With proper security procedures and technologies in place, institutions or enterprises can be sure that information systems are reasonably secured and are able to respond to incidents when they occur.

Security incidents can be caused intentionally or unintentionally by outsiders or inside individuals, or by malicious viruses. According to statistics, many security incidents are the work of insiders such as staff, previous employees, contractors, and others who have ready access to EPHI. Healthcare institutions or enterprises should involve the administration, IT department, public relations, and legal professionals to develop and implement policies and procedures for responding to and handling of security incidents. These procedures should be tested and used to formally respond to security incidents when they occur. Through risk assessment, various types of threats to EPHI and possible resulting incidents can be identified and corresponding solutions found. All relevant staff should be informed of the policies and procedures, and be made aware of the process of incident reporting. The policies and procedures should be periodically reviewed and revised to reflect an improvement in handling security incidents.

When a security incident occurs, it should be investigated to find out exactly what happened and have appropriate evidence collected. Then practical action can be taken to mitigate the effects of the incident, specific vulnerability causing the incident can be closed, and security safeguards can be strengthened to prevent similar incidents in future. The security incidents and their outcomes should be documented, and its occurrence should be reported to the administration. Additionally, in case of a security incident such as a lost unprotected laptop computer with sensitive EPHI, many states in the United States require the institution or enterprise to notify individuals whose personal identifiable information is exposed in the security incident, which has negative consequences because of damaging media publicity and incident-mitigation costs. As part of the American Recovery and Reinvestment Act of 2009, The Health Information Technology for Economic and Clinical Health (HITECH) Act requires the reporting by institutions in the event of certain protected health information-related security incidents.

12.5.9 Information System Activity Review

HIPAA requires healthcare institutions or enterprises to implement procedures to regularly review records of information system activities, such as audit logs, access reports, and security incident tracking reports. The procedures must specify what type of data are collected and how often they are reviewed, so that security incidents can be detected, appropriate remedial actions taken, and the record or log be retained as evidence for any legal purpose. An institution or enterprise should assign staff to be responsible for the review of information system activity. The frequency and amount of data to be reviewed should be reasonable and practical. If necessary, software utilities may be used to automatically process audit logs and access reports, and provide an alert to the staff responsible should any suspicious activities be detected.

12.6 DEVICE AND ACCESS CONTROL

Unlike other businesses that may have access to isolated pieces of sensitive personal information, data stored or accessible from computers of healthcare institutions or enterprises are much more comprehensive, including demographic, EPHI, and financial information. Although some people may not be concerned much about disclosure of their health information, almost everyone is concerned with disclosure of their demographic and financial information in public. Therefore, healthcare institutions and enterprises must ensure security and access control for workstations and electronic devices that store or access EPHI.

12.6.1 Workstation Security

According to HIPAA, workstations include laptop or desktop computers that perform similar functions, and the electronic media stored in its immediate environment. All PACS and EHR systems use workstations to access patients EPHI. Workstation security is to safeguard all workstations that access EPHI, to restrict access only by authorized users, and to protect against threats including accidental misuse, random abuse by individuals for personal curiosity, revenge, or gain which may lead to financial and legal consequences causing disruption to clinical operations. Therefore, healthcare institutions or enterprises must have accurate inventory information about all workstations, computers, laptop computers, and other mobile devices that contain or are capable of accessing EPHI, including who and where they are assigned to, the software installed, and where they are physically located. Policies and procedures for workstation use should identify capabilities and vulnerabilities of different workstations and devices that have access to patient's EPHI. Before a workstation is installed in a specific physical location, security risks should be evaluated, and it should be kept away from public areas, or face away from areas where displayed EPHI can be viewed. Workstations should also be turned off, logged out of, or applications suspended when users leave the workstations.

12.6.2 Laptop Computers

Laptop computers are increasingly used for accessing medical images and EHR by healthcare professionals. Because of the value of laptop computers and comprehensive personal data stored on them, naturally they are attractive targets for theft; thus, keeping data secure in such a mobile environment is a critical requirement that is very challenging. For laptop computers used for patient image and EHR access or storage, physical security is very important, because lost or stolen laptop computers make EPHI stored on them vulnerable to inadvertent disclosure. Therefore, laptop computers should be used responsibly and securely, and it is prudent to accurately track laptop computers, have protection to safeguard laptop computers from vandalism or unauthorized EPHI access, and plan for possible loss or theft. The security incident prevention mechanism for laptop computers may include:

- Inventory control and locks preventing unauthorized movement
- No user ID or password information attached
- Lock up when unattended
- Not to be used in public places
- Not to be reconfigured or changed by users
- Using regularly changed strong passwords
- Immediate report of any stolen or lost computers

In case laptop computers are lost, stolen, or in other suspicious circumstances, EPHI data stored on them are vulnerable to various threats including disclosure of stored EPHI and reconfiguration for other uses. Therefore, technical measures should be implemented as well to protect EPHI in case they are lost. There are antitheft technologies available to either quickly and remotely lock computers from reboot, or delete data on the hard drive. If the computers are returned or recovered, the data can be quickly restored. Such antitheft security measures not only protect EPHI on laptop computers, but also discourage vandalism, because these computers are deemed less valuable compared to those without such security measures.

12.6.3 Electronic Mobile Devices Usage

Electronic mobile devices such as cell phones capable of Internet access and image display are increasingly used by healthcare professionals for various purposes. These devices can increase efficiency and potentially decrease errors in patient care. With the emergence of Web-based PACS and EHR systems, these mobile devices become attractive for accessing, storing, or processing EPHI. However, they are inherently less secure than desktop workstations and even laptop computers, because their size and portability make them vulnerable to theft, loss, or misplacement, and they lack security measures compared to laptop computers. Therefore, they are prone to inadvertent EPHI disclosure. Even though they maybe owned by staff, institutions or enterprises must have policies and procedures concerning the use of such mobile devices, implement safeguards to prevent unauthorized use, and secure EPHI so that even when these devices fall in the hands of unauthorized users, EPHI cannot be accessed. Although HIPAA does not have requirements specifically for mobile devices, Security and Privacy Rules should still be applied and followed closely to prevent accidental EPHI disclosure. Policies and procedures should determine the extent of their use, a central inventory of types of EPHI stored or accessible from these devices, use of strong password, mandatory use of approved access control and data encryption, and allowing only approved applications on these devices, so that in case EPHI are potentially compromised because of loss or theft of the mobile devices, threats and risks can be mitigated accordingly.

12.6.4 Device and Media Control

According to HIPAA, electronic media include "memory devices in computers (hard disk drives) and any removable/transportable digital medium, such as magnetic tape or disk, optical disk, or digital memory card; or transmission media used to exchange information already in electronic storage media. Transmission media include, for example, the Internet (wide-open), extranet (using Internet technology to link a business with information accessible only to collaborating parties), leased lines, dial-up lines, private networks, and the physical movement of removable/transportable electronic storage media. Certain transmissions, including of paper, via facsimile, and of voice, via telephone, are not considered to be transmissions via electronic media, because the information being exchanged did not exist in electronic form before the transmission."

Device and media controls govern the receipt and removal of hardware and electronic media that contain EPHI into and out of a facility, and the movement of these items within the facility. Healthcare institutions or enterprises must implement policies and procedures to address security threats caused by disposal and re-use of media. Specific areas of security threats include unauthorized download of EPHI (including images from PACS and patients EHR) to widely available and easy to use large capacity flash memory devices, disposal of storage media still containing recoverable EPHI, and physical movement of storage media containing EPHI from the healthcare institution or enterprise. Therefore, policies and procedures must be implemented to address the final disposal of hardware or electronic media on which EPHI is stored, and to remove EPHI from the electronic media before it is made available for reuse. When data storage media is disposed when a legacy PACS system is replaced and data migration is completed, if data cannot be erased because they are stored on WORM devices, as discussed in Chapter 5, they should be destroyed. If EPHI is stored on re-writable media that can be erased such as magnetic tapes, hard disk drives, rewritable DVDs or MODs, they must be permanently and securely erased before disposal or reuse to prevent unauthorized recovery and disclosure of EPHI. It is also more cost effective to destroy rewritable storage media used by legacy systems than to erase them for re-use.

For workstations or laptop computers (especially leased ones) that were once used for EPHI access and storage, when they are decommissioned from such use, secure data deletion utilities should be used to permanently remove EPHI before they are discarded or released for general purpose use, because many easily obtainable utility software can recover simply deleted files.

12.6.5 E-mailing EPHI Consideration

E-mails are frequently used by physicians and other healthcare professionals to communicate with other staffs or patients. Users should always be aware of many privacy and security concerns that exist when using e-mail, because they can be easily misdirected by accidentally sending to unauthorized individuals. Although many institutions and enterprises have implemented encryption technologies to their e-mail systems, they can easily be forwarded, printed, circulated, or stored in numerous paper and electronic files, and distributed outside of the healthcare institution or enterprise. If intercepted, they are readable by anyone who receives the message. Since communications through plain e-mail are usually insecure and may lead to patient data exposure, using e-mails to exchange EPHI increases the risk of an accidental disclosure.

12.6.6 Authentication

Healthcare institutions or enterprises are required to have procedures to verify that a person or entity seeking access to EPHI is the one claimed to be. The authentication process is a necessary step to ensure that appropriate access control policies and procedures are properly carried out, and to confirm that an individual seeking to access patients' EPHI is actually the person who is authorized for such access. Typical authentication mechanisms include passwords, biometrics (such as fingerprint and iris patterns), and tokens for the verification of a user, so long as only the user knows the password and possesses the token. A user ID together with a password is the most common authentication method because of its convenience. HIPAA requires each individual to use a unique identifier to access applications or systems that contain EPHI. A unique login ID should be assigned to each individual, and the password should remain confidential. The password should be hard to guess in order to maintain the security of EPHI. Additionally, each individual is responsible for the security of their password, and it should not be shared with anyone. If it is suspected that a password has been exposed it should be changed immediately. If higher security is required, additional authentication mechanisms should be used with the trade-off of reduced convenience and increased cost. If any authentication mechanism is compromised, security staff should be notified immediately to reduce potential negative impact.

12.6.7 Data Integrity

To protect data integrity of EPHI from unauthorized alteration or destruction, healthcare institutions and enterprises must implement policies and procedures for the use of multiple appropriate technologies to ensure data integrity, so that equipment failure, unintentional incidents by authorized users, and intentional malicious attacks by unauthorized and authorized users can be prevented. These policies and procedures must be subjected to periodic review and revision to account for new technologies that can counter new threats.

12.6.8 Audit Control

Audit control is to implement hardware, software, and procedural mechanisms that record and examine activities in information systems that contain or use EPHI. The unique login ID and password assigned to a user allow recording and examining the activities of the user accessing EPHI. Audits can be performed on a routine or special case basis. The purpose of audit control is to monitor EPHI access by users and hold users accountable for their conduct. Although HIPAA requires audit control, it does not specify exactly what data should be audited. Healthcare institutions or enterprises must decide what information is to be collected, the technical means to log all user activities, the mechanism to review such logs, and how frequently the logs are to be reviewed to detect any unauthorized activities. Detected misconduct that continue can seriously jeopardize the mission, reputation, and legal status of the healthcare institution or enterprise. Staff should be informed of documented audit policies and procedures, and the audit program should be subjected to regular review and update as well.

12.7 DICOM AND IHE SECURITY MECHANISM

Having appropriate security policies in place is necessary for any level of security. Before anyone can access any image, security measures need to be implemented including local access control of a workstation, authorization at the level of application software with passwords or biometrics, authentication of remote workstations or devices, and the audit and logging mechanism, which logs any data access by anyone. Additionally, security is very important when patient images and reports on PACS can be accessed with Web distribution. When images are accessed with an unsecure Internet connection, data encryption with standard mechanisms and utilities need to be used, together with electronic signatures which can prevent original data from being altered. To achieve these goals, DICOM and IHE provide technical mechanisms that address security policies, with regard to the exchange of DICOM images or reports between two medical imaging devices.

12.7.1 DICOM Security Mechanisms

Dealing with private and confidential patient information, each imaging device must ensure that its local environment is secure before establishing secure communications with other medical imaging devices. It is assumed that all imaging devices involved in a DICOM image and report exchange have appropriate security policies. An imaging device can identify local users with their roles, whereas local users may be individual persons or pieces of equipment. DICOM also assumes imaging devices involved in the information exchange have mechanisms in place to determine if specific device users have access authorization from information owners such as healthcare institutions or enterprises, and the authorization may determine imaging device access control. Using access control as an example, DICOM does not set access control policies; however, it provides the imaging device technical means of information exchange for access control policy implementation. DICOM provides mechanisms for imaging devices to securely authenticate each other as well, so that any alteration of or tampering with messages exchanged can be detected. This is achieved when imaging devices exchange certificates during secure channel setup for DICOM information exchange, where a certificate is an electronic document identifying a party and the party's public encryption algorithm, parameters, and keys. The certificate also includes, among other items, the identity and a digital signature from the entity that creates the certificate. Imaging devices may use users information in the certificates for access control policy and audit trail generation implementation. The confidentiality of DICOM messages is protected during message transport in the communication channel through encryption. Depending on the level of trust imaging devices place in the communication channel, a combination of security technologies can be deployed.

12.7.1.1 DICOM Security Profiles

DICOM images or reports contain medical information, including patient demographic information. In addition to safekeeping activities on a computer where patient private and confidential data are processed or stored, security measures to be concerned about include traffic of patient data on the network and its storage on any digital storage media. Accordingly, the exchange of DICOM images and reports should be restricted to a secure network or securely protected media. As for security mechanisms, DICOM is a standard facilitating information exchange, which is only part of the overall information chain, and it only constitutes a relatively small part of the security measures a healthcare institution or enterprise needs to adopt for creating a secure environment. DICOM enables a wide variety of authentication and access control policies by specifying security-related profiles, which are all covered in the DICOM Part 15: Security and System Management Profiles. There are various profiles in the categories of Secure Use Profiles, Security Transport Connection Profiles, Digital Signature Profiles, and Media Storage Security Profiles. Security profiles are defined by using non-DICOM developed industry standard protocols such as TLS, which is derived from SSL. These non-DICOM specific industry standard protocols are utilized with attention to their use in a system for DICOM image and report exchange. DICOM defines an encryption method to encrypt the SOP instance and attributes, and uses security profiles to define protected attributes. In cases where only patient information is encrypted without image encryption, or only reports are encrypted without patient ID encryption, local profiles can be utilized for selective encryption.

12.7.1.2 Basic TLS Secure Transport Connection Profile

The Basic TLS Secure Transport Connection Protocol is derived from SSL 3.0, and is largely compatible with it. As discussed in Chapter 4, TLS protocol has a framework and a negotiation mechanism for secure transport connections. The purpose of TLS is to transfer data safely over the Internet, and it is used by DICOM implementation to support the basic TLS secure transport connection profile. The profile does not specify any certificate exchange during authentication and how to establish a TLS secure transport connection; thus, various mechanisms can be used to establish the TLS channel for secure DICOM information exchange. The secure transport channel can be used for DICOM upper layer association once it is established by the imaging device. In the profile there is no requirement for all security features of TLS such as authentication, data encryption, and integrity check to be supported in the implementation. Thus, each imaging device can follow its own site security policy to determine which features are to be implemented, and use the certificate owners' identities for audit log purpose and use access control framework to restrict access.

12.7.1.3 User Identity Plus Passcode Association Profile

The DICOM User Identity and Password Association Profile enables user identity of the application on the imaging device to be authenticated through a password, through either internal or external authentication systems. The user identity can then be logged for user identification in the audit log.

12.7.1.4 Secure Use of E-mail Transport Profile

Most countries have regulations requiring the use of data encryption or other protection when patients' private and confidential information is transported, and the DICOM profile meets most common regulatory requirements throughout the world. When a DICOM file set including images and reports is sent through e-mail, the DICOM Secure Use of E-mail Transport profile requires that the DICOM file set is an attachment to the e-mail body, and the entire e-mail including body, DICOM file set attachment, and any other attachment are encrypted. The e-mail is digitally signed by the sender, and the signing may be applied before or after encryption. The digital signature clearly states that the sender is attesting to his/her authorization to disclose the information in this e-mail to the recipient. By using a signature the e-mail composer and sender provides minimum sender information, confirms the integrity of the e-mail content and the attachments, and attests that he/she is authorized to transmit the e-mail with attached DICOM image and report data to the recipient. It should be noted that the e-mail signature is different from those presented in DICOM images and reports contained in the DICOM file set attached to the e-mail, because those signatures are used for clinical purposes. Additionally, all healthcare content attestations need to be encoded in digital form, not as the e-mail composer and sender signature, because the e-mail may be composed and sent by someone who does not make clinical attestations. It should be noted that there might be locally posed requirements which may include mandatory statements in the e-mail body and prohibiting patient private and confidential information content in the e-mail body to protect patient privacy. Finally, the e-mail body and attachments may be compressed, and any use of compression is separated from the profile.

12.7.2 IHE Security and Privacy Control

The IHE security and privacy control does not address security measures that protect health information systems against malicious attacks from network and virus infections. Instead, the purpose is to provide security measures related to information systems within the scope of healthcare applications, and define a standardized approach to use existing standards and technologies. In addition to the ATNA, EUA, XUA, and Basic Patient Privacy Consents (BPPC) integration profiles, there are other integration profiles such as Consistent Time (CT), Personal White Pages (PWP), Digital Signatures (DSG), Notification of Document Availability (NAV), and Cross-Enterprise Document sharing via Reliable messaging (XDR) that can be leveraged to meet security and privacy requirements.

12.7.2.1 Basic Patient Privacy Consents

The BPPC integration profile indicates the patient's consent or not to participate in the exchange of sensitive health information among healthcare enterprises, and is capable of processing several different policies covering most types of patients' privacy consent. It defines a mechanism to document patients' privacy consent by cross enterprise document sharing registries and data repositories, enables patient privacy consent to be marked with clinical documents published to XDS, which is discussed in Chapter 11, allows users of these XDS to enforce appropriate privacy consent during use, and is used to authorize the publication of patients' sensitive health information such as EHR for cross enterprise exchange. Although documents exchanged in XDS can be labeled with confidentiality codes, there is no clear guidelines as how to use this information to support patient privacy, and healthcare providers even from the same healthcare institution playing various roles during care delivery have different patient document access needs. Since the XDS integration profile itself does not clearly define privacy policies within an XDS affinity domain, which consists of multiple healthcare enterprises, BPPC complements XDS in this regard. It provides a way for XDS affinity domains to be able to develop and implement privacy policies, and integrate these policies with access control governed by EHR systems of healthcare enterprises, which are defined as XDS actors.

12.7.2.2 Enterprise User Authentication

User authentication is necessary for almost all healthcare information applications and data access processes. The EUA is an integration profile for establishing a connection of one name for each user, so that it can be used on all participating software and hardware devices implemented with the integration profile, and improve user workflow. This applies to a single healthcare enterprise governed by a single security policy, operating on a secure common network domain. Users with single sign-on can conveniently use services provided by these software applications or hardware devices with the convenience of centralized management for user authentication. It should be noted that other security features such as authorization management, audit trails, and access control are not addressed in this integration profile.

12.7.2.3 Cross-Enterprise User Assertion

The XUA integration profile enables identity claim communication of a user or system when patient healthcare information is exchanged among various healthcare enterprises. During these information transactions, it is necessary for the information provider to identify the information requesting user or system, so that appropriate access decisions can be made and proper audit entries logged. The XUA thus provides the user identity on the transactions so that the document registry and document repository know which systems are requesting information through strong network authentication, and enforce some level of access control by using the system identity to deny access to specific information. This integration profile allows enterprises to use either their own mechanisms of user authentication with their own user directories, or a third party to authenticate users. Unlike the EUA integration profile, whose function is for intraenterprise use, XUA is used for interenterprise applications where two involved enterprises each maintains its own independent user directory. It supports complex environments where different trust domains may operate under different procedures and policies with different technologies. For example, an independent physician's office very likely has a different patient clinical record access control model compared to a large healthcare enterprise with multiple hospitals. Although it is not necessary for them to have the same access control, mutually agreed processing rules at the policy level are needed. The XUA integration profile is thus effective for security auditing, but not very useful for access controls.

12.7.2.4 Audit Trail and Node Authentication

IHE enforces common basic security and privacy functionalities through the definition of the Audit Trail and Node Authentication (ATNL) integration profile. ATNA is one of the most important security integration profiles. It addresses system authentication and audit trail issues to meet security and privacy requirements imposed by regulations for healthcare information systems. It defines security measures such as user and node authentication, authorization, access control, audit transaction, and uses TLS to meet security requirements for node communications.

The authentication part of ATNA is composed of user authentication and network connection authentication. ATNA only requires local user authentication with either the access control technology of the local secure node's choice, or the EUA integration profile. For network connection authentication, ATNA mandates node authentication for both nodes involved in a network connection, instead of the actual user of the node. Certificate-based authentications are defined in DICOM and HL7 standards and in HTML protocols when Web access is used. When a user logs into a PACS workstation implemented with image display and secure node actors, he/she is authenticated to the PACS workstation according to the policy determined by the healthcare institution and the system manufacturer, and images can be requested from the image manager and secure node pair.

The audit trail of ATNA provides user accountability, tracking unacceptable conduct, and allows all activities to be audited by a security administrator of the healthcare institution or enterprise. In healthcare institutions and enterprises, users of information systems are mostly internal staff requiring different degrees of flexibility when interacting with these systems. They should understand that their activities on these systems are recorded and monitored, because auditing is an important part of a comprehensive security policy in addition to access control and authentication. The audited activities include secure domain policy compliance and behaviors such as improper access, alteration, deletion, and creation of patient healthcare information. The protected

PHRs include registration, order history, procedures, reports, and images, and so on. The ATNA audit trail also provides information about all patients' healthcare records accessed by a specific user, all users accessing a specific patient's health record, and failures of user and node authentications. IHE defines various actors and secure node pairs of other transactions to trigger audit record transactions. During the authentication process, the audit record repository actor accepts audit record transactions. In addition, IHE adopts the network time protocol as a transaction for various information system clock synchronization, so that there is a consistent date and time stamp for audit transactions from heterogeneous information systems.

Although data encryption is not required during data exchange, it is allowed if both nodes involved in the transaction are configured to support and request data encryption, and TLS is required by ATNA to ensure that all nodes are authorized secure nodes. Although the ATNA integration profile does not automatically make implementing healthcare institutions compliant with various national and international security regulations, it does provide a useful tool set for institutions or enterprises to achieve this goal. When used with security procedures and policies, confidentiality, integrity, and user accountability of patient information are ensured.

12.7.2.5 IHE Security Measures for Cross-Enterprise Exchange of Information

Most security and privacy integration profiles discussed are defined to be carried out within a healthcare enterprise. With the emerging cross-enterprise exchange of PHI by various healthcare enterprises, it is critical to establish appropriate security and privacy policies and procedures for the handling of EPHI when it crosses the borders of the enterprise.

Health information exchange across enterprises and relevant IHE integration profiles are discussed in Chapter 11. This requires all involved enterprises to implement security mechanisms in order to protect information exchange, and IHE only defines technical details that are necessary to ensure interoperability. Since there is no single IHE defined mandatory policy for each participating enterprise to ensure security and privacy of PHI, it distributes the responsibilities of security and privacy to the PACS and EHR systems that participate in the health information exchange. However, these participating enterprises need to establish common policies based on appropriate risk analysis, and the policies need to be harmonized with those policies of each individual participating enterprise. These policies and procedures must determine the acceptable level of risk, user authentication mechanisms, the acceptable types of documents and the length of time it is to be kept, who is allowed to publish documents, who has access to what type of documents, the type of sanctions imposed on those violating the policies, system and data backup, emergency mode of operation, and disaster recovery. With the establishment and implementation of these policies, the following security and privacy controls can be achieved:

- Access control to only allow authorized users or systems to access the information or system function

- Accountability control with security audit log, report, and warning

- Identification and authentication to verify the user or system is actually who they claim to be

- Patient privacy control to carry out patients' specific instructions about their sensitive health information

- Confidentiality and integrity control to ensure sensitive information is not disclosed or altered without authorization

- Nonrepudiation control to ensure the acknowledgment of actions by individuals or systems when they participate in information transactions

- Availability control to ensure data and information are available when they are needed

Healthcare institutions or enterprises have patient data repositories that must be protected against incidental and intentional disclosures. The type of information ranges from protected health information, patient demographics, to financial and insurance information. Although most information loss is caused by human error or improper information handling, many times it is caused by intentional malicious conducts.

Healthcare institutions and enterprises are increasingly relying on PACS and EHR systems for the clinical care of patients. When these systems become more complicated and government regulations more stringent, it is more important than ever before to analyze both the technical and business relations between these systems and the clinical operations they support. It is prudent that policies and procedures for quality control, fault tolerance, downtime planning, and disaster recovery processes be incorporated to ensure that patient health information including medical images are maintained confidential, free from unauthorized change, and readily available. Through careful planning and adopting appropriate technologies, these tasks can be accomplished, and ultimately the goal of improving the quality of patient healthcare with reduced cost can be achieved.

BIBLIOGRAPHY

Anderson, M. 2002. The toll of downtime: A study calculates the time and money lost when automated systems go down. *Healthcare Inform.* 19 (4): 27–30.

Avrin, D. E., K. P. Andriole, and L. Yin, et al. 2000. Simulation of disaster recovery of a picture archiving and communications system using off-site hierarchal storage management. *J. Digit. Imaging* 13 (Suppl 1): 168–170.

Absolute Software. Compliance, protection, recovery. A layered approach to computer security for healthcare Organizations. www.himss.org/content/files/Absolute%20Software%20-%20 Healthcare%20whitepaper.pdf. Accessed January 11, 2010.

Absolute Software. 2009. PC-disable delivers intelligent client-side protection for lost or stolen notebooks. www.absolute.com/resources/public/Whitepaper/20092704-Absolute_Intel_ Whitepaper.pdf. Accessed January 11, 2010.

Beaver, K. and R. Herold. 2004. *The Practical Guide to HIPAA Privacy and Security Compliance.* Auerbach. New York.

Bowers, G. H. What keeps CIOs awake at night: Information theft. *The Society for Imaging Informatics in Medicine 2008 Annual Meeting.* www.scarnet.org/WorkArea/showcontent.aspx?id=4470. Accessed January 11, 2010.

Buecher, A., P. Andreas, and S. Paisely, et al. 2008. Enterprise security architecture using IBM ISS security solutions. /www.redbooks.ibm.com/redbooks/pdfs/sg247581.pdf. Accessed January 18, 2010.

Burr, W., D. F. Dodson, and R. A. Perlner, et al. 2008. *Electronic Authentication Guideline.* National Institute of Standards and Technology, Information Security.

Carter, J. H. The legal health record in the age of e-discovery. www.himss.org/content/files/ LegalEHR_eDiscovery.pdf. Accessed January 11, 2010.

Caumanns, J., R. Kuhlisch, and O. Pfaff, et al. Access control, IHI IT infrastructure (ITI). www.ihe. net/Technical_Framework/upload/IHE_ITI_WhitePaper_XC_Dynamic_Data_2009-09-28.pdf. Accessed January 11, 2010.

Chenoweth, D., M. Perters, and B. Naremore. 2006. Hospital systems recovery in a worst case scenario. www.healthmgttech.com/features/2006_february/0206hospital_systems.aspx. Accessed January 11, 2010.

Code Green Networks. 2009. Preventing data loss in healthcare organizations. www. codegreennetworks.com/resources/downloads/CGN_Healthcare_Whitepaper_0309.pdf. Accessed January 11, 2010.

Disaster Recovery Journal. 1997. DRJ's sample plan. www.drj.com/index.php?option=com_cont ent&task=view&id=2069&Itemid=0. Accessed January 11, 2010.

Dreyer, K. J., D. S. Hirschorn, and J. H. Thrall. 2006. *PACS: A Guide to the Digital Revolution.* Springer, New York, NY.

Exploring HIPAA and HITECH ACT Definition 16. www.hipaa.com/2009/12/exploring-hipaa-and-hitech-act-definitions-part-16/. Accessed January 11, 2010.

Federal Emergency Management Agency. 1993. *Emergency Management Guide for Business and Industry, A Step-by-Step Approach to Emergency Planning, Response and Recovery for Companies of All Sizes.* FEMA 141. 1993. www.fema.gov/pdf/business/guide/bizindst.pdf. Accessed January 11, 2010.

Federal Register, Part VI 45 CFR Part 162. 2009. Health insurance reform; Modifications to the Health Insurance Portability and Accountability Act (HIPAA); Final rules. edocket.access.gpo.gov/2009/pdf/E9-740.pdf. Accessed January 11, 2010.

Hayes, J. 2008. PACS crash teaches administrator the hard way, http://www.diagnosticimaging.com/imaging-trends-advances/pacsweb/pacs-administration/article/113619/1181327. Accessed January 11, 2010.

HealthImaging, com. 2009. Cedars-Sinai endures more patient identity theft, fraud. www.healthimaging.com/index.php?option=com_articles&view=article&id=15711. Accessed January 11, 2010.

Huang, H. K. 2004. *PACS and Imaging Informatics: Basic Principles and Applications.* Wiley-Liss, Hoboken, NJ.

IHE. 2007. HIE security and privacy through IHE profiles. IHE IT infrastructure white paper. www.ihe.net/Technical_Framework/upload/IHE_ITI_Whitepaper_Security_and_Privacy_2007_07_18.pdf. Accessed January 11, 2010.

IHE. 2008. IHE IT Infrastructure (ITI) Technical Framework.

IHE. 2008. Cook book: Preparing the IHE profile security section (Risk Management in Healthcare IT white paper). www.ihe.net/Technical_Framework/upload/IHE_ITI_Whitepaper_Security_Cookbook_2008-11-10.pdf. Accessed January 11, 2010.

IHE. 2009. Access control. IHE IT infrastructure white paper.

Joint NEMA-MITA/COCIR/JIRA/Security and Privacy Committee. 2007. Management of machine authentication certificates. www.medicalimaging.org/documents/CertificateManagement-2007-05-Published.pdf. Accessed January 11, 2010.

Joint NEMA-MITA/COCIR/JIRA/Security and Privacy Committee. 2007. Information security risk management for healthcare systems. www.medicalimaging.org/documents/ISRM_for_healthcaresystems.pdf. Accessed January 11, 2010.

Joint NEMA/COCIR/JIRA Security and Privacy Committee. 2008. Remote services in healthcare-use cases and obligations for customer and service organizations. www.medicalimaging.org/documents/Remote-Service2008.pdf. Accessed January 11, 2010.

Kennedy, R. L. PACS policies and procedures. *The Society for Imaging Informatics in Medicine 2009 Annual Meeting.*

Khorasani, R. Communication of critical test results and the national patient safety goals: What you need to do. *The Society for Imaging Informatics in Medicine 2008 Annual Meeting.*

Leontiew, A. Radiology option for audit trail and node authentication. www.ihe.net/Participation/upload/rad13_ihe_wkshp07_atna_leontiev.pdf. Accessed January 11, 2010.

Meenan, C. D. 2008. Disaster recovery. *The Society for Imaging Informatics in Medicine 2008 Annual Meeting.* www.scarnet.org/WorkArea/showcontent.aspx?id=4448. Accessed January 11, 2010.

Moehrke, J. 2007. ITI security profiles—ATNA, CT. www.ihe.net/Participation/upload/iti7_ihewkshp07_atna_moehrke.pdf. Accessed January 11, 2010.

Moehrke, J. and L. Fourquet. 2007. Basic patient privacy consent. www.ihe.net/Participation/upload/pcc2_ihe_wkshp07_consents_moehrke.pdf. Accessed January 11, 2010.

National Institute of Standards and Technology. 2006. Minimum security requirements for federal information and information systems. Federal Information Processing Standards Publication, FIPS PUB 200.

National Institute of Standards and Technology. 2009. Digital signature standards (DSS). Federal Information Processing Standards Publication, FIPS PUB 186-3.

National Institute of Standards and Technology. 2010. Guide for applying the risk management framework to federal information systems: A security life cycle approach. csrc.nist.gov/publications/drafts/800-37-Rev1/SP800-37-rev1-FPD.pdf. Accessed January 11, 2010.

Navarro, D. 2009. Flames from the road: Remote system management tools for PACS administration and support. *The Society for Imaging Informatics in Medicine 2009 Annual Meeting*. Charlotte, NC.

Pyke, G. and L. Reed-Fourquet. Document digital signature (DSG). www.ihe.net/Participation/upload/2005-IHE-DGS-workshop-june262005.ppt. Accessed January 11, 2010.

Quinn, S., D. Waltermire, and C. Johnson, et al., *The Technical Specification for the Security Content Automation Protocol (SCAP): SCAP Version 1.1 (Draft)*. Recommendations of the National Institute of Standards and Technology. csrc.nist.gov/publications/drafts/800-126-r1/draft-sp800-126r1.pdf. Accessed January 11, 2010.

Ross, R., S. Katzke, and A. Johnson, et al. 2008. *Managing Risk from Information Systems: An Organizational Perspective*. National Institute of Standards and Technology, Information Security. NIST Special Publication 800-39.

Scholl, M., K. Stine, and J. Hash, et al. *An Introductory Resource Guide for Implementing the Health Insurance Portability and Accountability Act (HIPAA) Security Rule*. csrc.nist.go/publications/nistpubs/800-66-Rev1/SP-800-66-Revision1.pdf. Accessed January 11, 2010.

Skaer, K. 2009. Having the right stuff. *The Society for Imaging Informatics in Medicine 2009 Annual Meeting*. Charlotte, NC.

Tellis, W. M. 2008. Keeping your PACS afloat. *The Society for Imaging Informatics in Medicine 2008 Annual Meeting*. Seattle, WA.

Toland, M. 2008a. Things that go bump in the night: Frequent PACS failure modes. *The Society for Imaging Informatics in Medicine 2008 Annual Meeting*. Seattle, WA.

Toland, M. 2008b. PACS worst case scenarios: Understanding the implications of major downtimes and avoiding them. *The Society for Imaging Informatics in Medicine 2008 Annual Meeting*. Seattle, WA.

Index